METHODS IN MOLECULAR BIOLOGY™

Series Editor
John M. Walker
School of Life Sciences
University of Hertfordshire
Hatfield, Hertfordshire, AL10 9AB, UK

For other titles published in this series, go to
www.springer.com/series/7651

Cellular Programming and Reprogramming

Methods and Protocols

Edited by

Sheng Ding Ph.D.

Department of Chemistry, The Scripps Research Institute, La Jolla, CA, USA

Humana Press

Editor
Sheng Ding, Ph.D.
Department of Chemistry
The Scripps Research Institute
La Jolla, CA
USA
sding@scripps.edu

ISSN 1064-3745 e-ISSN 1940-6029
ISBN 978-1-4939-5714-9 ISBN 978-1-60761-691-7 (eBook)
DOI 10.1007/978-1-60761-691-7
Springer New York Dordrecht Heidelberg London

Printed on acid-free paper

Humana Press is a part of Springer Science+Business Media (www.springer.com)

Preface

Advances in stem cell biology are making possible new approaches to treat devastating human diseases, including cardiovascular disease, neurodegenerative disease, musculoskeletal disease, diabetes, and cancer. Such approaches may involve cell replacement therapy as well as the development of therapeutic drugs for stimulating the body's own regenerative ability to repair cells damaged by disease and injury. However, obstacles such as control of stem cell fate, immunorejection, and limited cell sources must be overcome before their therapeutic potentials can be realized. Recent studies have suggested that tissue-specific cells may overcome their intrinsic lineage-restriction to dedifferentiate or transdifferentiate upon exposure to a specific set of signals in vitro and in vivo. The ability to dedifferentiate or reverse lineage-committed cells to pluripotent/multipotent cells might overcome many of the obstacles (e.g., cell sources, immunocompatibility, and bioethical concerns) associated with using ES and adult stem cells in clinical applications. With an efficient dedifferentiation process, it is conceivable that healthy, abundant, and easily accessible somatic cells could be reprogrammed to become multipotent or pluripotent stem/progenitor cells, which can then be programmed to generate different types of functional cells for the repair of damaged tissues and organs. This series will cover the most recent technologies and their mechanistic understanding in cellular reprogramming and programming.

La Jolla, CA *Sheng Ding*

Contents

Contributors

ZOË D. BURKE • *Department of Biology & Biochemistry, Centre for Regenerative Medicine, University of Bath, Bath, UK*
LINZHAO CHENG • *Institute for Cell Engineering, Johns Hopkins University School of Medicine, Baltimore, MD, USA*
JOSH G. CHENOWETH • *Laboratory of Molecular Biology, National Institute of Neurological Disorders and Stroke, National Institutes of Health, Bethesda, MD, USA*
BIN-KUAN CHOU • *Institute for Cell Engineering, Johns Hopkins University School of Medicine, Baltimore, MD, USA*
JOSE B. CIBELLI • *Department of Animal Science, Michigan State University, East Lansing, MI, USA*
CÉSAR COBALEDA • *Departamento de Fisiología y Farmacología, Universidad de Salamanca, Salamanca, Spain*
CAROLINE DESPONTS • *Department of Chemistry, The Scripps Research Institute, La Jolla, CA, USA*
SHENG DING • *Department of Chemistry, The Scripps Research Institute, La Jolla, CA, USA*
JEONG TAE DO • *CHA Stem Cell Institute & CHA Biotech, Pochon CHA University, Seoul, Korea*
DANIEL EBERHARD • *Department of Biology & Biochemistry, Centre for Regenerative Medicine, University of Bath, Bath, UK*
TAKUMI ERA • *Division of Molecular Neurobiology, Institute of Molecular Embryology and Genetics, Kumamoto University, Kumamoto, Japan*
QIANG FENG • *Advanced Cell Technology, Worcester, MA, USA*
SARAH FERBER • *Endocrine Institute, Sheba Medical Center, Tel-Hashomer, Israel; Department of Human Genetics and Molecular Medicine, Sackler School of Medicine, Tel-Aviv University, Tel-Aviv, Israel*
MARTINE GERAERTS • *Stem Cell Institute Leuven, Catholic University of Leuven, Leuven, Belgium*
THOMAS GRAF • *Center for Genomic Regulation and ICREA, Barcelona, Spain*
JULIANE GUST • *Department of Biological Structure, University of Washington, Seattle, WA, USA*
BAO-YANG HU • *Department of Anatomy and Department of Neurology, Waisman Center, School of Medicine and Public Health, the WiCell Institute, University of Wisconsin-Madison, Madison, WI, USA*
WEI-SHOU HU • *Stem Cell Institute and Department of Chemical Engineering and Materials Science, University of Minnesota, Minneapolis, MN, USA*
YEUHUA JIANG • *Stem Cell Institute Leuven, Catholic University of Leuven, Leuven, Belgium*

CATHERINE V. LAIOSA • *Mount Sinai Hospital, New York, NY, USA*
DEEPAK LAMBA • *Department of Biological Structure, University of Washington, Seattle, WA, USA*
ROBERT LANZA • *Advanced Cell Technology, Worcester, MA, USA*
LINHENG LI • *Stowers Institute for Medical Research, Kansas City, MO, USA; Department of Pathology and Laboratory Medicine, Kansas University Medical Center, Kansas City, KS, USA*
WENLIN LI • *Department of Chemistry, The Scripps Research Institute, La Jolla, CA, USA*
GE LIN • *Stem Cell Institute and Department of Genetics and Developmental Biology, University of Connecticut Health Center, Farmington, CT, USA*
SHI-JIANG LU • *Advanced Cell Technology, Worcester, MA, USA; Stem Cell and Regenerative Medicine International, Worcester, MA, USA*
PRASHANT MALI • *Institute for Cell Engineering, Johns Hopkins University School of Medicine, Baltimore, MD, USA*
KRISTEN MARTINS-TAYLOR • *Stem Cell Institute and Department of Genetics and Developmental Biology, University of Connecticut Health Center, Farmington, CT, USA*
IRIT MEIVAR-LEVY • *Sheba Medical Center, Endocrine Institute, Tel-Hashomer, Israel; Department of Human Genetics and Molecular Medicine, Sackler School of Medicine, Tel-Aviv University, Tel-Aviv, Israel*
MANJA MUIJTJENS • *Stem Cell Institute Leuven, Catholic University of Leuven, Leuven, Belgium*
KATHLEEN M. MUNSON • *Developmental Biology Laboratory, Cardiovascular Research Center, Massachusetts General Hospital, Charlestown, MA, USA; Department of Medicine, Harvard Medical School, Boston, MA, USA*
KATHY O'NEILL • *Department of Biology & Biochemistry, Centre for Regenerative Medicine, University of Bath, Bath, UK*
D. JASON OWENS • *Stem Cell Institute and Department of Chemical Engineering and Materials Science, University of Minnesota, Minneapolis, MN, USA*
JENNIFER S. PARK • *Advanced Cell Technology, Worcester, MA, USA*
YONSIL PARK • *Department of Chemical Engineering and Materials Science and Department of Biomedical Engineering, University of Minnesota, Minneapolis, MN, USA*
KAREN A. PAUWELYN • *Stem Cell Institute Leuven, Catholic University of Leuven, Leuven, Belgium*
JOHN M. PERRY • *Stowers Institute for Medical Research, Kansas City, MO, USA*
THOMAS A. REH • *Department of Biological Structure, University of Washington, Seattle, WA, USA*
PABLO J. ROSS • *Department of Animal Science, Michigan State University, East Lansing, MI, USA*
HANS R. SCHÖLER • *Department of Cell and Developmental Biology, Max Planck Institute for Molecular Biomedicine, Münster, Germany*
YAN SHI • *Laboratory of Chemical Genomics, Shenzhen Graduate School of Peking University, Shenzhen, Guangdong, China*

KARTIK SUBRAMANIAN • *Stem Cell Institute and Department of Chemical Engineering and Materials Science, University of Minnesota, Minneapolis, MN, USA*
PAUL J. TESAR • *Department of Genetics, Case Western Reserve University, Cleveland, OH, USA*
DAVID TOSH • *Department of Biology & Biochemistry, Centre for Regenerative Medicine, University of Bath, Bath, UK*
EIRINI TROMPOUKI • *Stem Cell Program and Hematology/Oncology, Children's Hospital, Harvard Medical School, Howard Hughes Medical Institute, Harvard Stem Cell Institute, Boston, MA, USA*
FERNANDO ULLOA-MONTOYA • *Stem Cell Institute Leuven, Catholic University of Leuven, Leuven, Belgium; Stem Cell Institute and Department of Chemical Engineering and Materials Science, University of Minnesota, Minneapolis, MN, USA*
CATHERINE M. VERFAILLIE • *Stem Cell Institute Leuven, Catholic University of Leuven, Leuven, Belgium*
HUAFENG XIE • *Dana Farber Cancer Institute, Boston, MA, USA*
REN-HE XU • *Stem Cell Institute and Department of Genetics and Developmental Biology, University of Connecticut Health Center, Farmington, CT, USA*
ZHAOHUI YE • *Institute for Cell Engineering, Johns Hopkins University School of Medicine, Baltimore, MD, USA*
JING-RUEY J. YEH • *Developmental Biology Laboratory, Cardiovascular Research Center, Massachusetts General Hospital, Charlestown, MA, USA; Department of Medicine, Harvard Medical School, Boston, MA, USA*
JONATHAN YEN • *Institute for Cell Engineering, Johns Hopkins University School of Medicine, Baltimore, MD, USA*
SU-CHUN ZHANG • *Department of Anatomy and Department of Neurology, School of Medicine and Public Health, Waisman Center, the WiCell Institute, University of Wisconsin-Madison, Madison, WI, USA*
LEONARD I. ZON • *Stem Cell Program and Hematology/Oncology, Children's Hospital, Harvard Medical School, Howard Hughes Medical Institute, Harvard Stem Cell Institute, Boston, MA, USA*

Chapter 1

Human Embryonic Stem Cell Derivation, Maintenance, and Differentiation to Trophoblast

Ge Lin, Kristen Martins-Taylor, and Ren-He Xu

Abstract

Since the first report of derivation of human embryonic stem cell (hESC) lines in 1998, many progresses have been achieved to reliably and efficiently derive, maintain, and differentiate this therapeutically promising cell type. This chapter introduces some basic and widely recognized methods that we use in our hESC core laboratory. Specifically, it includes methods for (1) deriving hESC lines without using enzyme and antibody to isolate the inner cell mass; (2) sustaining hESC self-renewal under feeder-dependent, feeder-conditioned, and defined conditions as well as pluripotency validation and quality control assays; and (3) inducing hESC differentiation to trophoblast with BMP4.

Key words: hESCs, Derivation, Pluripotency, Differentiation, Trophoblast, BMP4, bFGF, TGFβ

1. Introduction

Human embryonic stem cells (hESCs) were originally derived and cultured on mouse embryonic fibroblast (MEF) cells as feeder cells in 80% DMEM medium (Invitrogen) and 20% fetal bovine serum (Hyclone) (1, 2). Alternatively, 80% DMEM medium and 20% KnockOut SR, a serum replacer optimized for mouse ES cells (Invitrogen), supplemented with 4 ng/mL bFGF can also support clonal as well as colonial growth of hESCs on MEF cells (3). Later, a feeder-independent system was reported, in which the SR-containing DMEM is conditioned by MEF, and then the conditioned medium (CM) is harvested and supplemented with 4 ng/mL bFGF to culture hESCs plated on Matrigel, a gelatinous protein mixture secreted by mouse tumor cells (BD Biosciences) (4).

We found that BMP4 induces hESC differentiation to the extra-embryonic lineage trophoblast (5). We further demonstrated that

S. Ding (ed.), *Cellular Programming and Reprogramming: Methods and Protocols*, Methods in Molecular Biology, vol. 636, DOI 10.1007/978-1-60761-691-7_1, © Springer Science+Business Media, LLC 2010

the SR contains BMP-like activity; MEF cells secret BMP inhibitors such as Noggin and Gremlin, and high-dose bFGF also represses BMP signaling in hESCs (6). Noggin and bFGF synergize to maintain hESC culture (6). When the bFGF concentration is increased to 100 ng/mL, Noggin is no longer necessary (7). We further demonstrated that TGFβ and FGF work concertedly to inhibit BMP signaling, and TGFβ-activated SMAD2/3 directly bind to the key pluripotency gene *NANOG* to regulate its expression (8). Others also reported that bFGF alone or together with ligands for TGFβ and WNT signaling supplemented in the SR-containing medium (9) or chemically defined media (10, 11) can support hESC culture on Matrigel or other matrices. This chapter will introduce the above basic and widely recognized methods that we use in our hESC core laboratory for hESC derivation, maintenance, and differentiation to trophoblast.

2. Materials

2.1. Reagents

1. Dulbecco's PBS, Without Ca^{++} and Mg^{++}.
2. Dulbecco's PBS with Ca^{++} and Mg^{++}.
3. 0.05% Trypsin in 0.53 mM EDTA.
4. Gelatin Solution (0.1%, 500 mL). Mix 0.5 g gelatin with 500 mL embryo-quality water (type 1, reagent grade), autoclave the solution for 30 min, store at room temperature, and keep sterile.
5. MEF Culture Medium (500 mL). Mix 445 mL DMEM, 50 mL FBS, and 5 mL Non-Essential Amino Acids (NEAA, 100×). Filter with 0.22 μm PES membrane. Store at 4°C, and use within 1 month.
6. MEF Cryopreservation Medium (2×, 50 mL). Mix 20 mL MEF Medium, 20 mL FBS, and 10 mL Dimethyl Sulfoxide (DMSO). Filter with 0.22 μm PVDF membrane. Use the same day, or store aliquots at −80°C.
7. Quinn's Advantage Thaw Kit (SAGE In Vitro Fertilization).
8. Quinn's Advantage Protein Plus Blastocyst Medium (SAGE In Vitro Fertilization).
9. OVOIL (Vitrolife).
10. hESC Medium (250 mL). Mix 200 mL DMEM/F-12, 50 mL Knockout Serum Replacer (KOSR), 2.5 mL NEAA 100×, 1.25 mL 200 mM l-glutamine solution, 1.75 μL beta-mercapto-ethanol, and 500 μL of 2 μg/mL bFGF solution. Filter with 0.22 μm PES. Store at 4°C and use within 2 weeks.

11. bFGF Solution (2 μg/mL, 5 mL). Dissolve 10 μg bFGF (Invitrogen) in 5 mL 0.1% Bovine Serum Albumin (BSA) in PBS. Mix with pipette. Aliquot into microfuge tubes. Store at −80°C.
12. 0.1% BSA Solution (150 mL). Mix 150 mL Dulbecco's PBS without Ca^{++} and Mg^{++} and 2 mL 7.5% BSA, Fraction V (Invitrogen). Filter with 0.22 μm PES membrane. Store at 4°C.
13. 1 mg/mL Collagenase solution (30 mL). Dissolve 30 mg collagenase in 30 mL DMEM/F12 (Invitrogen). Filter with 0.22 μm PES membrane. Store at 4°C for up to 1 month.
14. 0.5 mg/mL Dispase Solution (30 mL). Dissolve 15 mg Dispase in 30 mL DMEM/F12. Filter with 0.22 μm PES membrane. Store at 4°C for up to 1 month.
15. Matrigel (Becton Dickinson Labware).
16. hESC Cryopreservation Medium (2×, 50 mL). Mix 10 mL hESC medium, 30 mL FBS (Hyclone, Defined), and 10 mL Dimethyl Sulfoxide (DMSO). Filter with 0.22 μm PVDF membrane. Use it the same day, or store aliquots at −80°C.
17. Conditioned Medium (CM) plus 4 ng/mL bFGF (100 mL). Seed irradiated MEF cells at ~55,000 cells/cm^2 in MEF Culture Medium in T75 flask, incubate them at least 4 h or overnight to allow MEF attachment. Aspirate the MEF medium, add hESC Medium at 0.5 mL/cm^2, culture overnight for conditioning, and collect the CM. Mix 100 mL CM with 200 μL of 2 μg/mL bFGF Solution. Filter with 0.22 μm PES membrane. Store at 4°C and use within 7–10 days, or freeze at −80°C and use within a month.
18. mTeSR1 medium (Stem Cell Technologies).
19. BMP4 (R&D Systems).
20. Brefeldin A, Hoechst 33342 Solution, and mouse IgG (Sigma).
21. Primary Antibodies: Mouse anti-SSEA3 (Developmental Studies Hybridoma Bank), and mouse anti-SSEA4 (Developmental Studies Hybridoma Bank) antibodies. Mouse anti-TRA1-60 (Santa Cruz Biotechnology Inc), mouse anti-TRA1-81 (Santa Cruz Biotechnology Inc.), mouse anti-Oct-3/4 (C-10) antibody (Santa Cruz Biotechnology Inc.), and rabbit anti-NANOG antibody (Abcam Inc.). Mouse anti-human CGβ antibody (Abcam Inc.).
22. Secondary Antibodies: Alexa Fluor® goat anti-mouse IgG, goat anti-mouse IgM, and goat anti-rabbit IgG antibodies (Molecular Probes, Inc.).
23. Quantitative Alkaline Phosphatase ES Characterization Kit (Millipore).

24. Permeablization Buffer: Add 5% normal goat serum (Invitrogen, 10000C), 1% BSA, and 0.1% Triton X-100 in PBS.
25. Blocking Buffer: Add 5% normal goat serum and 1% BSA in PBS.
26. PBS-T: Add 0.4% Triton X-100 in PBS.
27. Fluorescence-activated cell sorting (FACS) Buffer: Add 2% fetal bovine serum (FBS), 0.1% Triton X-100, and 0.1% sodium azide to calcium- and magnesium-free PBS.
28. AxSYM Total hCGβ kit (Abbott, Lake Forest, IL).
29. PBS-T20: Add 0.2% Tween-20 in PBS.

2.2. Supplies

1. Watchmakers' Forceps (Fine Science Tools Inc.).
2. Iris Scissors (Fine Science Tools Inc.).
3. STEMPRO® EZPassage (Invitrogen) or Disposable Cell Scraper (Fisher Scientific).
4. Nalgene Freezing Box (Nalge Nunc International).
5. One mL U-100 Insulin Syringe 28G1/2 (Becton Dickson).
6. Four-well and six-well Culture Plates, 50, 75, and 90 mm Filter Units (Nalge Nunc International).
7. 0.22 mM Filter Unit (Millipore).
8. 40 mM Mesh (BD Labware).
9. Falcon (35/2054) 5 mL Polystyrene Round Bottom Tube (BD Labware).
10. TaqMan® Low Density Array Human Stem Cell Differentiation Panel (Applied Biosystems Inc.).
11. Glass Coverslips (Fisher Scientific).

3. Methods

3.1. hESC Derivation

hESC lines are traditionally derived from the inner cell mass (ICM) of early stage human embryos (days 5–6 after insemination) donated by patients of in vitro fertilization clinics. After zona pellucida digestion by pronase, the ICM is isolated by immunosurgery using anti-human serum antibody followed by exposure to guinea pig complement, and placed onto an irradiated MEF-coated plate for expansion (1, 2). hESC lines can also be derived by placing the ICM onto human fibroblasts (12–14), Matrigel or defined matrices (15, 16), or by isolating single blastomeres from cleavage-stage embryos and placing each of them onto a fluorescent hESC colony as a feeder (17). A surgical method that does not require the enzyme and antibody to isolate the ICM has been reported recently (18, 19). By using this method, we have derived four hESC lines, CT1-4. Below are the detailed procedures for the derivations (see Note 1).

3.1.1. Thawing and Culture of Human Embryos

1. One day prior to embryo thawing, prepare blastocyst culture dishes for thawed embryos as follows. Under a dissecting microscope that is installed within a UV-sterilized hood, make droplets of 50 μL Blastocyst Culture Medium onto a 60 mm culture dish, and cover the droplets with 6 mL paraffin oil. Note: Usually make two droplets on top for washing the embryo and make appropriate droplets on the bottom for embryo culture. Each culture droplet should not contain more than 3 embryos. Make appropriate number of droplets according to the number of embryos to be thawed. The embryos from one donor may not be mixed with those from another in the same plate unless ethical compliance requires so.
2. Equilibrate the dish in incubator at 37°C, 5% CO_2 and 95% humidity overnight or for at least 4–6 h.
3. On the day of embryo thawing, prepare four-well plates as follows 30 min prior to thawing. Add 1 mL 0.5 M Sucrose Solution from the Quinn's Advantage Thaw Kit into the top left well of the plates, 1 mL 0.2 M Sucrose Solution into the top right well, and 1 mL/well of Cryo Solvent into the bottom two wells.
4. Let the plates set at room temperature in a laminar flow cabinet for 30 min. Prepare one plate per thawed vial or straw.
5. In the meanwhile, prepare a flame-pulled open Pasteur pipette as follows. Pull a sterile Pasteur pipette over a flame with one hand holding the cotton end and forceps holding the open end. Bend the narrow part to break the tip to produce an open end approximately 150–250 μm in diameter, which is slightly larger than the diameter of a human embryo (110–120 μM). Check the cutting edge under the dissecting microscope to determine whether the diameter is appropriate. Keep the pulled Pasteur pipette sterile in the hood.
6. Using a pair of forceps, take a cryopreserved vial containing human embryo(s) out of liquid nitrogen. Place the lower part of the vial in a 30°C water bath, gently swirling the vial until ice crystals disappears. Be careful not to let the water emerge the cap of the vial during swirling and not to leave the vial in the water bath once the ice crystals have disappeared. Wipe off the water with paper towel. Spray the vial with 75% ethanol for sterilization, and let the vial air dry in the hood.
7. Based on ethical requirements, de-identify the information on the vial by assigning the thawed embryos with random codes.

8. Quickly remove the content from the vial with a sterile Pasteur pipette, put it onto a 60 mm culture dish, and check the embryos (often more than one embryos is stored per vial) under the dissecting microscope. Note: A healthy embryo often contains bright, similarly sized, and integral blastomeres. In contrast, an unhealthy embryo often contains gray, unevenly sized or fragmented blastomeres. Continue thawing the healthy embryos.
9. Using the pulled Pasteur pipette, transfer the embryos into the prepared four-well thawing plate for serial dilution of embryos in the cryopreservants. Place the embryos into the first well, which contains the 0.5 M sucrose solution, and incubate for 10 min at room temperature (RT). Move the embryos into the second well, which contains the 0.2 M sucrose solution, and incubate for 10 min at RT. Move the embryos into the third well, which contains the Cryo solvent, and incubate for 5 min at RT. Finally, move the embryos into the second well of Cryo solvent (well #4) and incubate for 10 min at 37°C (in a incubator). Note: During the dilution process, embryos are very fragile and easy to be damaged. Be cautious to manipulate them very gently. Do not repeatedly pipette the embryos up and down. If more than one vial is thawed, repeat the thawing process for each new vial.
10. After the final incubation step in the fourth well, use the pulled Pasteur pipette to transfer the embryos from the Cryo Solvent in the thawing plate to washing droplet #1 in the prepared blastocyst culture dish. Wash the embryos by moving them clockwise in the droplet and pipetting them up and down gently.
11. Transfer the embryos into washing droplet #2 and wash them as in droplet #1. Finally, move the embryos to the culture droplet containing Blastocyst Medium. Note: When moving the embryos from one droplet to another, try to carry as little medium as possible from the last droplet to the next.
12. Place the blastocyst culture dish into the incubator to culture the embryos, allowing them to develop to the blastocyst stage. After culturing for 48 h, transfer the embryos into freshly prepared Blastocyst Medium and continue culturing them for an additional 48 h. Note: The fresh media can only maintain the vigorous development of human embryos for a maximum of 48 h.

3.1.2. Preparation of MEF-Coated Plate

1. Harvest exponentially proliferating MEF cells (less than 3 passages in culture after thawing), irradiate them with gamma ray at 40 Gy, dilute them to 0.75×10^4 cell/mL in the MEF Culture Medium, and seed them at 1 mL/well into a gelatin-coated twelve-well plate (see Note 2).

2. Return the plate to the incubator to let the cells attach overnight. When placing the plate in the incubator, gently and straightly move it back and forth and left to right three times (do not swirl the plate) for even spreading. After plating, the MEF-coated plate should be used within 7 days.

3.1.3. Isolation of the Inner Cell Mass from Human Blastocysts

Usually, a human embryo will develop to the blastocyst stage on day 5 or 6 postfertilization. If there is no blastocyst formation on day 7, the embryo is considered to have arrested in development, which is often unlikely for hESC derivation.

1. Wait for the embryos to develop to the expanded blastocyst stage for easy evaluation of the inner cell mass.
2. Under an inverted microscope observe the number of blastocysts formed and record the quality of the blastocysts as described (20). Blastocysts with good quality often contains a clearly formed inner cell mass (ICM) with bright, similarly sized, and nonfragmented cells. Although the healthy blastocysts are more likely to become cell line, we still recommend continuing the derivation with any blastocysts that you may have.
3. Prepare dissection dishes by adding 5 mL prewarmed hESC Medium to a 60 mm culture dish, using one dish per blastocyst.
4. Aspirate the MEF Medium in the twelve well MEF-coated plate, add 1 mL PBS to each well to remove any traces of serum, aspirate the PBS, and add 1 mL hESC Culture Medium per well.
5. Transfer one blastocyst at a time from the blastocyst culture dish to the 60 mm dissection dish by using a 10 μL sterile pipette tip.
6. Locate the blastocyst in the dish under the dissecting microscope and use a 1 mL U-100 Insulin Syringe needle (28½ G) to make two slight, close, and paralleling scratches aside the blastocyst.
7. As shown in Fig. 1, use the needle to carefully push the blastocyst between the two scratches, orientating the embryo such that its ICM is to one side. The scratches will hold the blastocyst via its sticky outer zona pellucida in the right position for the following steps. Using the needle, cut open the blastocyst, which will also break the zona pellucida, and separate the ICM part from the zona pellucida.
8. Using a 10 μL sterile pipette tip, transfer the ICM to one well of the twelve-well MEF-coated plate. Return the twelve-well plate to the incubator. Repeat the above procedures for each blastocyst.

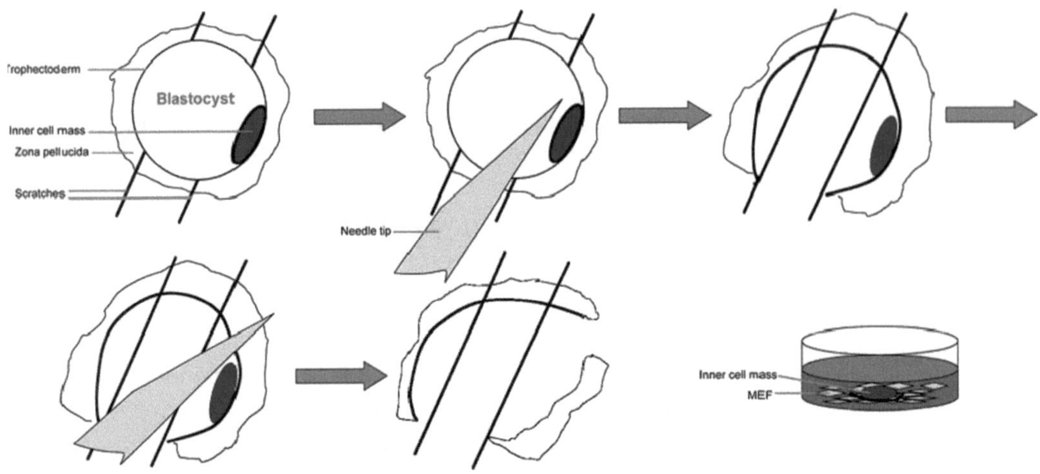

Fig. 1. Diagram for surgical isolation of the inner cell mass from human blastocyst by using a 28G1/2 needle linked to a 1-mL syringe

9. Leave the twelve-well plate undisturbed in the incubator for 48 h to allow the ICM to attach to the MEF layer (to facilitate the attachment, add 10 μM of the ROCK inhibitor Y27632).
10. Carefully refresh the medium every other day until the first passage.

3.1.4. Primary Culture of the ICM

In the primary culture, there are usually trophoblast surrounding the ICM, which may restrict the outgrowth and spread of the ICM cells, especially when the ICM is located in the center of the primary colony. In this case, the ICM may still proliferate but will form a ball- or cylinder-like structure and tend to differentiate. Thus, it is important to separate the entire ICM colony from the surrounding trophoblast prior to its differentiation and replate it as described as follows onto a new MEF-coated well for continued flat outgrowth.

1. Locate the primary ICM colony under dissecting microscope.
2. Use a 1 mL syringe needle to carefully cut and lift the entire ICM colony off from the primary culture dish, and be cautious not to damage the ICM colony.
3. Using a 10 μL sterile pipette tip transfer the ICM colony into a new MEF-coated well containing hESC Medium, and put the plate back into incubator for continued culture.
4. Refresh the medium every other day until a typical hESC colony appears.

3.1.5. Continuing Propagation of the hESC Cells

1. For the first passage, use a 1 mL syringe needle to cut the hESC colony into 2–4 clumps, gently lift the cell clumps, and transfer them into a new MEF-coated well.

2. For the next 2–3 passages, repeat the cutting/transferring steps into a MEF-coated six-well plate.
3. After the hESCs are scaled up to a full six-well plate, use routine hESC culture methods for continuous expansion, maintenance, and cryopreservation (see below).

3.2. hESC Culture

hESCs can be cultured in three kinds of conditions: feeder-dependent (1, 2), feeder-independent (4), and defined cultures (10, 11, 16, 21) (see Note 3). A typical hESC colony is often highly compacted with clear borders, and contains tightly contacted, tiny cells with unclear cell borders. Under high magnification, they have a high nucleus/cytoplasm ratio and clear nucleoli. On the other hand, a colony of differentiated hESCs often has diffused borders, containing loosely contacted and flattened cells with decreased nucleus/cytoplasm ratio. Spontaneous differentiation can occur in hESC culture. In this case, "weeding out" the differentiated cells or "picking up" the undifferentiated cells is often necessary to keep and expand the undifferentiated cells.

3.2.1. Culture in hESC Medium on MEF (1–3)

1. Determine when to split hESCs. In general, cells must be split when the MEF layer is 2-weeks old or when the hESC colonies are too dense or too large, whichever comes the first. These criteria often lead to a splitting frequency of about once a week.
2. Remove the six-well hESC culture plate from the incubator, aspirate the spent media, add 1 mL of 1 mg/mL Dispase Solution to each well of the plate, and incubate for 5–20 min or until the edges of the colonies begin to curl up.
3. Using a glass 5 mL pipette, blow the cell colonies off the surface of the plate. Transfer the cell suspension into a 15-mL conical tube. Gently pipette the cells up and down a few times to further break up the cell clumps. Be careful not to break up the clumps into a single cell suspension and not to cause foaming. Since the enzymatic splitting can promote cytogenetic instability of the cells (22), hESCs may be split mechanically by cutting and pasting or with a disposable stem cell passaging tool called STEMPRO® EZPassage, instead of enzymatically, to directly remove the cell colonies from the plates.
4. Pellet the cell clumps by centrifuging at $200\times g$ for 5 min. Aspirate the supernatant, and wash the cell pellet with 2–3 mL KnockOut DMEM/F12 medium in the 15-mL conical tube via centrifugation of the suspension and aspiration of the supernatants.
5. While the hESCs are centrifuging, aspirate MEF medium from fresh feeder plates, and add about 1 mL PBS to each well of the six-well feeder plate to remove any traces of

the serum. Note: Do not leave PBS on fibroblasts for more than 5 min.

6. Aspirate the supernatant from the hESC pellet, resuspend the pellet in 2–3 mL hESC Medium, and add a sufficient volume of hESC Medium to the 15-mL tube to ensure a total of 2.5 mL medium per well or 15 mL medium per six-well plate. Mix well by gently pipetting.
7. Aspirate the PBS from the wells of the feeder plate, add the cell suspension to each well of the plate as follows: (1) mix the suspension by gently pipetting up and down, (2) take only a portion of it to evenly add it drop-wise into each of the six wells, and (3) do not disperse the last several drops, instead, return them to the tube. Repeat the three steps until each well has 2.5 mL of the cell suspension. If the total volume is less than 2.5 mL/well, add fresh hESC Medium to fill the difference.
8. After plating hESCs, return the plate to the incubator. Move the plate in several quick, short, back-and-forth and side-to-side motions to disperse the cells across the surface of the wells. Note: Open and close incubator door carefully while cells are attaching. This will prevent disturbing the even distribution of cells to the surface of well.
9. Incubate the cells overnight to allow the cell clumps to attach. If cell attachment is a concern, 10 μM Y27632 may be added to the culture for the first night postsplit.
10. Refresh the medium daily until next split when the cell density is about 75% confluent, which usually takes about a week.

3.2.2. Culture in MEF-Conditioned hESC Medium on Matrigel (4)

1. Coat six-well culture plates by incubating with Matrigel (diluted 0.5 mg per six-well plate in cold KnockOut DMEM/F12 at 1 mL/well) and incubate at room temperature for 1 h. Plates may be stored and sealed at 4°C until use.
2. Add 1 mL of 1 mg/mL Dispase Solution to each well of hESCs that are at about 75% confluent in a MEF- or Matrigel-coated six-well plate, and incubate the plate at 37°C for 5–20 min or until the edges of the colonies start to curl up.
3. Using a glass 5 mL pipette, blow the cell colonies off the surface of the plate. Transfer the cell suspension into a 15-mL conical tube, gently pipette the cells up and down a few times to further break up the cell clumps. Be careful not to break up the clumps into a single cell suspension and not to cause foaming. As described above, hESCs may also be split mechanically, instead of enzymatically, with stem cell passaging tool to directly remove the cell colonies from the plates.

4. Wash each well with 1 mL KnockOut DMEM/F12 medium, transfer the cell wash to the 15-mL conical tube, and gently pipette to mix.
5. Pellet the cell clumps by centrifuging at 200 × *g* for 5 min. Aspirate the supernatant, and wash the cell pellet with 2–3 mL KnockOut DMEM/F12 medium in the 15-mL conical tube via centrifugation of the suspension and aspiration of the supernatants.
6. Aspirate the supernatant from the hESC pellet, resuspend the pellet in 2–3 mL CM supplemented with 4 ng/mL bFGF, and add a sufficient volume of the CM plus bFGF to the 15-mL tube to ensure a total of 2.5 mL medium per well or 15 mL medium per six-well plate. Mix gently by pipetting.
7. Aspirate excess Matrigel solution from the Matrigel-coated plates, and add the cell suspension to each well of the plate as described and place the plate in incubate as described in "Culture in hESC medium on MEF."
8. Incubate the cells overnight to allow the cell clumps to attach. If the cell attachment is a concern, 10 μM Y27632 may be added to the culture for the first night postsplit.
9. Refresh the CM plus bFGF daily until next split when the cell density is about 75% confluent, which usually takes about a week.

3.2.3. Culture in mTeSR1 Medium on Matrigel (16, 21)

1. To transit hESCs from other culture media to mTeSR1, replace the previous medium with prewarmed mTeSR1 at 2 mL/well about 3 days before passage. Refresh the medium daily.
2. When hESCs reach about 75% confluence – usually 5–7 days postsplit, aspirate the spent medium and incubate cells with prewarmed 1 mg/mL Dispase at 1 mL/well for 5–10 min at 37°C or until the edges of the cell colonies start to curl up.
3. Aspirate excess Matrigel solution and pipette 2 mL/well prewarmed mTeSR1 medium into the Matrigel-coated plate.
4. When the cell incubation with Dispase is complete, aspirate the Dispase solution and gently rinse the cells on the plate with 1 mL of warmed DMEM/F12 medium/well at least three times, instead of washing the cells by centrifugation. Note: Residual Dispase will reduce the cell attachment after split.
5. After the last rinse, add 2 mL of mTeSR1 medium to the well and gently scrape the cells from the plate, while slowly expelling the medium from a pipette. Plate cells into prepared Matrigel plates at a split ratio of 1:3 or 1:6 or from one original well to 3–6 fresh wells, distribute the cells evenly, and place the plate

into the incubator as described above in "Culture in hESC medium on MEF."

6. Incubate the cells overnight to allow the cell clumps to attach. If the cell attachment is a concern, 10 μM Y27632 may be added to the culture for the first night postsplit.

3.3. Validation and Quality Control of hESCs

Validation and quality control assays are necessary to establish a hESC line and should be started as soon as possible once a primary hESC culture is expanded. These assays typically include pluripotency tests, telomerase activity, DNA fingerprinting, karyotyping, HLA typing, and tests for viral pathogens and mycoplasma. We will mainly introduce pluripotency assays here, but also provide brief information for the other assays. Pluripotency assays often include detection of hESC markers, and both in vitro (e.g., embryoid body formation) and in vivo (e.g., teratoma formation) differentiation ability.

For hESC marker detection, alkaline phosphatase (AP) can be tested as a quick marker for pluripotent cells. Expression of stage-specific embryonic antigen (SSEA) 3 (SSEA-3), SSEA-4, and tumor recognition antigens (TRAs) 1-60 (TRA1-60) and TRA1-81 are widely used as cell surface markers for hESCs. These markers are often detected by immunocytochemistry for visualization and flow cytometry for quantification. The transcription factors OCT4 and NANOG are tested as nuclear markers for hESCs by immunocytochemistry, flow cytometry, and RT-PCR.

3.3.1. Alkaline Phosphatase Detection

hESCs are characterized by the high expression of alkaline phosphatase. Although there are several commercial kits available for the detection of alkaline phosphatase, we have found that the Quantitative Alkaline Phosphatase ES Characterization Kit works the best in our laboratory. Alkaline phosphatase detection is done per the manufacturer's recommendations. Alkaline phosphatase may also be detected by immunocytochemistry (See below).

3.3.2. Immunocytochemistry Detection for Cell Surface Markers in hESCs (Fig. 2)

1. Split hESCs on Matrigel-coated glass coverslips (Fisher Scientific, 12-545-80) that fit into four-well plates, ideally plating 2–5 colonies/well.
2. At 2–3 days post-cell split, aspirate the medium and cover the coverslips with 4% paraformaldehyde (freshly prepared or thawed from frozen aliquots). Fix the cells for 10 min at room temperature. Remove the fixitive, and wash with PBS.
3. Prepare the Permeablization Buffer containing 5% normal goat serum, 1% BSA, and 0.1% Triton X-100 in PBS. Note: Spin down the Permeabilizing Buffer for 10–15 min before use to remove insoluble aggregates to reduce background.
4. Cover the coverslips with the Permeabilization Buffer and incubate for 45 min at room temperature. Remove the

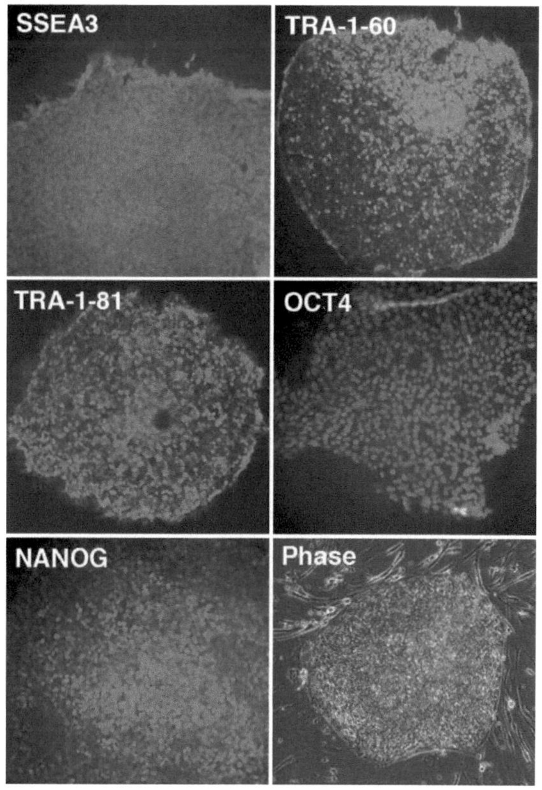

Fig. 2. Immunostaining for pluripotency markers in hESCs

Permeabilization Buffer, and wash each well with PBS three times.

5. Add Blocking Buffer containing 5% normal goat serum and 1% BSA in PBS, and incubate for 30 min at 37°C. Note: Spin down the Blocking Buffer for 10–15 min before use to reduce background.

6. Remove the Blocking Buffer, cover the coverslips with primary antibodies diluted in Blocking Buffer, and incubate at 37°C for 1 h or 4°C overnight, remove the antibodies, and wash with PBS containing 0.4% Triton X-100 (PBS-T) for three times. The primary antibodies include anti-SSEA3 (diluted 1:25), anti-SSEA4 (1:100), anti-TRA1-60 (1:100), anti-TRA1-81 (1:100), and anti-TRA-2-54/2 J (alkaline phosphatases, 1:10). Note: Spin down the antibody-containing Block Buffer for 10–15 min before use.

7. Remove the primary antibodies, and wash with PBS three times. Cover the coverslips with secondary antibodies diluted in Blocking Buffer, incubate them in the dark covered with aluminum foil at 37°C for 1 h, and wash with PBS three times. If using Alexa Fluor® secondary antibodies (Molecular

Probes), dilute the antibodies at 1:200 and spin down for 10–15 min before use.

8. Mount the cells with DAPI to counterstain the nuclei, seal the samples with clear nail polish, and observe under fluorescent microscope for phase and fluorescent images. Samples may be wrapped with aluminum foil and stored at 4°C for up to 1 week.

3.3.3. Immunocytochemistry Detection for Nuclear Markers in hESCs (Fig. 2)

1. Split hESCs on Matrigel-coated glass coverslips that fit into four-well plates, ideally plating 2–5 colonies/well.
2. At 2–3 days postcell split, aspirate the medium and cover the coverslips with 4% paraformaldehyde (freshly prepared or thawed from frozen aliquots). Fix the cells for 10 min at room temperature. Remove the fixitive, and wash with PBS.
3. Prepare PBS-T for cell permeablization.
4. Cover the coverslips with PBS-T and incubate for 10 min at room temperature. Remove the PBS-T, and wash each well with PBS three times.
5. Add Blocking Buffer (5% normal goat serum, 0.2% Tween-20, 0.2% fish skin gelatin) and incubate for 30 min at 37°C. Note: Spin down the Blocking Buffer for 10–15 min before use to reduce background.
6. Remove the Blocking Buffer, cover the coverslips with primary antibodies diluted in Blocking Buffer, and incubate at 37°C for 1 h or 4°C overnight, remove the antibodies, and wash with PBS-T20. The primary antibodies include mouse anti-Oct-3/4 (C-10) antibody (diluted 1:100), and rabbit anti-NANOG antibody (diluted 1:100). Note: Spin down the antibody-containing Block Buffer for 10–15 min before use.
7. Remove the primary antibodies, and wash with PBS-T20 three times. Cover the coverslips with secondary antibodies diluted in Blocking Buffer, incubate them in the dark covered with aluminum foil at 37°C for 1 h, and wash with PBS-T20 three times. If using AlexaFluor secondary antibodies (Molecular Probes), dilute the antibodies at 1:200 and spin down for 10–15 min before use.
8. Mount the cells with DAPI to counterstain the nuclei, and observe under fluorescent microscope for phase and fluorescent images. For later view, seal the samples with clean nail polish, wrap with aluminum foil, and store at 4°C.

3.3.4. Embryoid Body Formation (23, 24)

The in vitro pluripotency assay of embryoid body (EB) formation from hESCs is a simple way to test the pluripotency of the cells. Detection of differentiation markers for the three germ layers, as well as the extra-embryonic cell types such as trophoblast, in the EBs is a sign of the pluripotency of these cells. Although there are many versions of EB formation protocols, we have found that the following protocol used by Li and coworkers (23, 24) works the best in our laboratory.

1. Culture hESCs at about 75% confluence, similar to that prior cell passaging. Aspirate spent medium from hESC culture and add 1 mg/mL Dispase at 1 mL/well of a six-well plate. Incubate for 5–10 min and check for the edges of cell colonies to begin to curl off of the plate. Aspirate dispase.
2. Add 3 mL hESC Medium, gently blow off the hESC colonies from one entire six-well plate, and transfer the cell suspension into a 15-mL tube. Gently triturate three to five times to break the cell colonies into smaller clusters. Clusters should be roughly twice the size as clusters for passaging hESCs.
3. Allow the hESC clusters to settle to the bottom of the tube for 5 min. Aspirate the medium with caution so as not to aspirate the cell pellet.
4. Wash the cells once by adding 5 mL fresh hESC Medium and then centrifuge for 2 min at $200 \times g$.
5. Aspirate the supernatant and resuspend the cells in hESC Medium (for cells from six wells use 60 mL hESC Medium) and transfer them to low attachment T75 flasks. Note: Cell aggregates will initially look unhealthy from shock of separation from feeders. To speed cell recovery, feed for the first time within 12 h and replace most of the medium to remove debris. Switching cells to a new flask is also useful to remove MEF that may have attached during the first 12 h.
6. Continue feeding with hESC Medium every day for 4 days. Note: When feeding, use a 5 mL pipette to gently pull the cell aggregates up and then blow them back into the medium two to three times. This will help clean dead cells off the aggregate surface. Let the clusters settle to the bottom in a standing flask and aspirate off the medium.
7. EBs should be formed within 1 day. Healthy EBs often look bright and are in integral shape, whereas unhealthy EBs often look dark and are in irregular shapes with cell debris. Collect EBs at various times, e.g., 7 or 14 days for early or late differentiation markers or further differentiation purposes. The TaqMan® Human Stem Cell Pluripotency Low Density Arrays contain representative marker genes for pluripotency and various differentiation cell lineages and may be used to test mini-profiles of gene expression via quantitative RT-PCR of hESCs and EB cells differentiated at various times from hESCs (Fig. 3).

3.3.5. Teratoma Formation

The in vivo pluripotency assay of teratoma formation in immunodeficient mice remains the gold standard to test the ultimate

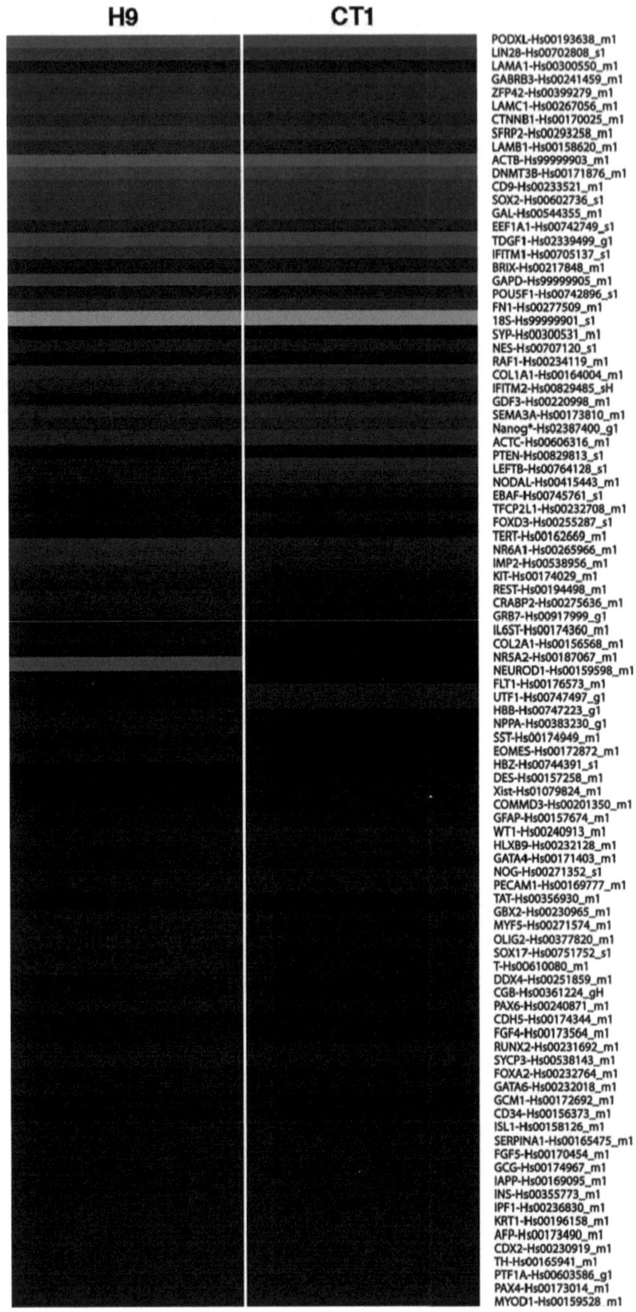

Fig. 3. Heat map for gene expression in hESC lines H9 and CT1 tested by the TaqMan® Low Density Array Human Stem Cell Differentiation Panel. The intensity of the colors correlates to normalized ΔCt values for transcripts of pluripotency and differentiation genes in hESCs. Displayed is the averaged ΔCt value from two biological replicates for each gene. Note: Ct for a referred gene is reversely related to its RNA level. Green color indicates high transcript levels, and red indicates low.

differentiation potential of both hESCs (1). As described below, immunodeficient (SCID-beige) mice are injected with hESCs (see Note 2). The mice will usually form teratomas 2–3 months postinjection. The resultant teratomas are collected and subject to histological examination to confirm the presence of cells from all the three germ layers and extra-embryonic cell types.

1. Grow hESCs on MEF-coated six-well plates with four wells of hESCs per mouse.
2. When the cells reach about 75% confluence, aspirate the spent medium, wash each well with 2 mL PBS.
3. Add 1 mL of 1 mg/mL Dispase to each well, incubate at 37°C for 15–20 min until entire colonies can be easily dislodged by tapping the plate, add 2 mL hESC Medium into each well, and transfer the cell suspension from each four wells into a 15-mL tube.
4. Centrifuge at 200 × *g* for 1.5 min, remove the supernatant, add 10 mL fresh hESC Medium into the tube, and pipette up and down of the cell clumps for eight times to break them down into small clumps.
5. Centrifuge at 12 rpm for 2 min, carefully remove the supernatant, add 100 μL of hESC Medium to resuspend the cell clumps, and transfer the cell suspension into a 1.5-mL microcentrifuge tube for injection. Note: 10 μM Y27632 may be added to the cell suspension before injection to increase the cell survival and terotoma formation efficiency.
6. Carry 1-mL U-100 Insulin Syringe 28G1/2, Sterile Alcohol Prep Pads, and the cell suspensions to the animal facility.
7. Tap the bottom of the cell-containing microcentrifuge tube to disperse the cell suspension before loading the cells to the syringe.
8. Suck the entire cell suspension into a 1-mL U-100 Insulin Syringe. Keep the suspension within the needle side and the air on the syringe side. Do not hold the needle side up to tap out the air bubble from the bottom.
9. Use the Alcohol Prep Pad to sterilize the skin of the left thigh of a SCID mouse (male, 4–6-weeks old), inject all of the cell suspension into the muscles of the left thigh until the air bubble reaches the link between the needle and the syringe.
10. Pull out the needle slowly, and make sure no blood or the cell suspension leaks out of the injection site.
11. Return the mouse to the cage and label the cage with detailed information.
12. Observe all the injected mice weekly, and about 5 weeks postinjection start to feel the injection site for tumor formation.
13. When the tumor grows to about 1 cm in diameter, dissect the tumor as follows.

14. Use the Alcohol Prep Pad to sterilize the skin above the tumor, use scissors to cut open the skin, use forceps to separate the surrounding tissues and expose the tumor, and use the scissors to cut out the tumor.
15. Place the tumor into a Petri dish containing 5 mL PBS, cut the tumor into two halves, transfer both pieces into a 15-mL tube containing 4% paraformaldehyde, and fix the samples at 4°C overnight.
16. Process the fixed samples for paraffin-embedded sections (5 μm thick) on glass slides, and develop the sections for hematoxylin–eosin staining.
17. Observe under a microscope for representative cell types differentiated from all the three germ layers, and take photographs (Fig. 4).

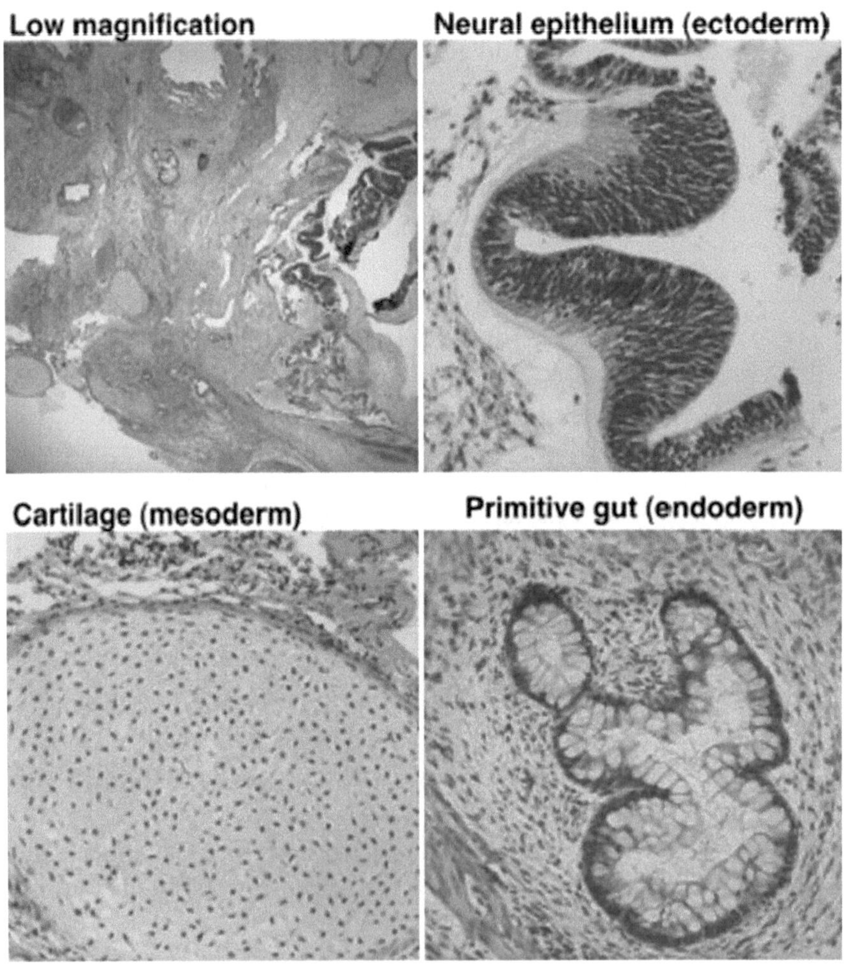

Fig. 4. Hematoxylin–eosin staining of sections from teratomas formed by CT2 hESCs injected into SCID-beige mice

3.3.6. Information for Other Quality Control Assays

1. Test for pathogenic viruses such as HIV, HTLV, HBC, HCV, CMV, EBV, and HSV: GlobelStem Inc. http://www.globlestem.com.
2. Mycoplasma test: MycoAlert Mycoplasma Detection Kit, Cambrex http://www.turnerbiosystems.com/doc/appnotes/PDF/S-0129.pdf.
3. Karyotyping: Cell Line Genetics http://www.clgenetics.com/.
4. DNA fingerprinting: PowerPlex® 1.2, Promega http://www.promega.com/tbs/tmd009/tmd009.html.
5. HLA typing: serology, PCR, or retrieval from transcriptomic database.

3.4. hESC Differentiation to Trophoblast (5)

hESCs can form derivatives of all three embryonic germ layers, as well as extra-embryonic tissues such as primitive endoderm and trophoblast. The use of hESCs to derive early human trophoblast is particularly valuable, because it is difficult to obtain these cells from other sources and they are significantly different from mouse trophoblast.

Here we describe a method by using BMP4, a member of the TGFβ superfamily, to induce the differentiation of hESCs to trophoblast. Immunoassays (as well as DNA microarray and reverse-transcription polymerase chain reaction analyses – data not shown) demonstrate that the differentiated cells express a range of trophoblast markers and secrete placental hormones. When plated at low density, the BMP4-treated cells form syncytia that express chorionic gonadotrophin β (CGβ). This technique underscores fundamental differences between human and mouse ES cells, which differentiate poorly, if at all, to trophoblast. Thus, hESCs are a novel tool for studying the differentiation and function of early human trophoblast and may provide a new understanding of some of the earliest differentiation events of human postimplantation development.

3.4.1. Trophoblast Induction

1. Split hESCs into a six-well plate as described above (see Note 4).
2. When the cells become ~30% confluent (usually on day 2–3 following split), treat the cells with or without 100 ng/mL BMP4.
3. Add fresh media and BMP4 every other day (see Note 5).
4. Observe morphological changes with following features (see Note 6):
    1. On day 2 (48 h) of BMP4 treatment, a synchronous wave of differentiation occurs at the edge of the colonies, and is characterized by flattened, enlarged cell types and reduced proliferation (Fig. 5).
    2. Gradually the differentiation continues inward to the center of the colonies.

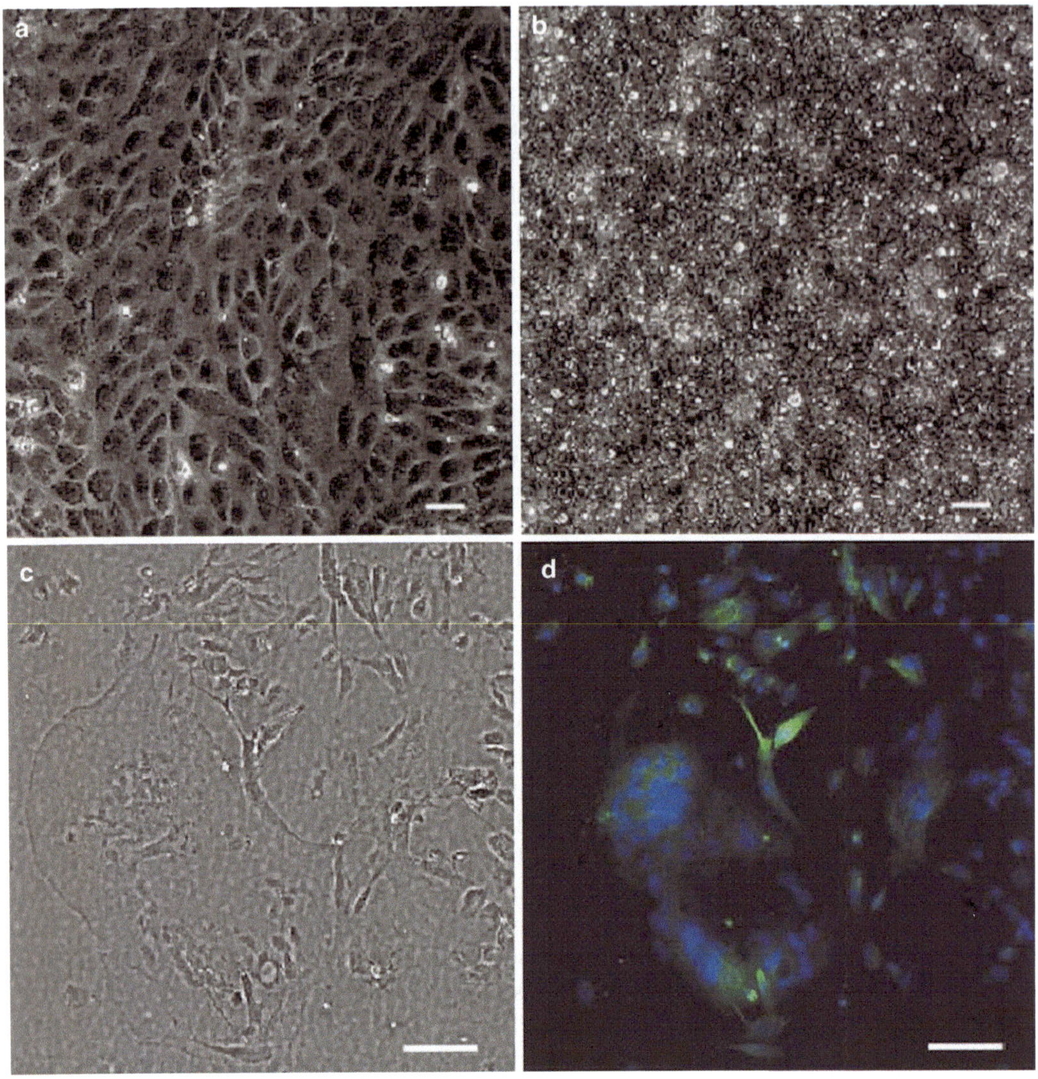

Fig. 5. Trophoblast differentiation from BMP4-treated hESCs. (**a**, **b**) H1 hESCs cultured in MEF-conditioned medium supplemented with 4 ng/ml bFGF were treated with (**a**) or without (**b**) 100 ng/ml BMP4 for 7 days. (**c**) A syncytial cell formed after 2 weeks of treatment of individualized hESCs by BMP4. (**d**) Immunofluorescence for CGβ (*green*) and Hoechst 33342 fluorescence for the nuclei (*blue*). Bars, 25μm.

3. The morphological changes become obvious by day 2 of the treatment for BMP4 at 100 ng/mL, days 3–4 for 10 ng/mL, and days 5–6 for 1 ng/mL.
4. BMP family members such as BMP2 (300 ng/mL), BMP7 (300 ng/mL), and GDF5 (30 ng/mL) induce similar morphological changes to that induced by BMP4 (100 ng/mL).
5. Addition of inhibitors of BMP signaling, such as the soluble BMP receptor IA (500 ng/mL) or the BMP-antagonizing protein Noggin (500 ng/mL), can block the morphological changes induced by the BMPs.

3.4.2. Induction of Syncytial Trophoblast

1. Aspirate spent medium from a well of hESCs cultured in a six-well plate, add 1 mL Trypsin/EDTA Solution to the well, and incubate at 37°C for at least 5 min, then add 1 mL MEF-conditioned medium (CM) to neutralize the Trypsin/EDTA Solution.
2. Scrape the cells from the plate with a glass pipette, transfer the cells to a 15-mL tube, break up the cell colonies by pipetting up and down several times, and add DMEM/F12 medium to a final volume of 10 mL.
3. Pellet the cells by spinning at 200 × g for 5 min, remove the supernatant and resuspend the pellet in CM, which contains mostly single cells.
4. Plate the cells at 10^5 cells/well, and incubate at 37°C overnight to allow the cells to attach.
5. Treat the cells with or without 100 ng/mL BMP4 next day, and add fresh media and BMP4 every other day.
6. Observe the formation of syncytial cells, which usually occurs within 1–2 weeks of the treatment, featuring giant and irregularly shaped cells containing more than two nuclei (Fig. 5). These differentiated phenotypes remain for a long time without obvious changes.

3.4.3. Characterization of the Induced Trophoblast

3.4.3.1. Immuno-cytochemistry

1. Culture hESCs in a four-well plate that contains 0.5 mL CM.
2. Treat the cells with or without 100 ng/mL BMP4 for 7 days as above.
3. On day 7 of the treatment, remove the spent medium from the treated cells, add fresh CM containing Brefeldin A, a Golgi blocker, at 1.25 g/mL, and incubate for 4 h at 37°C.
4. Remove the medium, wash the cells with PBS once, and fix them with 0.5% paraformaldehyde for 10 min at room temperature.
5. Remove the fixative and wash the cells with PBS once.
6. Block and permeabilize the cells by incubating them with PBS-T containing 5% milk powder at room temperature for 30 min.
7. Remove the solution and wash the cells with PBS-T.
8. Add 0.2 mL PBS-T containing mouse anti-human CGβ antibody at 1:100 dilution to the test well, and 0.2 mL PBS-T containing equal amount of mouse IgG to the control well.
9. Incubate the cells at 4°C overnight.
10. Remove the solution, and wash the cells with 0.5 mL PBS-T three times (5 min each time).
11. Add 0.2 mL PBS-T containing Alexa Fluor®-labeled goat anti-mouse IgG antibody at 1:200 to each well.

12. Incubate the cells at room temperature for 30 min.
13. Remove the solution, and wash the cells with 0.5 mL PBS-T three times (5 min each time).
14. Incubate the cells with the Hoechst 33342 solution for at least 5 min to stain the nuclei.
15. Aspirate the solution (to reduce light reflection) and photograph at 20× magnification to observe both phase and fluorescent images under a microscope (Fig. 5).

3.4.3.2. Placental Hormone Measurement

1. Collect 2 mL spent media daily from cultures of BMP4-treated hESCs.
2. Keep the media at –70°C or immediately test them for CGβ concentrations using the AxSYM Total hCGβ kit (see Note 7), and estradiol and progesterone concentrations by ELISA assays.

4. Notes

1. Before hESC derivation, appropriate ethical compliance documents must be obtained. For example, in the United States many institutions have set up embryonic stem cell research oversight committees. Investigators must obtain approvals from the committees regarding ethical compliances of their hESC research projects. To obtain informed consent from embryo donors, investigators must also apply to their institutional review board for its approval of the consent and authorization to use and disclose protected health information of the donors for research purposes.
2. For hESC culture on MEF or in MEF-conditioned medium, out-bred day-12 pregnant CF-1 strain mice are traditionally used to generate the MEF feeder cells. For in vivo pluripotency assay of hESCs, immunocompromised SCID-beige, or NOD-SCID mice are often used to produce teratomas from hESCs inoculated in the mice. For both purposes, investigators must obtain approval of their animal use protocols from institutional animal care committees.
3. All the three culture conditions, MEF, CM, and TeSR1, introduced above can be used to culture hESCs. In addition to economic and biosafety consideration, we use the MEF system for our stem cell core's hESC maintenance and training courses, the CM system for individual researcher's cell maintenance and general research, and the TeSR1 system for specific projects, e.g., study of signaling pathways.
4. For synchronous differentiation of hESCs to trophoblast, hESCs should be passaged as small colonies (about 200 μM

in size), and BMP4 added when the cells are ~30% confluent. Big colonies often end up with the cells in the middle of the colonies remaining undifferentiated.

5. We have observed that the potency of BMP4 added to hESC cultures twice on alternative days is equivalent to that of BMP4 added daily for 7 days, as evaluated by morphology and hCG secretion.
6. According to microarray and reverse-transcription polymerase chain reaction assays, the expression levels of pluripotency- and trophoblast-related genes in the hESCs change dynamically during BMP4 treatment (5). From 3 h through 7 days of BMP4 treatment, the expression of the following trophoblast-related genes are elevated: *TFAP2A, TFAP2C, MSX2, GATA2, GATA3, SSI3, HEY1, FZD, PIGF, CGB, CGA, LHB, GCM1, INSL4, PAEP, PAPPE, DEPP, MET, and HLA-G1*. At day 7, pluripotency marker genes *OCT4* and *TERT* are downregulated.
7. For detection of CGβ expression in trophoblast by immunocytochemistry, it is essential to enhance the signal by pretreating the cells with the Golgi blocker Brefeldin A for 4 h, permeabilizing the fixed cells with Triton X-100, and incubating the cells with anti-CGβ antibody at 4°C overnight.

Acknowledgments

The authors wish to thank the staff of Xu laboratory, former colleagues of Thomson laboratory at University of Wisconsin-Madison and WiCell Research Institute, and Dr. Xue-Jun Li for contributions to the protocols. This work was supported by Connecticut Stem Cell Research Grants #06SCB14 and 06SCD02 to R.X. The contents in this work are solely the responsibility of the authors and do not necessarily represent the official views of the State of Connecticut.

References

1. Thomson, J. A., Itskovitz-Eldor, J., Shapiro, S. S., Waknitz, M. A., Swiergiel, J. J., Marshall, V. S., Jones, J. M. (1998) Embryonic stem cell lines derived from human blastocysts. *Science* 282(5391), 1145–1147
2. Reubinoff, B. E., Pera, M. F., Fong, C. Y., Trounson, A., Bongso, A. (2000) Embryonic stem cell lines from human blastocysts: somatic differentiation in vitro. *Nat Biotechnol* 18(4), 399–404
3. Amit, M., Carpenter, M. K., Inokuma, M. S., Chiu, C. P., Harris, C. P., Waknitz, M. A., Itskovitz-Eldor, J., Thomson, J. A. (2000) Clonally derived human embryonic stem cell lines maintain pluripotency and proliferative potential for prolonged periods of culture. *Dev Biol* 227(2), 271–278
4. Xu, C., Inokuma, M. S., Denham, J., Golds, K., Kundu, P., Gold, J. D., Carpenter, M. K. (2001) Feeder-free growth of undifferentiated

human embryonic stem cells. *Nat Biotechnol* 19(10), 971–974
5. Xu, R. H., Chen, X., Li, D. S., Li, R., Addicks, G. C., Glennon, C., Zwaka, T. P., Thomson, J. A. (2002) BMP4 initiates human embryonic stem cell differentiation to trophoblast. *Nat Biotechnol* 20(12), 1261–1264
6. Xu, R. H., Peck, R. M., Li, D. S., Feng, X., Ludwig, T., Thomson, J. A. (2005) Basic FGF and suppression of BMP signaling sustain undifferentiated proliferation of human ES cells. *Nat Methods* 2(3), 185–190
7. Levenstein, M. E., Ludwig, T. E., Xu, R. H., Llanas, R. A., Vandenheuvel-Kramer, K., Manning, D., Thomson, J. A. (2006) Basic fibroblast growth factor support of human embryonic stem cell self-renewal. *Stem Cells* 24(3), 568–574
8. Xu, R. H., Sampsell-Barron, T. L., Gu, F., Root, S., Peck, R. M., Pan, G., Yu, J., Antosiewicz-Bourget, J., Tian, S., Stewart, R., Thomson, J. A. (2008) NANOG is a direct target of TGFbeta/activin-mediated SMAD signaling in human ESCs. *Cell Stem Cell* 3(2), 196–206
9. Wang, L., Li, L., Menendez, P., Cerdan, C., Bhatia, M. (2005) Human embryonic stem cells maintained in the absence of mouse embryonic fibroblasts or conditioned media are capable of hematopoietic development. *Blood* 105(12), 4598–4603
10. Yao, S., Chen, S., Clark, J., Hao, E., Beattie, G. M., Hayek, A., Ding, S. (2006) Long-term self-renewal and directed differentiation of human embryonic stem cells in chemically defined conditions. *Proc Natl Acad Sci USA* 103(18), 6907–6912
11. Lu, J., Hou, R., Booth, C. J., Yang, S. H., Snyder, M. (2006) Defined culture conditions of human embryonic stem cells. *Proc Natl Acad Sci USA* 103(15), 5688–5693
12. Hovatta, O., Mikkola, M., Gertow, K., Stromberg, A. M., Inzunza, J., Hreinsson, J., Rozell, B., Blennow, E., Andang, M., Ahrlund-Richter, L. (2003) A culture system using human foreskin fibroblasts as feeder cells allows production of human embryonic stem cells. *Hum Reprod* 18(7), 1404–1409
13. Richards, M., Fong, C. Y., Chan, W. K., Wong, P. C., Bongso, A. (2002) Human feeders support prolonged undifferentiated growth of human inner cell masses and embryonic stem cells. *Nat Biotechnol* 20(9), 933–936
14. Inzunza, J., Gertow, K., Stromberg, M. A., Matilainen, E., Blennow, E., Skottman, H., Wolbank, S., Ahrlund-Richter, L., Hovatta, O. (2005) Derivation of human embryonic stem cell lines in serum replacement medium using postnatal human fibroblasts as feeder cells. *Stem Cells* 23(4), 544–549
15. Klimanskaya, I., Chung, Y., Meisner, L., Johnson, J., West, M. D., Lanza, R. (2005) Human embryonic stem cells derived without feeder cells. *Lancet* 365(9471), 1636–1641
16. Ludwig, T. E., Levenstein, M. E., Jones, J. M., Berggren, W. T., Mitchen, E. R., Frane, J. L., Crandall, L. J., Daigh, C. A., Conard, K. R., Piekarczyk, M. S., Llanas, R. A., Thomson, J. A. (2006) Derivation of human embryonic stem cells in defined conditions. *Nat Biotechnol* 24(2), 185–187
17. Klimanskaya, I., Chung, Y., Becker, S., Lu, S. J., Lanza, R. (2006) Human embryonic stem cell lines derived from single blastomeres. *Nature* 444(7118), 481–485
18. Crook, J. M., Peura, T. T., Kravets, L., Bosman, A. G., Buzzard, J. J., Horne, R., Hentze, H., Dunn, N. R., Zweigerdt, R., Chua, F., Upshall, A., Colman, A. (2007) The generation of six clinical-grade human embryonic stem cell lines. *Cell Stem Cell* 1(5), 490–494
19. Zhou, J., Ou-Yang, Q., Li, J., Zhou, X. Y., Lin, G., Lu, G. X. (2008) Human feeder cells support establishment and definitive endoderm differentiation of human embryonic stem cells. *Stem Cells Dev* 17(4), 737–749
20. Bongso, A. (2006) Blastocyst culture for deriving human embryonic stem cells. *Methods Mol Biol* 331, 13–22
21. Ludwig, T. E., Bergendahl, V., Levenstein, M. E., Yu, J., Probasco, M. D., Thomson, J. A. (2006) Feeder-independent culture of human embryonic stem cells. *Nat Methods* 3(8), 637–646
22. Mitalipova, M. M., Rao, R. R., Hoyer, D. M., Johnson, J. A., Meisner, L. F., Jones, K. L., Dalton, S., Stice, S. L. (2005) Preserving the genetic integrity of human embryonic stem cells. *Nat Biotechnol* 23(1), 19–20
23. Li, X. J., Du, Z. W., Zarnowska, E. D., Pankratz, M., Hansen, L. O., Pearce, R. A., Zhang, S. C. (2005) Specification of motoneurons from human embryonic stem cells. *Nat Biotechnol* 23(2), 215–221
24. Li, X. J., Hu, B. Y., Jones, S. A., Zhang, Y. S., Lavaute, T., Du, Z. W., Zhang, S. C. (2008) Directed differentiation of ventral spinal progenitors and motor neurons from human embryonic stem cells by small molecules. *Stem Cells* 26(4), 886–893

Chapter 2

Isolation and Maintenance of Mouse Epiblast Stem Cells

Josh G. Chenoweth and Paul J. Tesar

Abstract

Epiblast stem cells (EpiSCs) are isolated from the postimplantation mouse embryo just after implantation but prior to gastrulation. EpiSCs are pluripotent and provide a tractable, in vitro system to study the processes that function during gastrulation to transition pluripotent cells to their differentiated derivatives. This chapter describes the methods for the isolation and maintenance of mouse EpiSCs. We also describe basic assays used to characterize new EpiSC lines.

Key words: Epiblast stem cells, Postimplantation epiblast, Egg cylinder, Pluripotency, Pou5f1 (Oct3/4)

1. Introduction

Prior to gastrulation, the mouse embryo must maintain a pluripotent population of cells while developing the extraembryonic tissues required for connection to the mother as well as support of the embryo proper. At 3.5 days postcoitus (dpc) the first overt differentiation event of the mouse embryo occurs where the trophectoderm and inner cell mass (ICM) become spatially distinct. The trophectoderm is an extraembryonic structure required for implantation in the female reproductive tract and will form most of the fetal portion of the placenta. The ICM is a small cluster of pluripotent cells that further segregates at 4.5 dpc into two cell types; the preimplantation epiblast and the primitive endoderm. The preimplantation epiblast is pluripotent and is the tissue source for mouse embryonic stem (ES) cell lines (1). After implantation the preimplantation epiblast undergoes morphological and molecular changes to form a radially symmetric, cup-shaped, pseudo-stratified epithelium called the postimplantation epiblast

S. Ding (ed.), *Cellular Programming and Reprogramming: Methods and Protocols*, Methods in Molecular Biology, vol. 636, DOI 10.1007/978-1-60761-691-7_2, © Springer Science+Business Media, LLC 2010

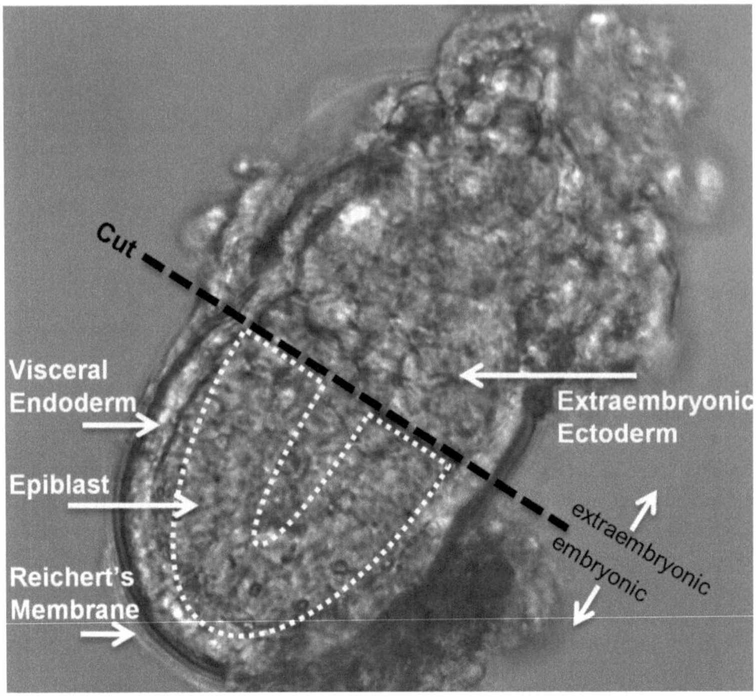

Fig. 1. 5.5 dpc mouse embryo. Tissues are labeled and indicated with *white arrows*. The location of the embryonic/extraembryonic boundary is shown (*black dashed line*). The postimplantation epiblast is outlined with *white dashes*

(also commonly referred to as the primitive ectoderm; see Fig. 1). Around 6.25 dpc the mouse embryo commences the process of gastrulation whereby the postimplantation epiblast differentiates into each of the three classical germ layers, ectoderm, mesoderm, and endoderm, as well as the primordial germ cells, the precursors to the gametes (2–4). The postimplantation epiblast, therefore, is the most proximal, pluripotent precursor to all the cell types of the embryo proper and is the ideal tissue in which to study the transition from pluripotency to the differentiated state. The molecular mechanisms regulating the postimplantation epiblast are currently not well defined. This is due to the difficulty accessing this early postimplantation stage of development and the lack of a robust in vitro model system.

Research on the biology of mammals has been transformed by the in vitro utilization of stem cell lines directly from the mouse embryo. These cell lines provide access to events that normally take place in vivo within the developing embryo and are not amenable to direct investigation. In addition, the early embryo contains very few cells and these derivative stem cell lines can provide the large source of cells necessary for biochemical and molecular analyses. The recent derivation of stable cell lines from the postimplantation epiblast, EpiSCs, has provided a unique system to interrogate the processes that function during gastrulation to generate all three germ layers as well as primordial germ cells (5, 6).

EpiSCs are isolated by direct culture of the postimplantation epiblast. These cells can be indefinitely propagated in vitro and maintained in an undifferentiated state. EpiSCs differ significantly from the more generally studied preimplantation-derived mouse ES cells (1, 7–9). These two pluripotent cell types have divergent global, transcriptional profiles illustrated by the restricted expression of Fgf5 in EpiSCs and Zfp42 (Rex1) in mouse ES cells. Both cell types express the pluripotency transcription factor Pou5fl, yet this is achieved through the use of distinct upstream enhancer elements. Related to this, mouse ES cells and EpiSCs are dependent upon different signal transduction pathways to maintain the pluripotent state.

EpiSCs recapitulate the molecular mechanisms functioning in the postimplantation epiblast, the most immediate precursors to the differentiated derivatives in the gastrulating mouse embryo. This trait makes them uniquely capable of serving as a robust model system to study the direct transition from pluripotency to differentiated derivatives. This chapter will detail the methods required to derive EpiSC lines as well as basic assays required for their characterization.

2. Materials

2.1. Mouse Embryonic Fibroblasts

1. Timed pregnant (see Note 1) CF1 strain mouse (Charles River) at 13.5 dpc.
2. Dissection tools: forceps (see Note 2) and scissors.
3. PBS: Phosphate-Buffered Saline without calcium and magnesium. PBS is stored at room temperature.
4. Dissection Medium (see Note 3): FHM Hepes-buffered medium. Dissection Medium is stored in aliquots at −20°C for up to 6 months. Once thawed, Dissection Medium can be used for up to 2 weeks.
5. Coating Solution: 0.1% gelatin Type A from porcine skin (w/v) in tissue culture grade water, dissolved and sterilized by autoclaving. Coating Solution is stored at room temperature and used within 3 months.
6. Mouse Embryonic Fibroblast (MEF) Culture Medium: Dulbecco's Modified Eagle Medium (DMEM) supplemented with 10% Fetal Bovine Serum (FBS; see Note 4), 2 mM L-glutamine, and 0.1 mM 2-mercaptoethanol. Antibiotics can also be added at final concentrations of 50 units/ml penicillin and 50 μg/ml streptomycin (see Note 5). MEF Culture Medium is filter sterilized using a 0.22 μm polyethersulfone Stericup-GP filtration unit (Millipore), stored at 4°C, and used for up to 2 weeks.

7. 2× Freezing Medium: 60% Knockout DMEM, 20% DMSO, and 20% FBS. The solution is filter sterilized with a 0.22 μm syringe filter. Excess 2× Freezing Medium can be stored at 4°C and used for up to 1 week.
8. Cryotubes: 1.8 ml.
9. Cryo-freezing Container: Nalgene "Mr. Frosty" (Nalgene). Container must be filled with isopropanol prior to use. Isopropanol should be changed after four uses. Between uses, store container with isopropanol at room temperature.
10. MEF Passaging Medium: 0.25% trypsin and 0.38 g/L EDTA-4Na in Hank's Balanced Salt Solution without $CaCl_2$ and $MgCl_2$. Trypsin/EDTA is thawed, aliquoted, and refrozen at –20°C for up to 1 year. Thawed aliquots can be used for up to 2 weeks when stored at 4°C.
11. Counting Chamber: Neubauer-improved, bright-lined (Marienfeld).
12. Tissue Culture Incubator: humidified and maintained at 37°C and 5% CO_2 in air.
13. Irradiator (see Note 6): cesium source or cabinet X-ray device (Faxitron).
14. Stereomicroscope: Leica MZ9.5 with transmitted light base.

2.2. Epiblast Isolation

1. Timed pregnant female mouse at 5.5 dpc (see Note 7).
2. Mouth pipette apparatus (see Note 8): tube assembly, 0.22-μm syringe filter, and latex tubing to adapt the end of the tube assembly to a Pasteur pipette.
3. Glass pipettes (see Note 9): siliconized Pasteur pipettes (Bilbate).
4. Glass scalpels (see Note 10): glass capillaries with an outer diameter of 1.00 mm and an inner diameter of 0.75 mm (World Precision Instruments). The central portion of each capillary is fused over a flame and a tapered cutting surface is created using an electrode puller (Sutter Instruments).
5. Dissociation Medium: 0.5% trypsin (w/v) and 2.5% pancreatin (w/v). Powders are stored at –20°C indefinitely. Combine 0.025 g of trypsin and 0.125 g of pancreatin in a 15-ml conical tube. Add 5 ml of PBS, lay the tube on its side and secure to a benchtop shaker with tape, and agitate at high speed for 1–2 h to dissolve the powders (see Note 11). Filter the solution through a 0.22-μm syringe filter and store at 4°C until use. Dissociation Medium can be stored for up to 3 days at 4°C or for 3 months at –20°C.
6. EpiSC Culture Medium: Knockout DMEM supplemented with 20% Knockout Serum Replacement (KSR), 5 ng/ml

recombinant human FGF2 (R&D Systems), 2 mM L-glutamine, 0.1 mM 2-mercaptoethanol, and 1× nonessential amino acids. Bottles of KSR are thawed upon arrival, aliquoted, and refrozen at −20°C for up to 6 months. FGF2 is dissolved in 0.1% Bovine Serum Albumin Fraction V (w/v) in PBS to a concentration of 10 ng/µl. FGF2 aliquots are stored at −80°C and used within 6 months. Once thawed, FGF2 aliquots are stored at 4°C and used within 2 weeks.

2.3. EpiSC Culture

1. EpiSC Passaging Medium: 1.5 mg/ml collagenase type IV dissolved in EpiSC Culture Medium (without FGF2) and filter sterilized. EpiSC Passaging Medium is stored at 4°C and can be used for up to 2 weeks.
2. Nunc culture plates and flasks with the Nunclon Δ-treated surface are used for the culture of EpiSCs and MEFs (six-well; 24-well; T175 flask).

2.4. EpiSC Characterization

2.4.1. RNA Extraction

1. TRIzol Reagent. Store at 4°C and use in a chemical safety fume hood. TRIzol contains phenol so treat waste as hazardous material.
2. Phase Lock Gel (PLG) Tubes. Store at 15–25°C and do not freeze.
3. Chloroform. Use in a chemical safety fume hood.
4. Isopropanol. Store in a flammable liquids cabinet below 37°C.
5. Diethyl pyrocarbonate (DEPC) water.
6. RNA Precipitation Salt Solution: 0.8 M Sodium Citrate, 1.2 M NaCl in DEPC Water. Store at room temperature in RNase-free tubes.
7. 75% Ethanol in DEPC Water (v/v).
8. DNA-*free* (Ambion).
9. NanoDrop ND-1000 Spectrophotometer (Thermo Scientific).
10. DNase-free\RNase-free\Sterile 1.6-ml microcentrifuge tubes and RNase-free filtered micropipette tips.

2.4.2. Semiquantitative Reverse Transcription Polymerase Chain Reaction

1. Superscript III Reverse Transcriptase 200 U/µl. This product also includes 0.1 M DTT and 5× First-strand Buffer: 250 mM Tris–HCL (pH 8.3), 375 mM KCL, 15 mM $MgCL_2$. Store the Reverse Transcriptase in a manual defrost freezer at −20°C.
2. Random Primers 3 µg/µl. Store at −20°C.
3. Taq DNA Polymerase 5 U/µl. This product also includes a 10× Polymerase Chain Reaction (PCR) Buffer. Store at −20°C in a manual defrost freezer.

4. PCR Nucleotide Mix 10 mM. Store at −20°C in a manual defrost freezer. Limit freeze/thaw cycles and aliquot the nucleotide mix if it will be used repeatedly.
5. PCR grade water.
6. Forward and reverse PCR Primers for mouse *Fgf5*, *Pou5f1* (*Oct3/4*), *Zfp42* (*Rex1*) and *Actin* genes. Prepare 10 μM working stocks in PCR grade water and store at −20°C. Primer sequences are as follows (5, 10): *Pou5f1*, forward 5′-CGT TCT CTT TGG AAA GGT GTT C-3′, reverse 5′-GAA CCA TAC TCG AAC CAC ATC C-3′; *Zfp42*, forward 5′-TGA AAG TGA GAT TAG CCC CGA G-3′, reverse 5′-GTC CCA TCC CCT TCA ATA GCA C-3′; *Fgf5*, forward 5′-CTG TAC TGC AGA GTG GGC ATC GG-3′, reverse 5′-GAC TTC TGC GAG GCT GCG ACA GG-3′; *Actin*, forward 5′-CTA GAC TTC GAGCAG GAG ATG GC-3′, reverse 5′-TCT GCA TCC TGT CAG CAA TGC C-3′.
7. RNase-free\DNase-free\Sterile 1.6-ml microcentrifuge tubes and 0.2-ml PCR tubes.
8. Thermal Cycler.

2.4.3. Oct3/4 Luciferase Reporter Assay

1. Plasmids: *Oct3/4* DE-SV40-Luc and *Oct3/4* PE-SV40-Luc (5), pGL3-Promoter (Promega), and pRL-TK (Promega).
2. Tris–EDTA (TE) pH 8.0: Tris base 1.21 g/L (10 mM); disodium EDTA $2H_2O$ 0.37 g/L (1 mM).
3. Mouse Embryonic Stem Cell Nucleofector Kit (Amaxa).
4. Nucleofector (Amaxa).
5. Dual-Glo Luciferase Assay System (Promega).
6. 96-well white Microlite microplates (Thermo Scientific).
7. Plate reader capable of quantitatively detecting luminescence such as the Wallac Victor^3 Plate Reader (Perkin Elmer).
8. RNase-free\DNase-free\Sterile 1.6 ml microcentrifuge tubes.

3. Methods

3.1. MEFs

3.1.1. MEF Isolation

1. Prewarm MEF Culture Medium at 37°C. You will need ~5 ml of medium per embryo.
2. Euthanize female mouse at 13.5 dpc according to local animal care requirements.
3. Cut open abdominal cavity, transfer the uterus to a 100 mm petri dish filled with PBS, and cut away excess fat and connective tissue from the uterus.

4. Transfer the uterus to a 100-mm petri dish filled with Dissection Medium and, using a stereomicroscope, carefully dissect out the embryos by tearing open the uterine muscle layers and extraembryonic membranes with forceps. Ensure embryos are at the correct developmental stage by consulting the Atlas of Mouse Development (7).
5. Transfer embryos to a clean dish of Dissection Medium. Pinch off the heads and remove spinal cords and all internal organ primordia.
6. Transfer the bodies to a clean dish of PBS and move to a sterile hood.
7. Transfer bodies to a 60-mm petri dish containing 5 ml of 0.125% trypsin/EDTA (0.25% trypsin/EDTA diluted 1:1 with PBS) and using a pair of sterile forceps and a sterile razor blade, mince the bodies thoroughly (see Note 12).
8. Use a 5-ml pipette to transfer minced tissue suspension to a 50 ml conical tube and triturate vigorously for 15–20 s.
9. Rinse the dish with an additional 5 ml of 0.125% trypsin/EDTA and add to the tube.
10. Incubate suspension at 37°C in a water bath for 8–10 min.
11. Triturate vigorously with a 5-ml pipette. Large clumps should be broken up leaving a mostly single-cell suspension.
12. Add an equivalent volume of MEF Culture Medium to the cell suspension to inactivate the trypsin. Do not try to pellet the cells at this point. The trypsin is sufficiently inactivated by the serum in the MEF Culture Medium.
13. Add 25 ml of prewarmed MEF Culture Medium to each T175 flask required (one flask for every five embryos).
14. Divide the cell suspension equally among the flasks and incubate in a Tissue Culture Incubator. This is the primary plating and is denoted as passage 0.
15. Change the medium the next morning to remove excess debris.

3.1.2. MEF Passaging, Freezing, and Thawing

1. Passage 0 MEFs require passaging when they are nearly confluent. This is initially 1–3 days after the primary plating and depends on the initial plating density and survival (see Note 13). Cells are passaged 1:4 (each passage 0 flask is split into 4). To passage, remove medium from each T175 flask, rinse with 5 ml PBS, add 3 ml MEF Passaging Medium, and incubate for 5 min at 37°C. Add 7 ml MEF Culture Medium to each flask and pipette to a single-cell suspension. Spin cells down at 200 × *g* for 5 min, resuspend the cell pellet in MEF Culture Medium, and distribute evenly among flasks containing 25 ml of prewarmed MEF Culture Medium.

2. When passage 1 cells are confluent, trypsinize cells as in Step 1 of Subheading 3.1.2, count using a Counting Chamber, pellet cells at 200 × *g* for 5 min, and resuspend in MEF Culture Medium at 14×10^6 cells/ml. Add an equal volume of 2× Freezing Medium, mix gently by inverting the tube three times, and distribute 1 ml of cell suspension into individual cryotubes (final concentration 7×10^6 cells/ml). Place tubes in cryo-freezing container and store at −80°C. The following day, transfer vials to a liquid nitrogen freezer for permanent storage.
3. Test thaw one vial from the batch of MEFs to determine survival percentage and to take samples for mycoplasma testing (see Notes 14–15). Usually 1 vial thawed into a T175 flask will be confluent (~25×10^6 cells) in 2–3 days. To thaw, remove vial from liquid nitrogen storage and thaw in 37°C water bath until almost completely melted. Gently add cell suspension to 8 ml of room temperature MEF Culture Medium. Rinse vial with 1 ml of MEF Culture Medium and combine with cell suspension (total of 10 ml). Spin down at 200 × *g* for 5 min, remove supernatant, and resuspend pellet in 5 ml MEF Culture Medium. Add cell suspension to a T175 flask containing 25 ml of prewarmed MEF Culture Medium. Place flask in a Tissue Culture Incubator.

3.1.3. Mitotic Inactivation of MEFs

1. Add enough Coating Solution to cover the bottom of each culture flask/plate that is required for your experiment. Incubate flasks/plates with Coating Solution at 37°C for 1–2 h (see Note 16).
2. Remove medium from a nearly confluent T175 flask of MEFs, rinse flask with 5 ml of PBS, add 3 ml of MEF Passaging Medium, and incubate at 37°C for 5 min.
3. Add 7 ml of MEF Culture Medium to inactivate the trypsin, triturate to a single-cell suspension, and spin down at 200 × *g* for 5 min.
4. Remove supernatant and resuspend pellet in 12 ml of MEF Culture Medium.
5. Transport the tube of cells to the irradiator and provide 30–60 Gy of radiation to mitotically inactivate the cells.
6. Resuspend suspension in case any settling has occurred during irradiation and count cells using a Counting Chamber.
7. Remove coating solution from plates/flasks and plate irradiated MEFs at 4×10^4 cells/cm² (see Note 17). We routinely plate primary epiblast explants onto irradiated MEFs in individual wells of a 24-well plate and culture our established EpiSC lines on irradiated MEFs in six-well plates.

3.2. Isolation and Culture of the Postimplantation Epiblast

3.2.1. Isolation of 5.5 dpc Postimplantation Mouse Embryos

1. Euthanize female mouse at 5.5 dpc according to local animal care requirements.
2. Cut open abdominal cavity and transfer the uterus to a 100-mm petri dish filled with PBS. Cut away excess fat and connective tissue from the uterus.
3. Transfer the uterus to a 60-mm petri dish filled with Dissection Medium and, under a stereomicroscope, carefully free each deciduum from the uterine muscle layers using properly sharpened forceps.
4. Carefully tease apart each deciduum to reveal each egg cylinder stage embryo (see Note 18). A representative 5.5 dpc embryo can be seen in Fig. 1.
5. Transfer each embryo to the center of a 60-mm petri dish filled with Dissection Medium.

3.2.2. Separation of Tissues from 5.5 dpc Mouse Embryos

1. Using properly sharpened forceps carefully reflect Reichert's membrane without damaging the embryo.
2. Using a glass scalpel, cut the embryo at the embryonic/extraembryonic boundary as shown in Fig. 1. Discard the extraembryonic fragment.
3. Transfer embryonic fragment (epiblast with overlying visceral endoderm) to a drop of cold Dissociation Medium and incubate for 5–8 min at 4°C.
4. Return embryonic fragment to Dissection Medium and allow it to rest for 5 min.
5. Hand-pull a glass pipette whose inner diameter is slightly smaller than the width of the embryonic fragment. Using a mouth pipette apparatus for fine control, carefully draw the embryonic fragment into the pipette (cut end first). The epiblast should enter the pipette and the visceral endoderm should peel away (see Note 19).

3.2.3. Plating and Culture of Primary Epiblast Tissue

1. Transfer each epiblast fragment to an individual well of a 24-well plate containing irradiated MEFs prepared the previous day (see Note 20). Primary epiblast fragments are grown in EpiSC Culture Medium (see Note 21).
2. Incubate overnight in a Tissue Culture Incubator and examine attachment and morphology of each colony the following day (see Note 22). Do not change medium on day 1.
3. Change medium on day 2. A representative image of a day 2 epiblast colony is shown in Fig. 2a.

3.3. EpiSC Culture

3.3.1. Passage of Primary Epiblast Outgrowth

1. After 2–3 days of growth, the epiblast outgrowth is ready for passage (see Fig. 2b). Under a stereomicroscope in a sterile hood, use a glass pipette or 30-gage needle to cut the colony into 4–6 equally sized pieces.

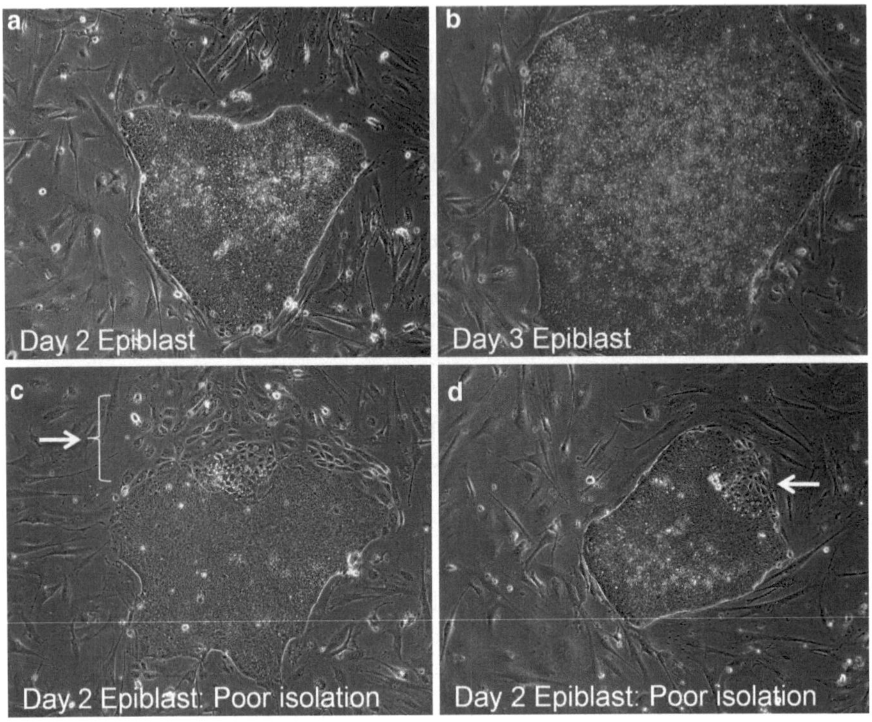

Fig. 2. Culture of primary postimplantation epiblast fragments. (*a*) Morphology of a primary, postimplantation epiblast colony after 2 days of culture on a layer of irradiated MEFs. Notice the flat, epithelial morphology and sharp, defined colony borders showing no evidence of differentiation. (*b*) Morphology of epiblast colony shown in (*a*) after 1 additional day of culture (3 days total). (*c–d*) Representative examples of cultured postimplantation epiblast explants where extra-embryonic tissues were not completely removed from the epiblast fragment prior to plating. *White arrows* denote areas of extraembryonic cell growth. It is much more difficult to derive EpiSC lines from epiblast explants that are not isolated cleanly as these colonies tend to differentiate rapidly

2. Transfer pieces to a fresh well of irradiated MEFs and culture in EpiSC Culture Medium for 2–3 days. Change medium daily.
3. Under optimal conditions, you can expect each primary epiblast explant to yield an EpiSC line. There is not a derivation process per se.

3.3.2. Routine Passaging of EpiSCs

1. After the first passage, EpiSCs can be passaged enzymatically. We culture our EpiSC lines in six-well plates. To passage, remove EpiSC Culture Medium from each well, add 1 ml of EpiSC Passaging Medium, and incubate at 37°C for 8–12 min (see Note 23).
2. Add 1 ml of EpiSC Culture Medium to each well and dislodge colonies away from irradiated MEFs by gentle pipetting with a 5-ml pipette. Combine colony suspensions from each plate into a 15-ml conical tube.

3. Separate colonies away from any remaining MEFs by centrifugation at 200×*g* for 15 s. The colonies will loosely pellet at the bottom of the tube while the individual MEFs will remain in the supernatant.
4. Aspirate the supernatant and discard. Gently resuspend the colony pellet in 5 ml EpiSC Culture Medium and perform another 15 s spin. Repeat rinse/spin cycle one more time for a total of three. FGF2 is omitted from EpiSC Culture Medium when used for rinsing.
5. Resuspend colonies in 100 μl of EpiSC Culture Medium and transfer to an individual well of a 96-well round bottom plate.
6. Triturate colonies with a P100 pipette set at 80 μl until only small clusters remain (see Note 24).
7. Add triturated suspension to prewarmed EpiSC Medium and plate 2.5 ml per well. EpiSCs are typically split 1:4–1:6 every 2–3 days. The density at which each EpiSC line is grown is very important and must be determined empirically.
8. EpiSC Culture Medium is changed daily.

3.3.3. Cryopreservation of EpiSCs

1. Isolate colonies free of MEFs as described in Subheading 3.3.2.
2. Triturate into small clusters just slightly larger than for passaging and spin down at 200×*g* for 5 min.
3. Resuspend pellet in MEF Culture Medium at a concentration of 0.5 ml for every 35 mm well being frozen (see Note 25). Add an equal volume of 2× Freezing Medium, mix gently by inverting the tube, and distribute 0.5 ml of cell suspension into individual cryotubes (final volume is 0.5 well/vial). Place tubes in cryo-freezing container and store at −80°C. The following day, transfer vials to a liquid nitrogen freezer for permanent storage.

3.3.4. Thawing of EpiSCs

1. Remove vial from liquid nitrogen storage and thaw in 37°C water bath until almost completely melted. Gently add cell suspension to 8.5 ml of room temperature MEF Culture Medium. Rinse vial with 1 ml of MEF Culture Medium and combine with cell suspension (total of 10 ml). Spin down at 200×*g* for 5 min, remove supernatant and gently resuspend pellet in 5 ml of prewarmed EpiSC Culture Medium. Remove MEF Culture Medium from MEFs irradiated the prior day, rinse with PBS, and aliquot suspension equally into 2, 35 mm wells. Place in a Tissue Culture Incubator.
2. If recovery is adequate, colonies should be ready to passage in 2–3 days.

3.4. Basic Characterization of EpiSCs

3.4.1. Total RNA Extraction from EpiSCs

1. Harvest colonies away from MEFs as described in Subheading 3.3.2. Transfer approximately 1×10^6 cells to a 1.6 ml RNase-free microcentrifuge tube and spin for 2 min at 1,000 × *g* to collect the cells. Remove the supernatant and resuspend the cells in 0.5 ml of TRIzol (see Note 26). These samples can be stored in the TRIzol at −80°C for later processing, or you can continue with the RNA isolation immediately (see Note 27).
2. Prepare one PLG tube for each sample by spinning each for 3 min at 13,000 × *g* in a microcentrifuge. The gel will collect at the bottom of the tube. Add 0.5 ml of sample-containing TRIzol to a prepared PLG tube followed by 0.1 ml of chloroform. Vortex the samples vigorously for 15 s and spin for 3 min at 13,000 × *g* in a microcentrifuge.
3. Transfer the aqueous phase to a new RNase-free microcentrifuge tube (see Note 28). Add 125 μl of isopropanol and 125 μl of the RNA Precipitation Solution. Mix by inversion and spin for 20 min at 22,000 × *g* in a microcentrifuge at 4°C. Thoroughly remove the supernatant from the resulting pellet with a micropipette (see Note 29). The pellet will be clear and may not stick well to the tube so avoid vacuum aspiration and take great care when removing the supernatant. Save the supernatants so the RNA can be recovered in case the pellet is accidentally transferred.
4. Add 500 μl of ice-cold 75% ethanol in DEPC water to the pellets and wash by inversion of the tube 4–6 times. Collect the pellet by spinning for 1 min at 22,000 × *g* in a microcentrifuge at 4°C. Remove the supernatant and repeat the wash once more. After removing the supernatant, allow the isolated RNA pellet in the microcentrifuge tube to air-dry for 5 min followed by resuspension in DEPC water. A yield of approximately 10–15 μg of RNA is expected from 1×10^6 EpiSCs depending upon cell line and extraction. A resuspension volume of 25–50 μl should result in a suitable working concentration (approximately 400 ng/μl).
5. DNase treat the RNA samples with the DNA-*free* reagent according to the manufacturer's protocol (see Note 30) and quantify the samples with a NanoDrop spectrophotometer (see Note 31). RNA can be stored at −80°C. Prepare aliquots to avoid the introduction of RNase contamination by repeated use of any one individual sample of RNA.

3.4.2. Reverse Transcription Using Total RNA

1. 200 ng–1 μg of total RNA is generally used for a 20 μl RT reaction to make cDNA. Dilute isolated RNA in DEPC water to a normalized working concentration of 100 ng/μl so RT master mixes can be prepared for the samples to be compared (see Note 32).

2. The RT reaction can be divided into two steps. The first step is target denaturation. The second step is cDNA generation. Master mixes should be made for each step.
3. For the first step, to a 0.2-ml PCR tube add 200 ng of RNA, 250 ng of random primers, 1 μl of the 10 mM dNTP mix, and an appropriate volume of DEPC water to bring the total volume to 13 μl. A master mix that contains the random primers, dNTPs, and water can be prepared if multiple samples are to be analyzed. Always prepare enough master mix for at least one additional sample.
4. Heat the samples in a thermal cycler for 5 min at 65°C. Immediately following the incubation, place the PCR tubes on ice for 1 min.
5. Spin the tubes briefly in a PCR tube centrifuge to collect the samples in the bottom of the tubes. For the second step, to each tube, add 4 μl of 5× First-Strand Buffer, 1 μl of 0.1 M DTT, 1 μl RNaseOUT and 1 μl Superscript III RT. A master mix of these components for multiple samples can be prepared while step one is incubating on ice. If including minus RT controls, prepare a second master mix where the Superscript III RT is substituted with DEPC water (see Note 33). Mix by pipetting and incubate at room temperature for 5 min. Transfer the tubes to a thermal cycler and incubate for 50 min at 50°C followed by reaction inactivation by heating at 70°C for 15 min.
6. Proceed to PCR-based target detection or store cDNA at −20°C for future use.

3.4.3. PCR Detection of Target cDNA

1. All PCR reactions should be prepared on ice in DNase-free 0.2-ml PCR tubes.
2. Two microliter of cDNA is used for each PCR reaction (see Note 34). Prepare a master mix for each primer set used for target detection. Each reaction consists of 5 μl of 10× PCR Reaction buffer containing 15 mM $MgCl_2$, 1 μl 10 mM dNTP mix, 1 μl of the 10 μM forward primer, 1 μl of the 10 μM reverse primer, 0.4 μl Taq polymerase and 39.6 μl PCR grade water. Add 2 μl of cDNA to the PCR tube followed by 48 μl of the reaction master mix. Mix by pipetting and spin briefly in a PCR tube centrifuge.
3. Program the thermal cycler with the following generic program:

   1 initial cycle of –

   95°C for 5 min (denaturation)

   30 cycles of –

   95°C for 30 s (denaturation)

60°C for 30 s (annealing)

72°C for 45 s (extension)

1 final cycle of –

72°C for 7 min (extension)

Hold at 4°C

This program is designed for the *Fgf5* primer set described in the materials section. Make the following modifications to the generic program for the indicated primer sets. *Pou5f1,* lower the annealing temperature to 54°C. *Actin,* lower the annealing temperature to 59°C. *Zfp42,* lower the annealing temperature to 56°C and increase the cycling extension time to 90 s.

4. Analyze 10 μl of each PCR reaction using agarose gel electrophoresis.
5. EpiSCs show a 320-bp band for Oct3/4, a 487 bp doublet for Fgf5, a 378 bp band for Actin, and no band for Zfp42 (see Note 35).

3.4.4. Oct3/4 Luciferase Reporter Assay

1. For each of the reporter constructs, *Oct3/4* DE-SV40-Luc, *Oct3/4* PE-SV40-Luc, and pGL3-Promoter, ~2.5×10^6 cells are used for nucleofection. Equimolar amounts of each construct will be cotransferred with the *Renilla* luciferase expression plasmid pRL-TK to normalize for nucleofection efficiency (see Note 36).
2. The approximate molecular weights of *Oct3/4* DE-SV40-Luc, *Oct3/4* PE-SV40-Luc, and pGL3-Promoter are 4.45×10^6 g/mol, 3.93×10^6 g/mol, and 3.30×10^6 g/mol, respectively. The optimal mass of plasmid DNA for the nucleofection is suggested to be from 2–20 μg. For EpiSCs, 4.5 μg of the largest reporter, *Oct3/4* DE-SV40-Luc, is a good starting point. Using the molecular weights and the concentrations of your reporter plasmid preparations, calculate the volume of each needed so that equal numbers of molecules are used for each reporter. Pipette equimolar amounts of each reporter construct into separate 1.6 ml microcentrifuge tubes.
3. Transfer an appropriate amount of the pRL-TK plasmid to each of the tubes so that a tenth of the number of molecules as compared to each reporter is cotransferred. The approximate molecular weight of pRL-TK is 2.67×10^6 g/mol. Bring the total volume of each tube up to 10 μl with TE.
4. Collect the EpiSCs colonies and isolate them from the MEFs as described in Subheading 3.3.2. Triturate into small clusters and resuspend ~2.5×10^6 cells in sterile 1.6 ml microcentrifuge tubes. Pellet cells at $1,000 \times g$ for

2 min, rinse with PBS, pellet cells again, and resuspend in 90 µl of Mouse ES cell Nucleofector Solution. Add the 10 µl DNA solution to the cell suspension and mix by gently pipetting three times. Transfer the colony suspension to an Amaxa certified cuvette and proceed with the nucleofection per manufacturer's protocol using program A-23 (see Note 37).

5. Following the nucleofection, add the 100 µl cell suspension to 3 ml of prewarmed EpiSC Culture Medium and plate 1 ml/well into three wells of a 24-well plate prepared with irradiated MEFs. Incubate for 48 h under standard EpiSC growth conditions and change the medium daily. This procedure produces three technical replicates for each sample as each of the three wells can be assayed individually. We recommend a minimum of three biological replicates (individual transfections) for each construct.
6. After 48 h the cells are ready to be harvested for analysis. Remove the EpiSC Culture Medium, rinse with PBS, add 250 µl of MEF Passaging Medium (0.25%trypsin/EDTA), incubate for 5 min at 37°C, add 250 µl of MEF Culture Medium, triturate with a P1000 pipette to a single-cell suspension, and transfer each sample to a 1.6-ml microfuge tube. Pellet the cells at 1,000 × *g* for 2 min, thoroughly aspirate off the supernatant, and resuspend each pellet in 75 µl of PBS. Add 75 µl of Dual-Glo Luciferase Reagent and mix by pipetting three times. Transfer all 150 µl of each sample to separate wells of a 96-well white microplate. Protect the plate from light and incubate at room temperature for 10 min. Measure the firefly luminescence using a luminometer to measure total luminescence for each sample for 1 s (see Note 38). This first measurement detects the firefly luciferase expressed from the Oct3/4 plasmids. After the firefly luminescence is recorded, to each well add 75 µl of the Dual-Glo Stop & Glo reagent that has been prepared according to the manufacturer's protocol. Protect the plate from light and incubate at room temperature for 10 min. Measure the luminescence using the luminometer to measure total luminescence for each sample for 1 s. This second measurement detects the *Renilla* luciferase expressed from the pRL-TK control plasmid.
7. Normalize the firefly luciferase activity for each sample to its corresponding *Renilla* luciferase activity.
8. In EpiSCs you should expect an approximate fivefold greater activity from the *Oct3/4* PE-SV40-Luc reporter versus the *Oct3/4* DE-SV40-Luc reporter (see Note 39).

4. Notes

1. Timed pregnant mice can be ordered from commercial suppliers (Charles River, Taconic, etc) or timed matings can be setup in house.
2. Forceps (Dumont #5) and scissors can be purchased from Fine Science Tools. Although often neglected, forceps should be diligently maintained (11). Properly sharpened forceps simplify the dissection of postimplantation embryos and result in less damage to the embryo.
3. Dissection medium can also be prepared in house (11). It is critical that the Dissection Medium is buffered for room air using a non-CO_2-based system and contains protein (BSA or FBS) to prevent sticking.
4. Different lots of FBS have distinct effects on the growth of MEFs. Since we also work extensively with mouse ES cells, we use the same lot of FBS for both MEFs and mouse ES cells. We screen 6–8 lots of FBS each year for mouse ES cell plating efficiency, mouse ES cell colony morphology, and maintenance of Oct3/4 expression. Serum is stored at −20°C or −80°C for up to 1 year.
5. We do not routinely use antibiotics for any of our cultures. Antibiotics are only added to primary cultures for the first 3 days after the initial plating.
6. If access to an irradiator is not feasible, chemical means such as mitomycin-c can be used to mitotically inactivate the MEFs.
7. Unlike mouse ES cells, the derivation of EpiSCs does not seem to be restricted to specific mouse strains.
8. Extreme care must be used when using a mouth aspirator. If you are uncomfortable with this technique or local safety regulations do not permit its use, alternatives such as the Stripper Pipette system (Mid-Atlantic Diagnostics) are available.
9. Pasteur pipettes can also be purchased and acid cleaned and siliconized in house.
10. A more detailed protocol for the production of glass scalpels can be found elsewhere (11).
11. Even after 1–2 h of shaking, not all the trypsin/pancreatin powder will have dissolved. Prior to filtering, the undissolved powder should be settled by gravity or a quick spin in a benchtop centrifuge. The cleared supernatant can be decanted prior to filtration to prevent clogging of the filter.
12. It is best to mince the bodies into small, uniform pieces. Spend no more than 1–3 min on this step to minimize the amount of cell death.

13. If the cells take longer than 3–4 days to reach confluence then your initial plating density was too low or proliferation rate is too slow. It is important not to plate the cells too sparse because then it takes them extra population doublings to reach confluence (so passage 1 cells would actually be more equivalent to passage 2 or 3 due to the excess number of population doublings).
14. All primary cells and cell lines should be mycoplasma tested immediately upon entry into the lab and on a regular basis thereafter. It is preferable to keep cells in a separate incubator until their mycoplasma status is known. We routinely use the MycoAlert Mycoplasma Detection Kit (Lonza).
15. Different batches of MEFs have distinct effects on the growth of EpiSCs (as well as ES cells). It is important to screen each batch of MEFs for the ability to support your EpiSCs. For EpiSCs, we only use MEFs up to Passage 3. In our experience, higher passage MEFs do not efficiently support the undifferentiated state of EpiSCs.
16. Pre-coating culture surfaces with gelatin prior to plating irradiated MEFs is important for EpiSCs. The MEFs tend to adhere to gelatin-coated surfaces more strongly and do not lift off the plate when treated with collagenase. This is vital when isolating EpiSCs away from the MEFs for passaging or analysis.
17. The density of the feeder layer is important to maintaining the undifferentiated state of EpiSCs. We recommend counting your MEFs just prior to plating (not before irradiation).
18. Isolation of early postimplantation mouse embryos is extremely difficult and requires practice and perseverance. Additional diagrams of this procedure can be found elsewhere (12).
19. Additional diagrams of this technique can be found elsewhere (13).
20. Although it is a laborious task, we only culture primary epiblast fragments and established EpiSC lines onto MEFs that were irradiated and plated the prior day.
21. Supplementation of Activin A (R&D Systems) at 10 ng/ml can be beneficial to primary epiblast fragments and established EpiSC lines. This is not typically required since the irradiated MEFs provide adequate stimulation of this signaling pathway. Supplementation with Activin A at 10 ng/ml is required if you decrease the plating density of your irradiated MEFs.
22. It is important that each epiblast fragment is completely isolated away from the extraembryonic tissues (see Fig. 2.2c, d). Any remaining extraembryonic tissue will induce differentiation of your epiblast colony.
23. A variety of factors can affect the amount of time required for the collagenase to free the EpiSC colonies from the plate.

Check the colonies every few minutes. They are ready when the edges begin to retract from the plate.

24. It is important to dissociate the colonies in a well of a 96-well plate so that you can monitor your dissociation. We typically triturate the colonies ten times and then check their size under the microscope. This is repeated until the colonies reach an optimal size. EpiSCs do not survive well as single cells so it is important not to triturate too much.
25. Since EpiSCs are not trypsinized to a single cell suspension, they are not counted for passaging or freezing. The passaging density and freezing density need to be determined empirically.
26. We find 0.5 ml of TRIzol works well to isolate total RNA from 1×10^6 EpiSCs. Too many or too few cells might lead to an inefficient extraction in 0.5 ml of TRIzol.
27. When working with RNA, take care to use RNase-free consumables and reagents and change gloves often. RNA is highly susceptible to degradation.
28. Avoid transferring any of the gel from the PLG tubes with the aqueous phase. Use a micropipette to transfer the RNA containing aqueous phase to a new microcentrifuge tube.
29. Depending on the efficiency of the extraction, the RNA pellet may be very small and loosely attached to the side of the microcentrifuge tube following centrifugation. Position the tubes in a common orientation in the microcentrifuge prior to spinning so you have an idea where the RNA pellet will be. Use a micropipette to carefully take off the supernatant. You might want to do a second quick spin to recollect any supernatant that may have attached to the sides of the tube and remove it prior to proceeding to the washing steps.
30. The DNA-*free* kit removes contaminating genomic DNA and divalent cations from the RNA preparation. Contaminating DNA could interfere with subsequent analysis and divalent cations promote nuclease activity. The kit includes recombinant DNaseI, a 10× DNaseI buffer, and a DNase inactivation reagent. Following a 37°C incubation of your RNA samples in 1× DNaseI buffer and 2 Units of DNaseI, the DNaseI inactivation reagent is added. The proprietary inactivation reagent removes the DNase without phenol\chloroform extraction or precipitation which helps avoid sample loss or degradation.
31. An $OD_{260}\backslash OD_{280}$ ratio of approximately 1.9–2.1 for total RNA is expected. A simple way to assess RNA quality is to run and analyze your sample on an ethidium bromide-stained 1% agarose gel. You should see two sharp high molecular

weight bands that are the 28S and 18S ribosomal RNA bands. Degraded RNA will look like a smear and will lack these sharp ribosomal RNA bands.

32. It is important to normalize RNA from different samples at the beginning of the protocol if they are to be compared using RT-PCR. This allows you to make master mixes of reagents that reduce experimental variability.

33. To control for DNA contamination, it is best to include a minus RT reaction for each sample to be analyzed by semi-quantitative RT-PCR. To do this, prepare duplicate RT reactions for each sample in step one of the RT reaction. In step two, only one of these will receive the reverse transcriptase enzyme, while the second will get just water and be termed the minus RT sample. Because the minus RT sample cannot be made into cDNA by definition, then it should not yield any PCR product in the PCR detection step. If it does, then you most likely have genomic DNA contamination of your RNA preparation.

34. We generally use 10% of the cDNA reaction, in this case 2 μl, for the template in the PCR reaction. This is sufficient to detect the desired targets from a starting amount of 200 ng of total EpiSC RNA. You may increase the amount of cDNA used in the PCR reaction to optimize the detection of additional targets. The Superscript III enzyme used in the reverse transcription (RT) reaction has been developed to prevent interference with the PCR reaction.

35. Consider a cDNA sample from blastocyst-derived mouse ES cells to serve as a positive control for the *Zfp42* primer set since EpiSCs do not express this gene and will give a negative result.

36. It is best to use equimolar amounts of each reporter and not just the same mass of each. The reporters are different sizes and later interpretation of the data will be facilitated by doing this.

37. Also prepare a control nucleofection sample of EpiSCs using the empty pGL-3 vector and analyze it in the same manner as the experimentals. This will give you an idea of background luminescence and the significance of your signal.

38. Consult your plate reader's manual to determine the proper operation to record luminescence.

39. You should perform the assay in a blastocyst-derived mouse ES cell line where you can expect the opposite result as compared to the EpiSCs. In the blastocyst-derived line, the *Oct3/4* DE-SV40-Luc reporter should have an approximate eight-fold greater expression level.

References

1. Brook, F. A., and Gardner, R. L. (1997) The origin and efficient derivation of embryonic stem cells in the mouse; *PNAS* **94**, 5709–12.
2. Diwan, S. B., and Stevens, L. C. (1976) Development of teratomas from the ectoderm of mouse egg cylinders. *J Natl Cancer Inst* 57, 937–42.
3. Lawson, K. A., Meneses, J. J., and Pedersen, R. A. (1991) Clonal analysis of epiblast fate during germ layer formation in the mouse embryo. *Development* **113**, 891–911.
4. Tam, P. P. L., and Zhou, S. X. (1996) The allocation of epiblast cells to ectodermal and germline lineages is influenced by the position of the cells in the gastrulating mouse embryo. *Developmental Biology* **178**, 124–32.
5. Tesar, P. J., Chenoweth, J. G., Brook, F. A., Davies, T. J., Evans, E. P., Mack, D. L., Gardner, R. L., and McKay, R. D. (2007) New cell lines from mouse epiblast share defining features with human embryonic stem cells. *Nature* **448**, 196–9.
6. Brons, I. G., Smithers, L. E., Trotter, M. W., Rugg-Gunn, P., Sun, B., Chuva de Sousa Lopes, S. M., Howlett, S. K., Clarkson, A., Ahrlund-Richter, L., Pedersen, R. A., and Vallier, L. (2007) Derivation of pluripotent epiblast stem cells from mammalian embryos. *Nature* **448**, 191–5.
7. Kaufman, M. H. (1992) The Atlas of Mouse Development, Elsevier Academic Press, London.
8. Martin, G. R. (1981) Isolation of a pluripotent cell line from early mouse embryos cultured in medium conditioned by teratocarcinoma stem cells. *Proc Natl Acad Sci U S A* **78**, 7634–8.
9. Tesar, P. J. (2005) Derivation of germ-line-competent embryonic stem cell lines from preblastocyst mouse embryos. *Proc Natl Acad Sci U S A* **102**, 8239–44.
10. Hirst, C. E., Ng, E. S., Azzola, L., Voss, A. K., Thomas, T., Stanley, E. G., and Elefanty, A. G. (2006) Transcriptional profiling of mouse and human ES cells identifies SLAIN1, a novel stem cell gene. *Dev Biol* **293**, 90–103.
11. Downs, K. M. (2006) In vitro methods for studying vascularization of the murine allantois and allantoic union with the chorion. *Methods Mol Med* **121**, 241–72.
12. Nagy, A. (2003) Manipulating the mouse embryo: a laboratory manual, Cold Spring Harbor Laboratory Press, Cold Spring Harbor, N.Y.
13. Beddington, R. S. (1987) Isolation, culture and manipulation of postimplantation mouse embryos, IRL Press, Oxford.

Chapter 3

Functional Assays for Hematopoietic Stem Cell Self-Renewal

John M. Perry and Linheng Li

Abstract

Stem cells are defined by the ability to self-renew. Specific functional assays have been developed for the rigorous identification and quantification of hematopoietic stem cells (HSCs), making these cells the benchmark in studies of self-renewal. Here, we review the theory behind these functional stem cell tests and discuss important considerations in choosing and designing these assays. Finally, we provide a basic protocol for the serial-dilution assay, a quantitative assay for HSCs, from which individual researchers can construct their own customized protocols utilizing the guidelines discussed.

Key words: Hematopoietic stem cell, Self-renewal, Transplantation, Reconstitution, Repopulation, Competitive repopulation assay, Serial-dilution assay, CRU

1. Introduction

Stem cells possess a combination of two key properties. First, they are multipotent, having the capacity to differentiate into multiple, distinct cell types. Secondly, they can undergo self-renewal, a process which regenerates stem cells (1). The hematopoietic system is maintained throughout an organism's life span by HSCs. Hematopoiesis is a particularly dynamic system containing stem cells, which have the remarkable ability to home from the peripheral circulation to their specific microenvironment or niche and there produce a vast hierarchy of progenitors and precursor cells which ultimately give rise to at least ten distinct lineages of cells (2, 3). Decades of study have produced an immense array of antibodies allowing for identification or enrichment of various hematopoietic cells including HSCs (4). In addition, the ability to isolate very specific populations using fluorescence-activated

S. Ding (ed.), *Cellular Programming and Reprogramming: Methods and Protocols*, Methods in Molecular Biology, vol. 636,
DOI 10.1007/978-1-60761-691-7_3, © Springer Science+Business Media, LLC 2010

cell sorting (FACS), transplant HSCs intravenously, and collect blood samples identifying donor and host cells periodically without sacrificing test animals has made HSCs the most intensively studied and best understood adult stem cell system. Due to these and other advantages, specific assays have been developed for the functional quantification of HSCs, making this system a gold standard in studies of self-renewal.

1.1. Phenotypic Identification of Putative HSCs

HSCs were first identified as being highly enriched in a relatively rare population of lineage marker negative (Lin−), Sca-1+ and Thy1.1^{Low} cells (5). This population proved to be of limited utility because commonly used strains of mice, such as C57BL/6, lack Thy1.1 expression, and the Thy1.2 allele in these strains is not expressed on HSCs. Later, the receptor for stem cell factor, c-Kit, was found to enrich for HSCs, thus identifying Lin−, Sca-1+, c-Kit+, or LSK cells as putative HSCs (6, 7). Further enrichments and hierarchical classifications have since been developed. Some important ones include the use of Flk2 for further purification and separation of long-term reconstituting HSCs (LT-HSC) in the Flk2− fraction and short-term reconstituting HSCs (ST-HSC) in the Flk2+ fraction (8). More recently, the use of CD34 has yielded further refinements with LT-HSCs classified as an even rarer population of LSK Flk2− CD34− cells (9). Many researchers now designate LSK Flk2− CD34+ cells as ST-HSC while their derivatives, LSK Flk2+ CD34+ cells, are termed multipotent progenitors (MPPs) (10). It is important to note that these designations are age-dependent as CD34+ cells contain LT-HSCs in mice younger than 10 weeks, for instance, and many researchers typically utilize 6–8-week-old mice (11). In general, ST-HSCs more rapidly reconstitute hematopoiesis biased toward myeloid differentiation. This is important for short-term survival (up to about 4 weeks), but ST-HSCs are incapable of longer-term reconstitution. Similarly, MPPs reconstitute only transiently but are biased toward lymphopoiesis (12). In addition, initial survival of animals conditioned for transplantation, typically by lethal-dose irradiation, require radioprotective cells. These cells have rapid and high proliferative capacity but no long-term potential. In contrast, while LT-HSCs are the only population with long-term potential, they are incapable of maintaining survival in the short-term. Consequently, both populations are necessary in transplantation assays involving severe myeloablative conditioning.

Numerous other schemes have been utilized for HSC identification. For example, various vital dyes such as rhodamine 123 or Hoescht can enrich for HSCs (13, 14). Efflux of Hoescht is utilized to identify "side-population" or SP cells, which are highly enriched in HSCs. More recently, CD201 (EPCR) and members of the SLAM family have been shown to aid in HSC enrichment, particularly CD150+, CD48− cells (15, 16). Complicated schemes utilizing multiple combinations of the above markers have yielded extremely high purity of functional HSCs from single cells with reports of greater than 50% of single cells yielding long-term, multilineage reconstitution. Regardless of this impressive progress, "HSCs" identified by FACS analysis can, at best, only be considered phenotypic or putative HSCs. Identification of *bona fide* HSCs requires functional testing in order to prove both multipotential and self-renewal capacity. In addition, while the variously identified phenotypic HSCs may meet the functional definition of HSC to different degrees, alternate populations may have previously unidentified variations in properties, such as lineage biases and differences in proliferative capacity, which require future functional investigations (17, 18).

1.2. Functional Assays for HSCs

In vitro functional assays have been developed for HSCs including the cobblestone area-forming cell (CAFC) assay and the long-term culture-initiating cell (LTC-IC) assay (19). In particular, the LTC-IC assay, originally developed for human HSC studies due to the previous lack of an adequate xenotransplantation model, can be quantitative and does not require radioprotective or competitor cells. However, the use of these assays remains controversial. They clearly identify relatively immature cells, but it is unclear whether or not they identify true HSCs. Thus, more rigorous standards involve an in vivo functional assay.

The first functional, in vivo assay for presumptive HSCs involved the colony forming unit-spleen or CFU-S (20). These cells give rise to large hematopoietic colonies which can be observed on the spleen between 1–3 weeks after irradiation. While the CFU-S was long thought to be a stem cell, these cells only have very short-term potential and are progenitors, not HSCs (21, 22). To test for self-renewal and thus stem cell capacity, long-term assays are required and have since been developed.

1.3. Long-Term Repopulation Assay

The long-term repopulation assay involves transplantation of test cells into irradiated or otherwise compromised hosts. Donors and hosts are chosen ideally so that each can be distinguished by different alleles of the hematopoietic cell marker CD45, with host cells expressing the CD45.1 allele and donor cells expressing the CD45.2 allele, for instance. Blood samples are collected at various intervals, and donor vs. host cells are analyzed by FACS for their potential presence and contribution to major hematopoietic lineages, typically including myeloid as well as B and T lymphoid lineages. A test population is considered to contain LT-HSC(s) when a recipient contains donor-derived cells of all these lineages even after at least 16 weeks following transplantation. Modifications of this basic scheme include the use of competitor cells.

1.4. Competitive Repopulation Assay

In this assay, a standard source of HSCs, typically a specific number of whole bone marrow cells derived from mice congenic with the host, are included with the test cells. This allows for the quality of the test cells to be assessed relative to the standard (23, 24). However, it does not allow for rigorous quantification of HSC frequency and number in the test population. For this, a further modification is required.

1.5. Limiting-Dilution Assay

The limiting-dilution assay utilizes a titration of varying doses of test cells (23, 25, 26). The percentage of hosts which fail to engraft donor cells is determined for each dose. With a minimum of three different doses in which both positive and negative recipients are present, a best-fit line is generated by Poisson statistics. This is utilized to calculate the frequency of competitive repopulating units (CRUs) and thus the number of HSCs.

1.6. Serial Transplantation Assay

The most stringent test for stemness is the serial transplantation assay (27). It can utilize any of the above assays. Here, test cells are transplanted into primary recipients as above but then harvested from these primary recipients and transplanted into secondary recipients and further from secondary into tertiary hosts. Only the most primitive HSC can yield long-term, multilineage repopulation in this assay.

There are a number of important considerations when designing the above assays (10). The first major consideration involves the test cells and competitor cells. The number of cells utilized is critical. Typically, $1–2 \times 10^5$ whole bone marrow competitor cells are utilized. It is estimated that 1×10^5 whole bone marrow cells contain about 3–5 long-term multilineage repopulating HSCs. When the number of test cells is very low, a high number of competitors could result in false negatives (19). However, when competitors are also required as rescue cells for myeloablated hosts, there must be an adequate number of these cells for survival of recipients. In this case, the use of compromised competitor cells may be preferred. Originally, bone marrow which had undergone two rounds of serial transplantation was utilized in the competitive repopulation assay in order to reduce the number of CRUs without reducing the radioprotective effect (28). This or similar schemes, such as transplantation of sorted radioprotective cells (without HSCs), can substantially improve engraftment of the test population. Of course, this also compromises the quality readout; however, it can be useful if the purpose is to demonstrate that a rare population possesses the capacity of a true HSC even if it does not compete well with a known standard.

A second major consideration involves the choice and treatment of recipients. Histocompatible mice should be utilized whenever possible; otherwise, immunocompromised recipients are necessary. Numerous stains of immunocompromised mice are available; the more severely compromised ones naturally tend to yield the best engraftment. In addition, myeloablative therapy such as lethal-dose irradiation is necessary in order to empty the HSC niche and thus allow for donor cell engraftment (29, 30). This myeloablative therapy necessitates the use of rescue/competitor cells. However, mutant mouse models exist which allow for engraftment without this preconditioning. Mice with mutations in c-Kit, W/W^V for instance, allow for engraftment of donor cells without irradiation (31, 32). Alternatively, administration of ACK2, an antibody that blocks c-Kit function, can transiently remove endogenous HSCs from their niches allowing for engraftment without myeloablative therapy (30). Generally, there are both advantages and disadvantages to both competitive and noncompetitive assays which should be considered according to the purposes of the experiment. Uncompetitive assays are actually the most clinically relevant; however, they lack quantification relative to a known standard. In contrast, competitive assays allow for rigorous quantification but may yield false negatives, particularly if limited test cell numbers are utilized (19). Theoretically, an HSC could be out-competed in this assay in spite of its capacity as an HSC under normal physiological circumstances in the original donor animal.

The third major consideration involves the time following transplantation at which donor engraftment should be measured. If engraftment is analyzed too early, the assay may be measuring progenitors rather than primitive HSCs. Theoretically, an HSC should be able to repopulate the host throughout its life span; however, this is impractical since hosts can live for several years. In practice, engraftment rates tend to stabilize by about 16 weeks posttransplantation (17). Current standards consider long-term engraftment to be achieved by 16 weeks at minimum (10).

The fourth important consideration involves the threshold for defining positive and negative engraftment. This is particularly important for the limiting-dilution assay where mice are considered either positive or negative and level of engraftment is not considered in calculating CRU frequency. With very low engraftment, it can be difficult to distinguish true engraftment from nonspecific staining. Some researchers somewhat arbitrarily set the threshold for engraftment at 1%. Others have considered true engraftment at 0.1%. Arbitrary thresholds should not be utilized; all thresholds should be empirically based on each experiment. In particular, a control group should be treated as similarly as possible to the experimental group but without injection of any donor-derived cells. Blood from the control group should be collected and processed alongside the experimental group. The average "engraftment" from this control group should be subtracted from the engraftment of the experimental group.

Here, we provide a basic structure for designing and carrying out a limiting-dilution assay for HSCs. Certain details and modifications should be considered by each individual investigator depending on the goals of the particular study. The considerations discussed above as well as in the following Notes section should aid in this process.

2. Materials

2.1. Recipient Mice/ Competitor Cell Donors

1. B6.SJL (CD45.1) mice, 4–8-weeks old.

2.2. Donor Mice for Test Cells

1. C57BL/6 (CD45.2) mice, 6–12-weeks old.

2.3. Reagents/ Equipment

1. Cell counter.
2. 15-ml sample tubes.
3. 5-ml flow tubes.
4. DMEM without phenol red for resuspension of test/competitor cells.
5. 29-gage and 22-gage needles with syringe.
6. Phosphate-buffered saline (PBS) +2% fetal bovine serum (FBS) for staining medium.
7. Ethylenediaminetetraacetic acid (EDTA)-coated blood collection tubes.
8. Red blood cell (RBC) lysis buffer (150 mM NH_4Cl, 10 mM $KHCO_3$, 10 mM EDTA).
9. 40 μM cell strainers/nylon mesh.
10. Anti-mouse antibodies:
    FITC CD45.2.
    PE-Cy5 CD45.1.
    APC CD3.
    PE B220.
    APC-Cy7 Gr1.
    PE-Cy7 conjugated Mac-1.
    IgG Isotype controls.
11. FACS analyzer.
12. L-Calc software™ (StemCell Technologies, Vancouver).
13. Irradiator.
14. Heat lamp.

3. Methods

3.1. Preparation of Recipient Mice

1. Place recipient mice on acidified or antibiotic-containing water 3 days prior to irradiation and continue for 2 weeks. Be sure to have enough mice for at least three doses typically separated by three to fivefold dilutions. Each dilution should have at least eight mice. Be sure to include a group which is treated as equivalently as possible but is not transplanted with any test cells. This group will be utilized as the base-line for establishing true engraftment (see discussion in the Introduction). Pilot experiments may be needed to gage the optimal doses. Note that each of the 3+ doses will need to contain at least one animal which is negative for use in the CRU calculation.
2. Identify mice by institutionally approved protocols.
3. Irradiate recipient mice with 10 Grays the morning before transplant.

3.2. Preparation of Test and Competitor Cells

1. Harvest bone marrow from mice by flushing marrow from tibias and femurs using a 22-gage needle with syringe and gently draw the solution through the needle to produce a single-cell suspension (see Note 1).
2. Obtain highly accurate cell counts.
3. Add test cells to competitor cells such that the concentration of competitor cells is 2×10^5 + desired number of test cells per 100 µl in DMEM for each recipient (see Note 2).

3.3. Transplantation

1. Place recipient mice under heat lamp for about 2 min. Do not overheat.
2. Place mouse in restraint with tail exposed. Carefully inject cells into the lateral tail vein (see Note 3).

3.4. Evaluation of Engraftment

1. Collect about 50–100 µl of peripheral blood by institutionally approved methods into EDTA-coated tubes (see Note 4).
2. Add cold RBC lysis buffer to blood collection tube and pipette out diluted sample. Add to 15-ml tube containing 5 ml of RBC lysis buffer (see Note 5). Mix gently and place on ice. Continue with additional samples to a maximum of 10–15 samples per group, depending on speed of work. It is important that lysis is relatively consistent, and that samples are not overlysed. To aid in this, all materials should be set up and arranged prior to beginning lysis procedure.
3. Place samples in 37°C water bath for 7 min. Remove and centrifuge at $400 \times g$ for 5 min at 4°C. Immediately aspirate

lysis buffer without disturbing cell pellet. Wash with 5 ml PBS + 2% FBS and resuspend in 100 μl of PBS + 2% FBS. Any remaining blood clots should be removed or filtered out.

4. Transfer samples to 5-ml flow tubes. Add repopulation antibody cocktail to each tube (cocktail should contain 0.25 μg FITC CD45.2; 0.05 μg PE-Cy5 CD45.1, PE B220, and PE-Cy7 Mac-1; 0.25 μg APC CD3 and APC-Cy7 Gr1) (see Note 6). Place on ice for 25–30 min.
5. Wash with 4–5 ml of PBS + 2% FBS.
6. Aspirate leaving approximately 100 μl above cell pellet and resuspend by quickly vortexing.
7. Analyze on FACS cytometer (see Note 7).
8. Calculate CRU frequency using L-Calc™ software (available from StemCell Technologies).

4. Notes

1. Due to normal differences in HSC frequency of individual mice, bone marrow pooled from at least three mice as the source of competitor cells may reduce variation between different experimental groups.
2. Fewer or compromised competitor cells can be added, especially when very limited numbers of test cells are utilized (see further discussion in the Introduction). It may be necessary to first ensure that at least 90% of mice receiving only competitor cells survive. Also, while larger volumes can be injected into adult mice, 100 μl allows for relatively rapid injection whereas lower volumes may compromise accuracy and precision of volume delivered.
3. This is a critical and challenging step and should only be performed by trained personnel. A small amount of blood can be drawn from the vein to ensure that the needle is actually within the vein immediately prior to injection. The fluid should enter smoothly without being forced.
4. Puncture of the submandibular plexus is the preferred method of blood collection. Mixing the blood within the tube by immediately vortexing for a few seconds helps reduce clotting which otherwise requires time-consuming filtration steps and loss of usable material.
5. If blood clots are present, they can often be removed by simply pipetting them out. If numerous small clots are present, entire sample can be filtered through 40-μm nylon mesh or cell strainer.

6. This staining scheme allows for host/competitor and donor myeloid as well as B and T lymphoid cells to be analyzed in a single sample but requires six-color analysis. Four-color analysis can be utilized by splitting each sample into two aliquots and staining host/competitor and donor myeloid cells in one tube, and host/competitor and donor lymphoid cells in the second tube. An alternative scheme is to utilize one color for myeloid cells, a different color for T cells, and both these colors for B cells. This allows four-color, single tube staining as well (B cells are the double positive population) (33).
7. Be sure to collect enough events to allow proper statistical analysis of potential low engraftment accounting for the fact that certain lineages may further represent a low percentage of the donor population.

Acknowledgments

We would like to thank Jeff Haug and the Flow Cytometry facility staff at Stowers Institute for help in designing the repopulation assay.

References

1. Weissman, I. L., Anderson, D. J., and Gage, F. (2001) Stem and progenitor cells: origins, phenotypes, lineage commitments, and transdifferentiations. *Annu Rev Cell Dev Biol* 17, 387–403.
2. Orkin, S. H., and Zon, L. I. (2008) Hematopoiesis: an evolving paradigm for stem cell biology. *Cell* 132: 631–44.
3. Watt, F. M., and Hogan, B. L. M. (2000) Out of Eden: stem cells and their niches. *Science* 287, 1427–30.
4. Weissman, I. L., and Shizuru, J. A. (2008) The origins of the identification and isolation of hematopoietic stem cells, and their capability to induce donor-specific transplantation tolerance and treat autoimmune diseases. *Blood* 112, 3543–53.
5. Spangrude, G. J., Heimfeld, S., and Weissman, I. L. (1988) Purification and characterization of mouse hematopoietic stem cells. *Science* 241, 58–62.
6. Ikuta, K., and Weissman, I. L. (1992) Evidence that hematopoietic stem cells express mouse c-kit but do not depend on steel factor for their generation. *Proc Natl Acad Sci U S A* 89, 1502–6.
7. Okada, S., Nakauchi, H., Nagayoshi, K., Nishikawa, S., Miura, Y., and Suda, T. (1992) In vivo and in vitro stem cell function of c-kit- and Sca-1-positive murine hematopoietic cells. *Blood* 80, 3044–50.
8. Christensen, J. L., and Weissman, I. L. (2001) Flk-2 is a marker in hematopoietic stem cell differentiation: a simple method to isolate long-term stem cells. *Proc Natl Acad Sci U S A* 98, 14541–6.
9. Osawa, M., Hanada, K., Hamada, H., and Nakauchi, H. (1996) Long-term lymphohematopoietic reconstitution by a single CD34-low/negative hematopoietic stem cell. *Science* 273, 242–5.
10. Purton, L. E., and Scadden, D. T. (2007) Limiting factors in murine hematopoietic stem cell assays. *Cell Stem Cell* 1(3), 263–70.
11. Matsuoka, S., Ebihara, Y., Xu, M., Ishii, T., Sugiyama, D., Yoshino, H., Ueda, T., Manabe, A., Tanaka, R., Ikeda, Y., Nakahata, T., and Tsuji, K. (2001) CD34 expression on long-term repopulating hematopoietic stem cells changes during developmental stages. *Blood* 97, 419–25.
12. Yang, L., Bryder, D., Adolfsson, J., Nygren, J., Mansson, R., Sigvardsson, M., and

Jacobsen, S. E. (2005) Identification of Lin(-)Sca1(+)kit(+)CD34(+)Flt3- short-term hematopoietic stem cells capable of rapidly reconstituting and rescuing myeloablated transplant recipients. *Blood* 105, 2717–23.

13. Goodell, M. A., McKinney-Freeman, S., and Camargo, F. D. (2005) Isolation and characterization of side population cells. *Methods Mol Biol* 290, 343–52.
14. Li, C. L., and Johnson, G. R. (1992) Rhodamine123 reveals heterogeneity within murine Lin-, Sca-1+ hemopoietic stem cells. *J Exp Med* 175, 1443–7.
15. Balazs, A. B., Fabian, A. J., Esmon, C. T., and Mulligan, R. C. (2006) Endothelial protein C receptor (CD201) explicitly identifies hematopoietic stem cells in murine bone marrow. *Blood* 107, 2317–21.
16. Kiel, M. J., Yilmaz, O. H., Iwashita, T., Terhorst, C., and Morrison, S. J. (2005) SLAM family receptors distinguish hematopoietic stem and progenitor cells and reveal endothelial niches for stem cells. *Cell* 121, 1109–21.
17. Dykstra, B., Kent, D., Bowie, M., McCaffrey, L., Hamilton, M., Lyons, K., Lee, S. J., Brinkman, R., and Eaves, C. (2007) Long-term propagation of distinct hematopoietic differentiation programs in vivo. *Cell Stem Cell* 1, 218–29.
18. McKenzie, J. L., Gan, O. I., Doedens, M., Wang, J. C., and Dick, J. E. (2006) Individual stem cells with highly variable proliferation and self-renewal properties comprise the human hematopoietic stem cell compartment. *Nat Immunol* 7, 1225–33.
19. van Os, R., Kamminga, L. M., and de Haan, G. (2004) Stem cell assays: something old, something new, something borrowed. *Stem Cells* 22, 1181–90.
20. Till, J. E., and McCulloch, E. A. (1961) A direct measurement of the radiation sensitivity of normal mouse bone marrow cells. *Radiat Res* 14, 213.
21. Jones, R. J., Wagner, J. E., Celano, P., Zicha, M. S., and Sharkis, S. J. (1990) Separation of pluripotent haematopoietic stem cells from spleen colony-forming cells. *Nature* 347, 188–9.
22. Ploemacher, R. E., and Brons, R. H. (1989) Separation of CFU-S from primitive cells responsible for reconstitution of the bone marrow hemopoietic stem cell compartment following irradiation: evidence for a pre-CFU-S cell. *Exp Hematol* 17, 263–6.
23. Szilvassy, S. J., Humphries, R. K., Lansdorp, P. M., Eaves, A. C., and Eaves, C. J. (1990) Quantitative assay for totipotent reconstituting hematopoietic stem cells by a competitive repopulation strategy. *Proc Natl Acad Sci U S A* 87, 8736–40.
24. Harrison, D. E., Jordan, C. T., Zhong, R. K., and Astle, C. M. (1993) Primitive hemopoietic stem cells: direct assay of most productive populations by competitive repopulation with simple binomial, correlation and covariance calculations. *Exp Hematol* 21, 206–19.
25. Taswell, C. (1981) Limiting dilution assays for the determination of immunocompetent cell frequencies. I. Data analysis. *J Immunol* 126, 1614–9.
26. Szilvassy, S. J., Lansdorp, P. M., Humphries, R. K., Eaves, A. C., and Eaves, C. J. (1989) Isolation in a single step of a highly enriched murine hematopoietic stem cell population with competitive long-term repopulating ability. *Blood* 74, 930–9.
27. Lemischka, I. R., Raulet, D. H., and Mulligan, R. C. (1986) Developmental potential and dynamic behavior of hematopoietic stem cells. *Cell* 45, 917–27.
28. Harrison, D. E., and Astle, C. M. (1982) Loss of stem cell repopulating ability upon transplantation. Effects of donor age, cell number, and transplantation procedure. *J Exp Med* 156, 1767–79.
29. Micklem, H. S., Clarke, C. M., Evans, E. P., and Ford, C. E. (1968) Fate of chromosome-marked mouse bone marrow cells tranfused into normal syngeneic recipients. *Transplantation* 6, 299–302.
30. Czechowicz, A., Kraft, D., Weissman, I. L., and Bhattacharya, D. (2007) Efficient transplantation via antibody-based clearance of hematopoietic stem cell niches. *Science* 318, 1296–9.
31. Waskow, C., Madan, V., Bartels, S., Costa, C., Blasig, R., and Rodewald, H. R. (2009) Hematopoietic stem cell transplantation without irradiation. *Nat Methods* 6, 267–9.
32. Boggs, D. R., Boggs, S. S., Saxe, D. F., Gress, L. A., and Canfield, D. R. (1982) Hematopoietic stem cells with high proliferative potential. Assay of their concentration in marrow by the frequency and duration of cure of W/Wv mice. *J Clin Invest* 70, 242–53.
33. Challen, G. A., and Goodell, M. A. (2008) Promiscuous expression of H2B-GFP transgene in hematopoietic stem cells. *PLoS ONE* 3, e2357.

Chapter 4

Isolation Procedure and Characterization of Multipotent Adult Progenitor Cells from Rat Bone Marrow

Kartik Subramanian, Martine Geraerts, Karen A. Pauwelyn, Yonsil Park, D. Jason Owens, Manja Muijtjens, Fernando Ulloa-Montoya, Yeuhua Jiang, Catherine M. Verfaillie, and Wei-Shou Hu

Abstract

Multipotent adult progenitor cells (MAPCs) are adult stem cells derived from the bone marrow of mouse and rat and were described for the first time in 2002 (Jiang et al., Nature 418:41-49, 2002), and subsequently (Breyer et al., Exp Hematol 34:1596-1601, 2006; Jiang et al., Exp Hematol 30:896-904, 2002; Ulloa-Montoya et al., Genome Biol 8:R163, 2007). The capacity of rodent MAPC to differentiate at the single-cell level into some of the cell types of endoderm, mesoderm, and neuroectoderm germ layer lineages makes them promising candidates for the study of developmental processes. MAPC are isolated using adherent cell cultures and are selected based on morphology after a period of about 8–18 weeks. Here, we describe a step-by-step reproducible method to isolate rat MAPC from fetal and adult bone marrow. We elaborate on several aspects of the isolation protocol including, cell density and medium components, and methods for selecting and obtaining potential MAPC clones and their characterization.

Key words: Adult stem cells, Bone marrow, Pluripotency, Differentiation

1. Introduction

Multipotent adult progenitor cells (MAPCs), derived from rodent bone marrow, were described for the first time in 2002 (1). In addition to bone marrow, murine MAPC have also been isolated from muscle and brain (2). MAPC can undergo extensive proliferation without the loss of potential to differentiate into several cell types, such as, endothelial cells and blood (3–5). Since the isolation of MAPC a number of other groups reported the isolation of

S. Ding (ed.), *Cellular Programming and Reprogramming: Methods and Protocols*, Methods in Molecular Biology, vol. 636, DOI 10.1007/978-1-60761-691-7_4, © Springer Science+Business Media, LLC 2010

cells with broader differentiation capacity than classical adult stem cells from somatic tissues such as (a) bone marrow, including human bone marrow-derived stem cells (hBMSCs) (6), marrow-isolated adult multilineage inducible cells (MIAMI cells) (7), pre-mesenchymal stem cells (pre-MSCs) (8), (b) umbilical cord, including unrestricted somatic stem cells (USSCs) (9), and very small embryonic-like cells (VSELs) (10, 11), (c) amniotic fluid, amniotic fluid stem cells (AFSs) (12), and from (d) liver, heart, and bone marrow, multipotent adult stem cells (MASCs) (13). The relationship between these different "more pluripotent" populations of stem cells and whether they exist in vivo or are created in vitro is yet to be determined conclusively. Regardless of their origin, the isolation of cells with pluripotent capacity from postnatal somatic tissues yields another source of cells to study and compare self-renewal and differentiation mechanisms in stem cells, and for potential applications, such as cell therapy and in vitro drug toxicity screening.

All isolations should be performed following approval from an ethical commission. Rat MAPC are isolated from bones from the entire hind limbs in fetal (E18–E19) and newborn rats and from the tibia and femur of 4–6 weeks old rats. Six to ten million cells per well are plated on six-well tissue culture plates (9.6 cm^2 of surface/well) in medium optimized for the process of MAPC isolation and growth (*see* Subheading 2.1, item 7). The plates are incubated in a humidified incubator at 37°C with 5% oxygen and 5–6% CO_2 gas. After 4 weeks of culture, cells are depleted of hematopoeitic cells using magnetic microbeads and the remaining cells are seeded into 96-well plates at 5–10 cells/well. Morphology of cells in the different wells of 96-well plates is very heterogeneous and only wells with mainly small cells are subsequently expanded. After 3–14 weeks, cells with the typical spindle-shaped MAPC morphology appear in some of the selected wells (cells are 10–15 μm size, Fig. 1). To demonstrate that these clones are indeed MAPC, the level of Oct4 mRNA is determined using quantitative RT-PCR and the presence of

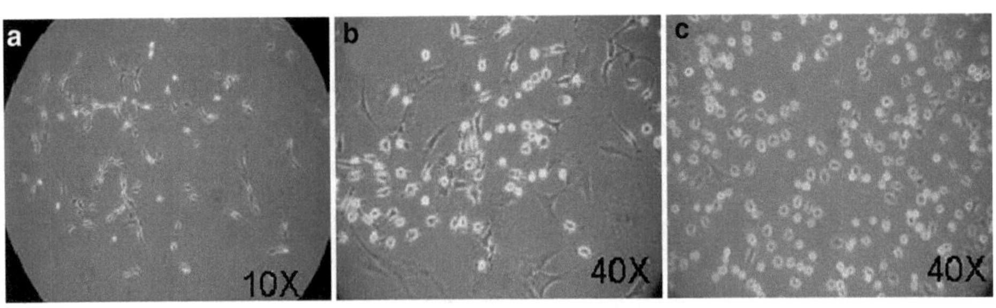

Fig. 1. Cell morphology of rat MAPC lines. Cell morphology of (**a**) selected clones from 96-well plates, (**b**) rat MAPC cells in heterogeneous cultures, and (**c**) rat MAPC line after subcloning

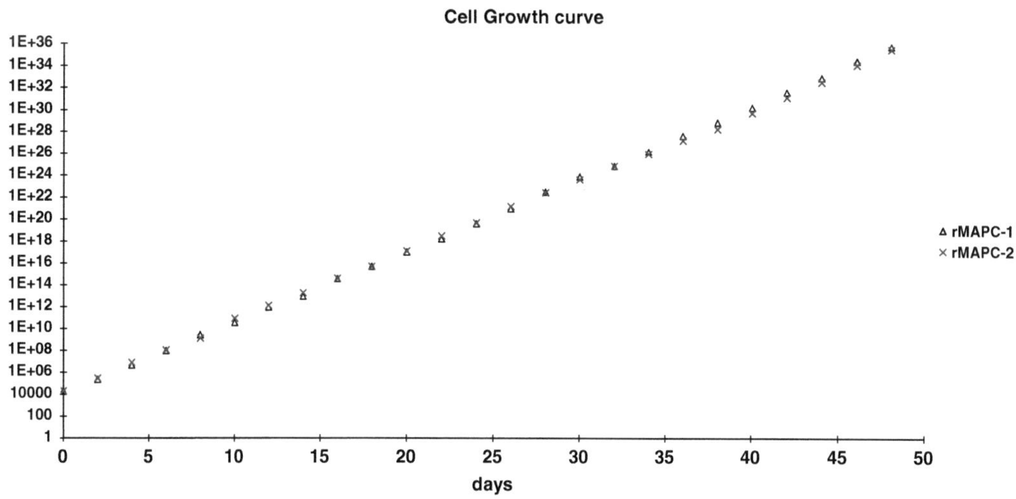

Fig. 2. Cell expansion curve for rat MAPC lines. Cell growth curve of two rat MAPC lines, the doubling time is about 12 h

CD31, typical for rat MAPC is analyzed by fluorescence-activated cell sorting (FACS) (14). Once a cell line with MAPC phenotype (Oct4 mRNA at ΔCt of 4–6, and CD31 positive) is derived, a large frozen stock of master and working banks is prepared. The characteristics of the putative MAPC are then evaluated (endothelial, hepatocyte, and neural precursor-like differentiation; and further transcriptome phenotype) as described by Ulloa-Montoya et al. (14) (Figs. 3–5), and cytogenetic analysis is done as well. During subsequent cell expansion/maintenance of an established line, quality control studies are performed on a weekly to monthly basis. Routine quality control includes (a) determination of the phenotype by FACS, (b) evaluation of the expression level of the pluripotency transcription factor Oct-4, at the mRNA level, (c) monitoring of karyotype by chromosome counting weekly and intermittent G-banding, (d) exclusion of mycoplasma contamination, and (e) proof of multilineage differentiation potential. Importantly, and different from what we initially described in Nature 2002, undifferentiated MAPC are derived and grown in low oxygen conditions (5%).

The process of isolating MAPC from the heterogeneous mixture of cells of the BM is a lengthy procedure (8–18 weeks). The exact frequency of wells that will yield MAPC is still unknown. This chapter aims to provide a detailed description of the isolation protocol of rat MAPC that has enabled multiple investigators in different laboratories to obtain rat MAPC lines with very similar phenotype and differentiation capabilities.

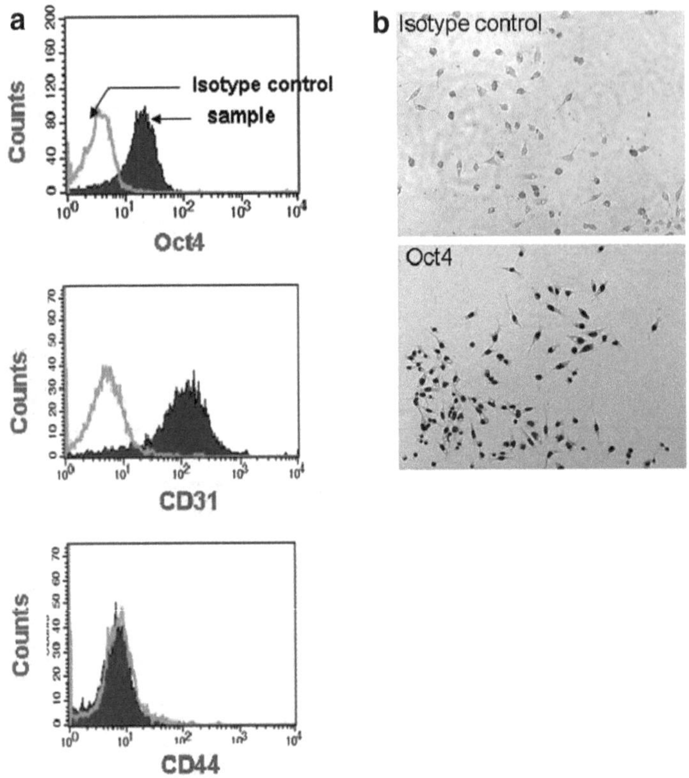

Fig. 3. Characterization of rat MAPC by flow cytometry and immunocytochemistry. Upon generation of new rat MAPC lines, typical MAPC markers are checked by (**a**) FACS analysis for Oct4, CD31, and CD44 and (**b**) Oct4 immunocytochemistry

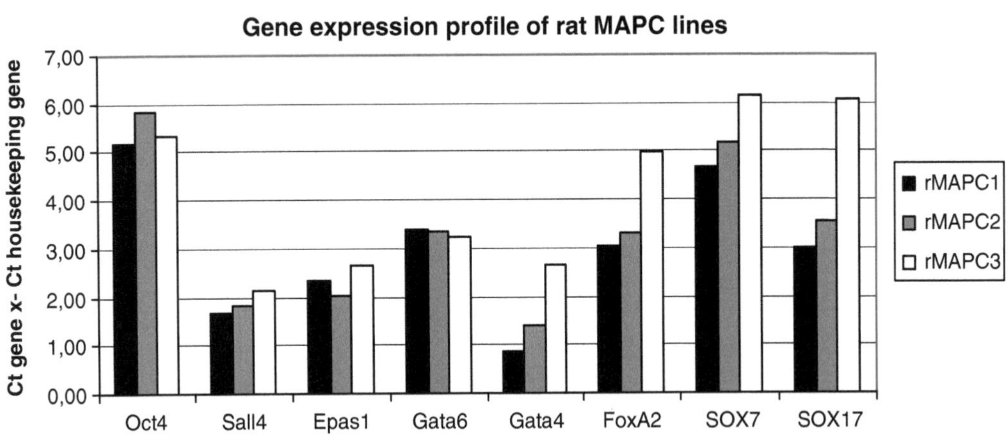

Fig. 4. Typical gene expression profile of undifferentiated rat MAPC lines. The gene expression profile of typical rat MAPC genes as evaluated by qRT-PCR shows high levels of expression of pluripotency genes Oct4 and Sall4 and primitive endodermal genes Gata4, Gata6, Sox7, and Sox17

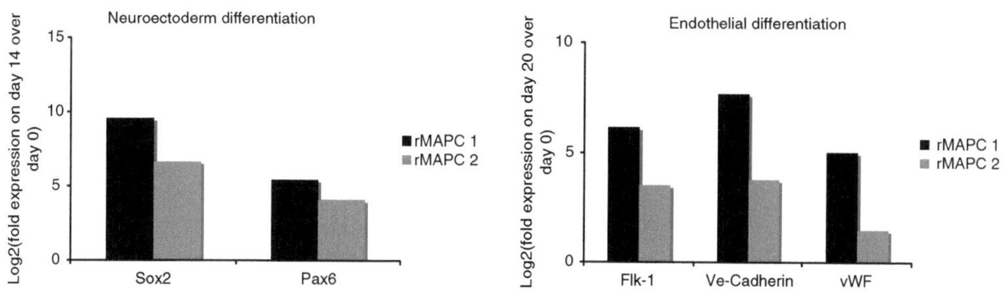

Fig. 5. Directed differentiation of rat MAPC lines to hepatocytes, endothelial, and early neuroectoderm-like cell lineages. Directed differentiation of the rat MAPC lines to hepatocyte, endothelial, and neuroectoderm-like progenitors, respectively. Upregulation of genes expressed in cells of hepatic endoderm (will be published separately), endothelial mesoderm (Flk-1, Ve-Cadherin, vWF), and in neuroectoderm-like cells (Sox2, Pax6) are observed demonstrating the multilineage differentiation capability of the rat MAPC lines in cell culture

2. Materials

2.1. Isolation of Rat MAPC from Bone Marrow and Cell Culture

1. PBS (w/o Calcium/Magnesium) with 0.2% Collagenase D and 0.02% DNAse.
2. PBS with 2% fetal bovine serum (FBS).
3. 23-G needle, a 10-ml Luer-Lok syringe, 40-μm nylon cell mesh strainer.
4. MCDB-201: Dilute 1 vial (17.7 g) of MCDB-201 in 1,000 ml of MilliQ distilled water by stirring with magnetic stirrer at room temperature. Adjust the pH to 7.2 and then filter using a 0.22-μm filter. Store at 4°C for a maximum of 1 month before use.
5. Recombinant human platelet-derived growth factor (PDGF-BB): Prepare 4 mM HCl with 0.1% bovine serum albumin (BSA) by adding 0.1 ml HCl to 300 ml water, and subsequently 0.1 g BSA in 100 ml of this 4 mM HCl solution. Filter the solution with a 0.22-μm filter. For vials containing 0.5 mg of PDGF-BB, dissolve the contents in 50 ml of 4 mM HCl + 0.1% BSA and mix well. Aliquot 500 μl/vial and store in −80°C. The stock concentration is 10 μg/ml (use 1 vial for 500 ml of MAPC medium for a final concentration of 10 ng/ml).
6. Mouse epidermal growth factor (EGF): Prepare 0.3% BSA–PBS solution by adding 0.3 g of BSA to 100 ml of PBS. Add 100 ml of this 0.3% BSA–PBS solution to 1 mg of EGF. Mix and aliquot 500 μl/vial and store in −80°C freezer. The stock concentration is 10 μg/ml (use 1 vial for 500 ml of MAPC medium for a final concentration of 10 ng/ml).

7. MAPC medium: Mix 60% Dulbecco's Modified Eagle medium (DMEM) low (1 g/l) glucose, 40% MCDB-201 solution at pH 7.2, 1× Insulin–transferrin–selenium (ITS), 1× Linoleic acid–Bovine serum albumin (LA-BSA), 100 IU/ml Penicillin and 100 μg/ml Streptomycin, 10^{-4} M l-Ascorbic acid (add 256 mg of l-Ascorbic acid to 100 ml PBS), 2% qualified (see Note 2) FBS, 10 ng/ml human PDGF-BB, 10 ng/ml mouse EGF, 0.05 μM dexamethasone, 10^3 units/ml mouse leukemia inhibitory factor (LIF), 55 μM 2-mercaptoethanol (to be added freshly) . Sterilize the medium using a 0.22-μm filter and use within a month.
8. 6-well cell culture plates, 24-well cell culture plates, 96-well cell culture plates, 10-cm cell culture dishes.
9. Fibronectin coating solution: Prepare 0.5 mg/ml stock solution of rat fibronectin by adding 4 ml of PBS to 2 mg fibronectin. Do not shake or pipet solution. Put in incubator for 30 min, then aliquot, and store at −80°C. Dilute rat fibronectin to working solution (100 ng/ml) by adding 100 μl of stock solution to 500 ml of PBS. Filter sterilize through a 0.22-μm filter and store at 4°C.
10. Trypsin–EDTA 0.05%.
11. MACS separation CS columns, PE anti-rat CD45 antibody, anti-PE microbeads.
12. Cell freezing: Prepare freezing medium A: 80% (v/v) MAPC medium and 20% (v/v) FBS (qualified for MAPC) and freezing medium B: 60% (v/v) MAPC medium, 20% (v/v) FBS, and 20% (v/v) DMSO. Cryovials, Nalgene Cryo 1°C freezing container, must change isopropanol at least once every 4–5 times of use.

2.2. Characterization of MAPC and Quality Control

2.2.1. mRNA Expression of Pluripotency Markers and Lineage-Specific Genes

1. RNA mini/micro kit.
2. Cell lysis buffer from RNA mini/micro kit: Mix 990 μl RLT buffer with 10 μl β-mercaptoethanol.
3. Nuclease-free microcentrifuge tubes (1.5 ml) with flat cap to collect cell lysates.
4. Ethanol, Depc-treated water: Prepare 70% and 80% (v/v) Ethanol in Depc-treated water.
5. DNAse turbo kit.
6. Superscript III first-strand synthesis system for RT-PCR.
7. PCR microtubes and RT-PCR grade water.
8. Platinum SYBR green qPCR Supermix-UDG, primer stocks at 5 μM (see sequence Table 1), Thermo-fast 96 Detection plate (Thermo Scientific), Ultra-Clear Cap Strips (Thermo Scientific).

Table 1
Forward and reverse primer sequences for pluripotency and differentiation markers

Tissue/lineage	Gene	Forward primer	Reverse primer
Housekeeping	Gapdh	AAGGGCTCATGACCACAGTC	GGATGCAGGCATGATGTTCT
Pluripotency	Oct4	CTGTAACCGGCGCCAGAA	TGCATGGGAGAGCCCAGA
	Sall4	AGAACTTCTCGTCTGCCAGTG	CTCTATGGCCAGCTTCCTTC
	Epas1	CAGTCCTCCAGGAGCTCA	CTCCCCTGCAGGAGTGTAGA
Primitive endoderm	Gata6	GTCTGGATGGAGCCACAGTT	ATCATCACCACCCGACCTAC
	Gata4	CTGTGCCAACTGCCAGACTA	AGATTCTTGGGCTTCCGTTT
	Foxa2	GCAGAACTCCATCCGTCATT	TCGAACATGTTGCCAGAGTC
	Sox7	CAAGGATGAGAGGAAACGTC	CTCTGCCTCATCCACATAGG
	Sox17	GCCAAAGACGAACGCAAGCGG	TCATGCGCTTCACCTGCTTG
Hepatic endoderm	AFP	ACCTGACAGGGAAGATGGTG	GCAGTGGTTGATACCGGAGT
	Albumin	TCTGCACACTCCCAGACAAG	AGTCACCCATCACCGTCTTC
	AAT	CAAACAAGGTCAGCCATTCTC	CAGCATCATTGTTGAAGACCC
Endothelium	Flk-1	CCAAGCTCAGCACACAAAAA	CCAACCACTCTGGGAACTGT
	Ve-Cadherin	GGCCAACGAATTGGATTCTA	GTTTACTGGCACCACGTCCT
	vWF	CCCACCGGATGGCTAGGTATT	GAGGCGGATCTGTTTGAGGTT
Neuroectoderm	Sox2	GGCCAACGAATTGGATTCTA	GTTTACTGGCACCACGTCCT
	Pax6	GTCCATCTTTGCTTGGGAAA	TAGCCAGGTTGCGAAGAACT

2.2.2. Flow Cytometry

1. Antibodies: mouse anti-rat CD31-PE (TLD-3A12, BD Pharmingen), mouse anti-rat CD44-FITC (OX-49, BD Pharmingen), goat Oct3/4 (Santa Cruz).
2. Isotype controls: Mouse IgG_1-PE (BD Pharmingen), Mouse IgG_{2a}-FITC (BD Pharmingen), Goat IgG (Jackson), and Donkey anti-goat-Cy3 (Jackson) as secondary antibody for Oct4 staining.
3. FACS round bottom tubes with cell strainer cap.
4. Prepare formaldehyde fixation solution by diluting 400 ml of 10% Ultrapure Formaldehyde (Polysciences) in 600 ml PBS.
5. SAP buffer – 0.1% (w/v) saponin, 0.05% (w/v) sodium azide in PBS, SAP serum buffer – SAP buffer with 10% donkey serum (Jackson).

2.2.3. Immunocytochemistry

1. Fish skin gelatin (in water).
2. 0.3% (v/v) Hydrogen peroxide solution (30% w/w) in methanol, prepare just before use.
3. 0.2% Triton X-100 solution for permeabilization: add 100 μl Triton X-100 to 500 ml of PBS.
4. Antibodies: Goat Oct3/4 (Santa Cruz) and Biotin-labeled donkey anti-goat secondary antibody (Jackson) in Dako Real Antibody diluent (Dako).
5. Vectastain ABC kit (Vector Laboratories).
6. Dako Liquid DAB+ substrate chromogen system containing Substrate Buffer and DAB+ Chromogen (Dako).

2.2.4. Hepatocyte Differentiation (Endoderm)

The precise current protocol will be published shortly in a separate manuscript; hence, readers are referred to that manuscript for the detailed protocol.

2.2.5. Endothelial Differentiation (Mesoderm)

1. Basal differentiation medium: 60% Dulbecco's Modified Eagle Medium, low glucose , 40% MCDB-201, 1× ITS, 1× LA-BSA, 100 IU/ml Penicillin, and 100 μg/ml Streptomycin (Cellgro), 10^{-4} M L-Ascorbic acid, 1 μM Dexamethasone, 55 μM β-mercaptoethanol (Gibco) (to be added freshly). Combine and filter through a 0.22-μm steriflip (Millipore) and store at 4°C no longer than 4 weeks.
2. Recombinant Human $VEGF_{165}$ (R&D): Dissolve 50 μg of lyopholized cytokine in 10 ml of PBS + 0.3% BSA and aliquot in 100 μl vials and store at −80°C.
3. Endothelial differentiation medium step 1: To prepare 50 ml of medium, add 47.5 ml of basal differentiation medium, 0.1 ml of rh-$VEGF_{165}$ 5 μg/ml, 2.5 ml of qualified FBS. Combine and filterize through 0.22 μm 50 ml steriflip and store at 4°C no longer than 2 weeks.
4. Endothelial differentiation medium step 2: To prepare 50 ml of medium, add 48.75 ml of basal differentiation medium, 0.1 ml of rh-$VEGF_{165}$ 5 μg/ml, 1.25 ml of qualified FBS. Combine and filterize through 0.22 μm 50 ml steriflip and store at 4°C no longer than 2 weeks.

2.2.6. Neuroectoderm Differentiation

1. N2B27 medium: 50% Neurobasal-A medium, 50% DMEM/F12, 0.5× B27, 0.5× N2 plus (R&D Systems), 0.05 mM β-mercaptoethanol, 2 mM Glutamine.
2. NSE medium: Euromed-N medium (Annovum/Euroclone), 2 mM L-Glutamine, 10 ng/ml bFGF, 10 ng/ml EGF, 1× N2 plus supplement, 2 μg/ml heparin, 0.05 mM 2-mercaptoethanol.
3. 0.1% gelatine solution (Chemicon).

2.3. Cytogenetics by G-Banding

1. Colcemid 10 μg/ml (Irvine Scientific).
2. Hanks' Balanced Salt Solution.
3. Trypsin–EDTA 1× solution (0.25%).
4. Hypotonic solution: 0.075 KCl (pH 6.0–6.7) (Mallinckrodt): Dissolve 11.2 g KCl in 2 L MilliQ water, stir for 2–3 h and store at 4°C until needed. No expiration date for this solution.
5. Methanol/Acetic Acid Fixative: Mix 1 part glacial acetic acid with three parts absolute methanol immediately before use. Store at room temperature.
6. Pancreatin: Mix 2.5 ml trypsin–EDTA 10× (Biochrom) and 50 ml of 1× Hanks balanced salt solution.
7. Wright's stain: In a 6-L flask, dissolve 12.5 g of Wright's stain (Richard-Allan Scientific) into 5 L of methanol and then add 0.225 g of Giesma stain (Fisher). Cover the flask with foil to protect from light and mix on a stirrer for at least 30 min (preferably overnight). Filter the solution into 1-L brown glass bottles. For best results, let the stain age for at least 1 month by storing in dark at room temperature.
8. Phosphate Buffer (pH 6.8): Prepare solutions A (0.06M KH_2PO_4 – dissolve 16.32 g KH_2PO_4 in 2 L distilled water) and solution B (0.06 M Na_2HPO_4 – dissolve 17.04 g Na_2HPO_4 in 2 L distilled water) in separate flasks. Pour solution A into a large clean plastic container and then add most of solution B to solution A and mix well. Check pH and adjust to 6.8 with the remainder of solution B.

3. Methods

3.1. Isolations of Rat MAPC from Bone Marrow

3.1.1. Bone Marrow Aspiration from E18 and 3/4-Week-Old Rats

1. Sacrifice one timed-pregnant rat (E18-E19) or 3–4-weeks-old Fischer or Sprague-Dawley rats by CO_2 inhalation (see Note 1).
2. Take out the whole litter of embryos and dissect the skin and muscle from the tibiae and femur, under sterile conditions. Collect all material in 10-cm dishes in PBS 2% FBS.
3. Cut of the ends of the bones to be able to insert the needle. It is important to flush the end of the bones as well. Flush the bones thoroughly with 15–20 ml of PBS containing 2% FCS using a 23-G needle, until the bones are more or less transparent. Collect the flushed fluid in sterile 10-cm cell culture dishes.
4. Triturate cells by passing the cell suspension 5–10 times through a 23-G needle attached to a 10-ml syringe to obtain a single-cell suspension.

5. After flushing of fetal bones, the bone is crushed and incubated for 20–30 min with PBS containing 0.2% collagenase and 0.02% DNAse at 37°C. After incubation, cells are pooled with the flushed cells.
6. Filter the total cell suspension through a 40-µm nylon cell mesh strainer.
7. Centrifuge cells at 600 × *g* for 6 min.
8. Wash the cells three times with regular MAPC expansion medium by centrifugation at 600 × *g* for 6 min.
9. Count the cells.
10. Plate cells in the fibronectin-coated wells (at 6–10 × 10^6 cells/well in 2 ml of MAPC medium).

3.1.2. High-Density Culture of Cells for 1 Month

1. Three days after plating of the cells, add 1-ml prewarmed medium to each well.
2. For the rest of the first week only add 1 ml of prewarmed medium every other day. Since many of the cells will not yet be attached to the plate, do not remove any medium.
3. From the second week (days 7–14), change half of the medium (2 ml/well) every other day. Add the medium slowly from the side wall of each well to minimize the chances of detachment of cells from the bottom of the plate.
4. On day 14 (beginning of the third week), trypsinize and replate at 80% confluence (about 2 × 10^4 cells/cm^2), which can be achieved by doing a 1–1.5 split.
5. For the next 2 weeks (days 14–28), trypsinize cells 1–2 times/week when the plate is 100% confluent and replate again at 80% confluence (1–1.5 split).

3.1.3. Column Depletion of CD45+ Population

Column depletion is a negative selection strategy whereby the unwanted cells are magnetically labeled and eliminated from the cell mixture (see Note 3).

1. *Column preparation*: Attach the three-way stopcock to the column, with a 20-G needle attached below and a 10-ml syringe on the side. Attach column with three-way stopcock to MACS triangle. Wash the CS column by pushing 30 ml of MACS buffer through the 10-ml syringe from bottom to top of the column and let the fluid evacuate through the 20-G needle. *Never let the column dry out!*
2. *Cell preparation*: Trypsinize the cells, centrifuge them and count. Use about 3–5 million cells for each run through the column. Wash and centrifuge the cells again in MACS buffer. Remove supernatant and resuspend in 80 µl of MACS buffer. Add anti-rat CD45-PE antibody. Incubate solution

for 20 min at 4°C with intermittent shaking. Wash the cells twice in 10 ml of MACS buffer. Incubate in a second step with anti-PE microbeads for 20 min at 4°C with intermittent shaking. Wash cells again and following the second wash, resuspend the cells in 500 µl of MACS buffer.

3. *Cell depletion*: Attach a 23- to 25-G needle to the three-way stopcock. Apply the cell suspension to the MACS column. The cells positive for CD45 will be attached against the magnetic column and the negative cells will migrate through the column. Let cells flow through the column at a slow pace (1 drop every 4–5 s). Collect in 15-ml tubes. Wash the column with an additional 30 ml of MACS buffer. Spin down the cells in different fractions with each fraction consisting of 10 ml of eluted buffer. Collect 3–4 fractions (see Note 3). Count the cells and plate them at 5–10 cells/well of 96-well fibronectin-coated plates for expansion. To obtain several cell lines with high levels of Oct4 expression, plate five 96-well plates per fraction.

3.1.4. Subcloning, Culture, and Screening of Potential MAPC Clones

1. Change half of the medium every other day. Inspect the plate under the microscope daily from the second week after plating to identify wells with clones consisting of small, triangular to round cells (see Note 4).
2. Once the colony consists of 30–50 cells, trypsinize the cells (50 µl of 0.05% trypsin/well of 96-well plate) and replate in progressively larger fibronectin-coated plates; a well of a 24-well plate for about 3–4 days, followed by 6-well plates for another 7–14 days and eventually a 10-cm dish. From then, trypsinize and maintain cells in 10-cm dishes passaging every 2–3 days at ratios between 1:3 and 1:6 depending on cell density. Make sure to keep the cells at low density, without forming cell–cell contacts.
3. Continue with maintenance of cells till small clusters of cells begin to appear and/or the large cells begin to disappear from the culture. Once a homogeneous population of small, spindle/triangular cells with MAPC morphology appear, passage every 2 days and seed each 10-cm dish at 300 cells/cm^2 (see Note 4). Within the first week of obtaining MAPC-like cells by morphology, expand cells to obtain a frozen stock (master bank of early passages) and for characterization of cells for MAPC phenotype. After some initial characterization and confirmation of new MAPC lines using quantitative real-time polymerase chain reaction (qRT-PCR) and immunohistochemistry for Oct4, create another frozen stock (working bank) within the following week. Complete all the characterizations before using or distributing the MAPC lines for experiments.

3.2. Expansion or Maintenance Culture of MAPC Lines (see Note 5)

3.2.1. Coating of Wells and Dishes

1. Cell Culture dishes (Nunc) or wells (Corning) are coated with 100 ng/ml rat fibronectin (3 ml/10 cm dish).
2. Dishes or wells are precoated for 2 h at room temperature, 1 h at 37°C or overnight at 4°C. Upon plating of the cells, fibronectin is removed and the rat or mouse growth medium containing the cells is added. Typically 20,000 cells are seeded per 10-cm dish (300 cells/cm^2).

3.2.2. Procedure for Thawing MAPC

1. Remove the vial from the liquid nitrogen tank and place on dry ice.
2. Partially thaw in a 37°C water bath.
3. When about half of the liquid in the vial is thawed, collect the medium and cells and add to 10 ml of fresh MAPC expansion medium.
4. Spin down at 300 × *g* for 8 min (freshly thawed cells are more fragile, hence g-force should be lower).
5. Count and plate the cells as described on fibronectin-coated dishes.
6. Change full medium the next day and passage the following day.

3.2.3. Passaging of MAPC Lines

1. Precoat 10-cm dishes with fibronectin.
2. Prepare fresh rat MAPC medium.
3. Aspirate the expansion medium. Collect the removed medium in 50-ml tubes.
4. Rinse the dishes with PBS without calcium and magnesium (3 ml/dish) by rigorous swirling and aspirate the PBS.
5. To deattach the cells, add ~20 μl/cm^2 trypsin; 0.05% for rat MAPC. Cover the surface of the plate by swirling. Incubate for 1–2 min at room temperature. Dislodge the cells by tapping the plates gently. Check microscopically if all cells are deattached.
6. Add the collected expansion medium to neutralize the trypsin. Harvest the cells in a 15-ml or 50-ml centrifuge tube.
7. Centrifuge at 600 × *g* for 6 min.
8. Aspirate the supernatant and resuspend the pellet in fresh growth medium.
9. Count the cells with a hemocytometer.
10. Remove fibronectin solution from the new 10-cm dishes. Plate rat cells at 300 cells/cm^2 (about 20,000 cells/10 cm dish) use 6 ml of rat MAPC medium.

3.2.4. Procedure for Freezing MAPC

1. Place freezing vials on ice.
2. Trypsinize and count cells. Freeze about 0.5×10^6 cells/vial.

3. Centrifuge at $600 \times g$ for 6 min and then discard the supernatant.
4. Resuspend 0.5×10^6 cells in 500 µl freezing medium A for each vial. Gently add 500 µl freezing medium B for each vial. Add drop by drop and tap or shake tube. Do not repeatedly pipet solution.
5. Immediately place the vial in Nalgene freezing box (containing 100% isopropanol) and store at −80°C for a minimum of 6 h or overnight.
6. Move the vials to the liquid nitrogen freezer (−180°C) for long-term storage.

3.3. mRNA Expression of Pluripotency Markers

Rat MAPC lines express some pluripotency genes as Oct4 and Rex1 but do not express Nanog and Sox2 as described by Ulloa-Montoya et al. (14). Upon isolation of new lines with characteristic MAPC morphology (Fig. 1), the cells are expanded (Fig. 2) and the levels of Oct4 and primitive endodermal genes are evaluated by qRT-PCR (Fig. 4). Generally rat MAPC lines express Oct4 at a ΔCt between 2 and 6, with GAPDH as housekeeping gene. Only lines with these levels of Oct4 are further characterized for MAPC phenotype and function (Figs. 3–5).

3.3.1. Preparation of Cell Lysates

1. After 2 days of culture, trypsinize cells in 10-cm cell culture dishes as described before, centrifuge 6 min $600 \times g$.
2. Remove medium and add 350 µl (for $<5 \times 10^6$ cells) or 600 µl (5×10^6–1×10^7 cells) of RNA lysis buffer to the cell pellet. Cells expand tenfold after seeding, yielding 200,000 cells from 1 10-cm cell culture dish. Leave lysis buffer for 30 s on the cells. Pipet the lysis buffer up and down to recover most of the cells.
3. Collect the lysed cells into a nuclease-free 1.5-ml microcentrifuge tube. Vortex shortly. Store at −80°C until further use. Extract RNA within a month to prevent loss in the quality of the sample.

3.3.2. Total RNA Isolation with Mini Kit

1. Thaw the samples on ice.
2. Clean the designated working space for RNA preparation with RNAse Zap Wipes. In order to avoid cross-contamination with (c) DNA or nucleases, it is advisable to use a separate set of micropipettes and to wear gloves during the entire procedure.
3. RNA extractions are performed according to the manufacture's protocol: Briefly, add 350 µl of sample and 350 µl of 70% ethanol to a 1.5-ml nuclease-free tube and mix solution before applying to loading column, placed in a collector tube. Spin in a table top microcentrifuge for 30 s at 13,200 rpm (maximum speed).

4. Apply subsequently RW1 washing buffer, RPE buffer and 80% ethanol to the column. Centrifuge between every step for 30 s to 2 min at 13,200 rpm.
5. Spin an additional minute to remove all of the RPE buffer/ethanol traces before eluting RNA from column with 20 μl of nuclease-free water and centrifugation for 2 min at 13,200 rpm.
6. The RNA samples are now ready to be processed further or can be stored at –80°C until further use.

3.3.3. DNAse Treatment

1. The kit contains nuclease-free water, 10× buffer, 10× inactivation reagent, and turbo DNAse. Thaw the components on ice. All the steps of the DNAse treatment can be performed at room temperature.
2. Add 2 μl 1× buffer and 0.4 μl of DNAse to 18 μl of samples.
3. Incubate the samples for 30 min in a water bath or heating block at 37°C.
4. Spin down shortly.
5. Add 2 μl 1× inactivation mix to each sample. Once added to every sample, incubate for 2 min at room temperature and mix the samples with the inactivation reagent by flicking the tubes regularly.
6. Centrifuge for 2 min at 13,000 rpm. Transfer the supernatant to a new Eppendorf tube without disturbing the white pellet at the bottom of the tube.
7. The RNA is now ready for further processing or can be stored at –80°C.

3.3.4. Prepare cDNA with Superscript III First-Strand Synthesis System of RT-PCR

1. Thaw all reagents necessary for the reverse transcription procedure on ice, vortex solutions and briefly centrifuge each component before use.
2. For the first step, combine the following in 0.2-ml nuclease-free tubes to obtain a total volume of 10 μl: use 1 μl of random hexamers (50 ng/μl) and 1 μl of dNTP mix (10 mM) per sample and add the mixture to the amount of volume of sample corresponding to 1 μg of RNA. Thus, a maximum of 8 μl of RNA can be used. Therefore, if the RNA concentration is higher than 1 μg, dilute RNA in the required volume of RNAase-free water to make up the volume.
3. Incubate the tubes at 65°C for 5 min and bring back to 4°C, using a thermocycler. Keep the samples at 4°C for at least 1 min.
4. Prepare a second reaction mix as follows per sample: 2 μl 10× RT buffer, 4 μl $MgCl_2$ (25 mM), 2 μl DTT (0.1 M),

1 μl RNAseOUT (40 IU/μl), and 1 μl Superscript III RT (200 IU/μl), in this order to avoid precipitation of certain components.

5. Add 10 μl of this reaction mixture to each sample of step 3. Mix and collect by brief centrifugation.
6. cDNA synthesis reaction is performed on a thermocycler as follows: 10 min at 25°C, 50 min at 50°C, 5 min at 80°C, and return to 4°C for thermal hold.
7. Add 1 μl of RNAse H to each tube and incubate for 20 min at 37°C and cool the samples down to 4°C.
8. Add 79 μl of nuclease-free water to every sample to obtain a final volume of 100 μl. If the whole procedure is done appropriately and pure RNA was obtained, the Ct value for the housekeeping gene GAPDH should be 16–19.
9. The cDNA samples are now ready for quantitative PCR or can be stored at –20°C until further use.

3.3.5. Quantitative Real-Time Polymerase Chain Reaction with Eppendorf Mastercycler ep Realplex

1. Set up individual PCRs in individual wells of a thermo-fast 96 detection plate.
2. PCRs are performed in a total volume of 10 μl per well per sample by adding 2.0 μl cDNA to a 8 μl of a master mixture prepared for each primer (volume of master mixture depends on the number of samples) as follows: 5.0 μl SYBR Green (Platinum SYBR green qPCR Supermix-UDG), 0.5 μl forward primer, 0.5 μl reverse primer (primer stock concentrations of 5 μM), and 2 μl water. Primers used are shown in Table 1.
3. Run each sample-primer combination at least in duplicates. For each sample, also determine the expression of a housekeeping gene to correct for variations in cDNA concentrations. Run all reactions in duplicate and include a no-template control (in which the cDNA is replaced by nuclease-free water) for each particular PCR mix. For every primer set, include an adequate positive control.
4. Cover the plates with ultra-clear cap strips and then vortex the plate and centrifuge at 100 × *g* for 1 min before setting up the plate for the PCR reaction in the Eppendorf Mastercycler ep realplex real-time PCR system.
5. Run the PCR using the following program: 50°C for 2 min, 95°C for 10 min, and 40 cycles at 95°C for 15 s and 60°C for 1 min. In the end, follow a dissociation protocol to obtain a melting curve by adding the following steps: 95°C for 15 s, 60°C for 10 min (stepwise temperature changes for melting curve), and 95°C for 15 s.

6. Results of gene expression can be calculated using different methods: ΔCt ($Ct_{gene\ of\ interest} - Ct_{GAPDH}$) and then represented as % of expression of GAPDH ($2\text{–}\Delta^{Ct} \times 100$). It is also sometimes represented as ΔΔCt (ΔCt (sample 1) – ΔCt (sample 2)). For example, in analyzing the outcome of differentiation with time, the expression of a lineage marker is represented as Log_2 (relative expression over day 0 undifferentiated cells) (or ΔΔCt calculated for an expression in a sample at a given time point along with the that at day 0) (see Note 8).

3.4. Flow Cytometry

The cell surface phenotype of a rat MAPC is $CD31^+$ and $CD44^-$.

3.4.1. Expression of Surface Markers by FACS Analysis (Fig. 4.3a)

1. Collect cells by trypsinization and wash with PBS containing 3% FBS.
2. Block the cells for 10 min in PBS containing 3% FBS.
3. Distribute cells into three FACS tubes (about 100,000 cells/tube) per sample. Centrifuge at 450 × *g* for 5 min. Discard the supernatant.
4. Dilute antibodies for CD31-PE and CD44-FITC and isotype controls 1:20 times in PBS containing 3% FBS.
5. To the first tube, add 100 μl of PBS + 3% FBS without any antibodies (unstained control). In the second tube, add 5 μl of 1:20 diluted isotype controls and in the third tube, add 5 μl of 1:20 diluted CD31-PE and CD45-FITC antibodies and make to a total volume of 100 μl in PBS + 3% FBS.
6. Run the samples on a flow cytometer (e.g., BD Canto) and follow the manufacturer's instructions to obtain data on the samples.

3.4.2. Intracellular Staining for Oct4 by Flow Cytometry (Fig. 4.3a)

1. Harvest cells by trypsinization and wash with PBS + 3% FBS (twice) and count the number of cells to divide 100,000 cells per FACS tube in 500 μl. For each sample, divide into following tubes: (a) Unstained control, (b) Isotype control, and (c) samples.
2. Spin down for 6 min at 450 × *g* at 4°C and fix the cells in 4% formaldehyde using 1 ml/tube for 15–20 min at RT.
3. Wash with 2 ml SAP serum buffer and block in SAP serum buffer for 1 h at RT.
4. Wash and add goat Oct4 antibody diluted in SAP buffer to a concentration of 1 μg/ml. Add 100 μl per tube and gently pipet to mix the cells and leave at RT for 1 h. Add 1 μg/ml of isotype to the isotype control tube.
5. Add 1 ml SAP serum buffer and spin down (1,400 rpm, 6 min, 4°C). Add Cy3 anti-goat antibody at 1 μg/ml in SAP buffer in 100 μl per tube and incubate for 30 min at RT.

6. Wash twice in SAP buffer and then once in PBS and strain the contents through the filter before analysis on the FACS.

3.5. Immunocytochemistry

Immunocytochemistry is performed to confirm nuclear protein expression of Oct4 (Fig. 3b).

1. Remove culture medium and rinse cells gently once with PBS at room temperature.
2. Fix cells for 15 min with 4% formaldehyde solution.
3. Rinse once with PBS, proceed with staining or store slides in PBS at 4°C (3–4 weeks).
4. Permeabilize cells 15 min with PBS + Triton for intracellular staining for Oct4.
5. Block endogenous peroxidase activity by incubating cells for 30 min with 0.3% H_2O_2 in 100% methanol.
6. Rinse twice with distilled water and transfer to PBS–Triton solution.
7. Block aspecific binding sites for 30 min with 0.4% fish skin gelatin in PBS.
8. Remove blocking buffer and incubate overnight in humid conditions at 4°C with primary antibody goat Oct4 (Santa Cruz) diluted 1:200 in Dako antibody diluent.
9. Rinse thrice (for 5 min each time) with PBS + Triton. Incubate with secondary biotinylated donkey anti-goat antibody diluted 1:1,500 in Dako antibody diluent for 30 min at RT. Immediately, prepare ABC complex according to the manufacture's protocol by incubating solution A and solution B for 30 min at room temperature in PBS–Triton.
10. Remove secondary antibody solution and rinse thrice (for 5 min each time) with PBS–Triton.
11. Incubate 30 min at room temperature with ABC complex (after 30 min of complexation).
12. Rinse thrice (for 5 min each time) with PBS–Triton.
13. Develop around 5 min with DAB chromogen (watch for overdevelopment of isotype control under microscope).
14. Rinse thrice with distilled water and transfer to PBS for analysis for nuclear staining under light microscope.

3.6. Multilineage Differentiation Capacity

To conclusively determine if one has isolated MAPC, it is necessary to perform differentiation toward the three main lineages (mesoderm, endoderm, neuroectoderm). Low density is required for neuroectodermal differentiation, while endodermal differentiation is performed at 100% confluency and for endothelial differentiation 80–90% confluency is required. Precise cell numbers/cm^2

are given below. For instructions on the preparation of growth/differentiation medium, coating material and stock solutions of the recombinant cytokines, we refer to the materials section (see Notes 8–10 and Fig. 5 for representative example of results).

3.6.1. Hepatocyte Differentiation (Endoderm)

The precise current protocol will be published shortly in a separate manuscript; hence, readers are referred to that manuscript for the detailed protocol.

3.6.2. Endothelial Differentiation (Mesoderm)

1. Prepare 24-well culture plates (Corning) by coating them with 650 μl/well of 1 μg/ml fibronectin.
2. Trypsinize and count the undifferentiated MAPC.
3. For endothelial differentiation, the cell seeding density is 60,000 cells/cm^2.
   Note: It is particularly important to count the cells accurately, since the seeding density is one of the critical parameters determining the result of the differentiation.
4. Resuspend the MAPC in expansion medium. Aspirate the fibronectin solution just before seeding and add 1 ml of cell suspension per well (after carefully mixing the cell suspension to homogeneity). Incubate the wells in a humidified incubator (5% O_2 – 5.8% CO_2 – 37°C) and let the cells attach.
5. After about 16 h after seeding, remove the expansion medium and add 1 ml of PBS to each well. When finished with all the wells, take off the PBS and at this point, add the endothelial differentiation step 1 medium containing 5% serum. Incubate the wells in a humidified incubator (5% O_2 – 5.8% CO_2 – 37°C). Remove 650 μl of the medium and add 700 μl of fresh step 1 medium on day 2.
6. On day 4, remove 500 μl from each well and add 500 μl fresh differentiation medium. Add the endothelial differentiation step 2 medium with 2.5% FBS for the remaining time of the differentiation process. Repeat medium change of step 6 every 2 days from day 6 to day 20.
7. Finish the differentiation process on day 20. Collect samples for RNA with 350 μl of cell lysis buffer. Perform RNA isolation, DNAse treatment and cDNA synthesis. Check for expression of endothelial markers (Flk-1).
8. See Fig. 5 for a representative example of endothelial transcripts levels obtained with this protocol.

3.6.3. Neuroectoderm Differentiation

1. Precoat six-well plates with 0.1% gelatine solution for 30 min at room temperature.
2. Trypsinize and count cells (see Subheading 3.2.3).

3. Plate rat MAPC at 1,500 cells/cm^2 in N2B27 medium for 2 days.
4. Change medium completely to NSE medium on day 2 of differentiation.
5. Finish the differentiation process on day 9, when small clusters of cells detach from the plastic and start forming neurosphere-like structures. Collect samples for RNA with 350 μl of cell lysis buffer. Perform RNA isolation (RNA mini kit), DNAse treatment and cDNA synthesis (Superscript III). Check for expression of early neuroectodermal markers (Sox2 and Pax6) (*see* Subheading 3.3.1, Table 1).
6. See Fig. 5 for a representative example of neural precursor transcripts levels obtained with this protocol.

3.7. Cytogenetics by G-Banding

1. Look at cells. Observe rounded-up mitotic cells. Ensure that there are enough cells (about 60% confluency) in the flask for cytogenetic analysis.
2. Add 60 μl Colcemid to each culture (10 ml medium).
3. Incubate for about 3 h at 37°C until about 50% of the cells are floating.
4. Remove medium to conical tubes.
5. Rinse dish twice with 4–5 ml of Hanks' balanced salt solution and pour off into the cell waste container.
6. Add 1–2 ml trypsin and incubate at 37°C for 2 min. Tap the flask or plate to lift off the cells and then add 2 ml of fresh complete medium with serum to stop the action of trypsin.
7. Centrifuge at 300 × *g* for 5 min.
8. Pour off supernatant, transfer pellet to a 15-ml conical tube with PBS.
9. Centrifuge at 300 × *g* for 5 min.
10. Add 10 ml of Hypotonic solution – 0.075 M KCl, prewarmed to 37°C, let stand for 16 min in a 37°C water bath.
11. Centrifuge at 300 × *g* for 5 min.
12. Pour off supernatant. Invert tube and blot on paper towel to remove most of the supernatant. Use remaining supernatant to resuspend the pellet by gently flicking the tube.
13. Add 10 ml fresh methanol/acetic acid fixative and flick gently to mix and let stand at room temperature for 20 min.
14. Centrifuge at 300 × *g* for 10 min.
15. Remove supernatant, resuspend in 200–300 μl of fresh fixative.

16. Label slide. Put 50 μl of cell suspension on slide. Wait ~2 s until drop has spread, then dry with the blow-dry method: place slide on slide warmer and blow to quickly evaporate methanol.
17. Store excess cell suspension at 4°C for a few months or at -20°C indefinitely. Before reuse, dilute with fixative to 10 ml, repellet, discard supernatant, resuspend in 200–300 μl of fixative, and continue with step 16.
18. Leave slides on slide warmers in 90°C oven for 2 h to dry the remaining fixative.
19. Dip slide in 50% pancreatin solution for 1–3 s and rinse in cool water and drain onto a paper towel.
20. Place the slide on staining rack and stain with a mixture made by mixing one part of Wright's stain with three parts of phosphate buffer (stain for an optimal time, determined daily by using one or more test slides).
21. Dip twice in distilled water then place slides on slide warmer until dry.
22. Use oil-immersion lens (40×) to collect at least 20 countable spreads. If banding does not appear optimal, try varying the time of dipping in trypsin and restaining or try destaining the slide in methanol for 2 min and restaining as above.

3.8. Screening of Mycoplasma Contamination

1. Trypsinize cells and wash once in PBS.
2. Discard the supernatant and store the cell pellet at -20°C until further analysis.
3. Isolate genomic DNA (Invisorb spin cell mini kit) according to kit instructions and dilute DNA to 50 ng/2 μl.
4. Perform real-time PCR, using platinum SYBR green qPCR Supermix-UDG. Fw Primer (used at 50 nM): ACACC-ATGGGAGCTGGTAAT; Rv Primer (used at 50 nM): CCTCATCGACTTTCAGACCCAAGGCAT. PCR program (Eppendorf Realplex): 2 min at 50°C, 2 min at 95°C. Forty cycles with 30 s of denaturation at 94°C, 2 min annealing at 64°C and 2 min of extension at 72°C, followed by a melting curve (15 s at 95°C, 10 min at 60°C, and 15 s at 95°C).

4. Notes

1. Rat MAPC lines were successfully isolated from Fisher rats and Sprague-Dawley rats. So far, it is not known whether MAPC can be obtained from other rat strains.

2. A critical component of the MAPC medium is the lot of the serum that is used. It is essential to screen different batches of serum for their ability in supporting the isolation and maintenance of MAPC. Once a new line is generated, this line can be used for screening of the new batches of serum.
3. For the column depletion, cells are labeled with antibodies directed against hematopoietic cell-specific cell surface antigens, coupled to magnetic particles (MACS microbeads are ~50 nm sized superparamagnetic particles). A strong magnetic field is then created by placing an MACS column into a permanent magnet (MACS separator). When the cell-loaded column is rinsed with buffer, the unwanted (nontarget) cells are retained on the column whereas all the unlabeled target cells pass through the column, and are collected as a highly pure (~99.99%) fraction. We observed differences in the size of cells between the fractions (10 ml units of collection of cells that have passed through the column). In addition, a higher frequency of MAPC appears to grow out from the second and third fractions than the first fraction.
4. The subcloning step is very important. When maintained as bulk populations, isolation of MAPC is less efficient, possibly because other cells within the mixed population prevents the proliferation of MAPC clones. Hence, subcloning is performed after column depletion. Cells are selected from the 96-well plates 2 weeks later. The wells selected for further propagation contain cells with small morphology and which do not form colonies, i.e., proliferate without contact with neighboring cells. Subcloning immediately following CD45 depletion at the single-cell level is difficult. Therefore, we commonly perform the initial subcloning at 5–10 cells/well, and repeat the subcloning at 0.5 cells/well when an MAPC line is generated. It is important to evaluate these clonal populations for cytogenetic stability since cytogenetic abnormalities have been observed.
5. Although the phenotype of established rat MAPC lines is known, i.e., $CD31^+$ and expression of pluripotency genes as Oct4 and primitive endodermal genes as Sox7 and Sox17 (see Figs. 3 and 4), we have not yet been able to use this information to prospectively isolate MAPC from rat BM. This could mean that the cell that gives rise to MAPC is phenotypically different and potentially undergoes reprogramming under specific culture conditions at a particular point during the isolation (this will be addressed in a forthcoming manuscript).
6. As the process of derivation of MAPC involves culture of cells from the bone marrow for multiple weeks, it is necessary to evaluate the karyotype of newly isolated lines. In addition, intensive follow-up of the karyotype has to be done as well.

7. We have evidence that rat MAPC expressing *Oct4* at a ΔCt level of 3–6, once isolated under low-density conditions can be maintained at higher cell densities. Culture for >40 days at high density or in 3D aggregates in complete MAPC medium does not induce loss of Oct4 or loss of differentiation ability. However, if cultured at high density in the absence of EGF, PDGF, and LIF, spontaneous differentiation is seen, with acquisition of early endodermal and mesodermal markers (papers being submitted).
8. In our protocol, Sybr Green is used for real-time detection of PCR products. Sybr Green is less specific compared to systems with gene-specific fluorescently labeled probes. Since Sybr Green also can bind nonspecific to dsDNA, it can give more aspecific readings (i.e., primer-dimers). Therefore, it is recommended to include a dissociation curve analysis in the end of the PCR reaction. For each set of primers, it is advised that the obtained amplified product is send the first time for sequencing to assure that the primers amplify the correct gene. Some investigators also run the amplified products on a gel to assure that a band of the correct size is obtained with each qRT-PCR reaction, aside from performing dissociation curve analysis. Results of gene expression can be calculated using different methods. We mostly use the ΔCt method by calculating the $Ct_{\text{gene of interest}} - Ct_{\text{GAPDH}}$. The closer the gene of interest is to the housekeeping gene, the higher the Ct value of this gene and the lower the ΔCt value.
9. At the start of a differentiation, always freeze an RNA sample of the cells for determination of pluripotency and differentiation gene expression before induction of differentiation. In general, it is important to handle the cells gently; never let the wells with the cells dry out during rinsing or changing medium. To avoid dislodging of the cells, it is advisable not to add the medium straight on the cells, but instead add the medium slowly from the side of the well, avoiding contact with the cell monolayer, also when removing medium. It is particularly important to count the cells accurately, since the seeding density is one of the critical parameters determining the result of the different differentiations.
10. In contrast to previous published papers with mouse MAPC (15), the neuroectoderm differentiation of rat MAPC with the described mesendodermal phenotype is the most difficult to perform. So far, we only obtain induction of early neuroectodermal genes Sox2 and Pax6. Further optimization of the protocol is being carried out in our lab to obtain functional neurons in cell culture.

Acknowledgments

The authors would like to thank Dr Hirsch and LeAnn Oseth of the University of Minnesota cytogenetics core facility for accurate performance and analysis of G-banding experiments. Tine Decuyper, Sarah Mertens, Nathalie Feyaerts, Annelies Deman, April Breyer, Tineke Notelaers, Pieter Berckmans, and Sumitha Jacob are acknowledged for their excellent technical support with cell culture, stainings and qRT-PCR. Vick Vanduppen is acknowledged for his FACS support. This work was supported by grants to CMV: NIH-T15 HL076653, KUL CoE, FWO Odysseus Fund; FP6-STREP: STROKEMAP. KS was funded by Doctoral dissertation fellowship (DDF), Graduate School, University of Minnesota. MG was funded by FWO; DJO was supported by the NIH biotechnology Training Grant (GM08347); KP was funded by the FWO, BAEF, and VVGE.

References

1. Jiang, Y., Jahagirdar, B. N., Reinhardt, R. L., Schwartz, R. E., Keene, C. D., Ortiz-Gonzalez, X. R., et al. (2002) Pluripotency of mesenchymal stem cells derived from adult marrow. *Nature.* **418,** 41–49.
2. Jiang, Y., Vaessen, B., Lenvik, T., Blackstad, M., Reyes, M. and Verfaillie, C. M. (2002) Multipotent progenitor cells can be isolated from postnatal murine bone marrow, muscle, and brain. *Exp. Hematol.* **30,** 896–904.
3. Luttun, A., Ross, J. J., Verfaillie, C., Aranguren, X. L. and Prosper, F. (2006) Differentiation of multipotent adult progenitor cells into functional endothelial and smooth muscle cells. *Curr. Protoc. Immunol.* **Chapter 22,** Unit 22F 29.
4. Aranguren, X. L., McCue, J. D., Hendrickx, B., Zhu, X. H., Du, F., Chen, E., et al. (2008) Multipotent adult progenitor cells sustain function of ischemic limbs in mice. *J. Clin. Invest.* **118,** 505–514.
5. Serafini, M., Dylla, S. J., Oki, M., Heremans, Y., Tolar, J., Jiang, Y., et al. (2007) Hematopoietic reconstitution by multipotent adult progenitor cells: precursors to long-term hematopoietic stem cells. *J. Exp. Med.* **204,** 129–139.
6. Yoon, Y. S., Wecker, A., Heyd, L., Park, J. S., Tkebuchava, T., Kusano, K., et al. (2005) Clonally expanded novel multipotent stem cells from human bone marrow regenerate myocardium after myocardial infarction. *J. Clin. Invest.* **115,** 326–338.
7. D'Ippolito, G., Diabira, S., Howard, G. A., Menei, P., Roos, B. A. and Schiller, P. C. (2004) Marrow-isolated adult multilineage inducible (MIAMI) cells, a unique population of postnatal young and old human cells with extensive expansion and differentiation potential. *J. Cell Sci.* **117,** 2971–2981.
8. Anjos-Afonso, F. and Bonnet, D. (2007) Nonhematopoietic/endothelial SSEA-1+ cells define the most primitive progenitors in the adult murine bone marrow mesenchymal compartment. *Blood.* **109,** 1298–1306.
9. Kogler, G., Sensken, S., Airey, J. A., Trapp, T., Muschen, M., Feldhahn, N., et al. (2004) A new human somatic stem cell from placental cord blood with intrinsic pluripotent differentiation potential. *J. Exp. Med.* **200,** 123–135.
10. Kucia, M., Halasa, M., Wysoczynski, M., Baskiewicz-Masiuk, M., Moldenhawer, S., Zuba-Surma, E., et al. (2007) Morphological and molecular characterization of novel population of CXCR4+ SSEA-4+ Oct-4+ very small embryonic-like cells purified from human cord blood: preliminary report. *Leukemia.* **21,** 297–303.
11. Kucia, M., Reca, R., Campbell, F. R., Zuba-Surma, E., Majka, M., Ratajczak, J., et al. (2006) A population of very small embryonic-like (VSEL) CXCR4(+)SSEA-1(+)Oct-4+ stem cells identified in adult bone marrow. *Leukemia.* **20,** 857–869.
12. De Coppi, P., Bartsch, G., Jr., Siddiqui, M. M., Xu, T., Santos, C. C., Perin, L., et al.

(2007) Isolation of amniotic stem cell lines with potential for therapy. *Nat. Biotechnol.* **25,** 100–106.

13. Beltrami, A. P., Cesselli, D., Bergamin, N., Marcon, P., Rigo, S., Puppato, E., et al. (2007) Multipotent cells can be generated in vitro from several adult human organs (heart, liver, and bone marrow). *Blood.* **110,** 3438–3446.
14. Ulloa-Montoya, F., Kidder, B. L., Pauwelyn, K. A., Chase, L. G., Luttun, A., Crabbe, A., et al. (2007) Comparative transcriptome analysis of embryonic and adult stem cells with extended and limited differentiation capacity. *Genome Biol.* **8,** R163.
15. Jiang, Y., Henderson, D., Blackstad, M., Chen, A., Miller, R. F. and Verfaillie, C. M. (2003) Neuroectodermal differentiation from mouse multipotent adult progenitor cells. *Proc Natl Acad Sci U S A.* **100 Suppl 1,** 11854-11860.
16. Breyer, A., Estharabadi, N., Oki, M., Ulloa, F., Nelson-Holte, M., Lien, L., et al. (2006) Multipotent adult progenitor cell isolation and culture procedures. *Exp. Hematol.* **34,** 1596–1601.

Chapter 5

Generation of Functional Insulin-Producing Cells from Human Embryonic Stem Cells In Vitro

Yan Shi

Abstract

Human pancreatic islet transplantation at present is the preferred therapeutic option for type I diabetes treatment. However, this therapy is not widely utilized because of the severe shortage of donor islets. The capacity for self-renewal and differentiation of human embryonic stem (hES) cells makes them a potential new source for generation of functional pancreatic islet cells for treating type I diabetes mellitus. Here, we report a simple and effective protocol, carried out in a serum-free system, which could induce human ES cells to differentiate into functional insulin-producing cells. Activin A was first used in the initial stage to induce definitive endoderm lineage differentiation from human ES cells. And all-trans Retinoic Acid (RA) was then utilized to promote pancreatic differentiation. After maturation in the final induction stage with bFGF and Nicotinamide, the differentiated cells expressed islet specific markers. The secretion of insulin and C-peptide by these cells corresponded to the variations in glucose levels. Our method provides a promising in vitro differentiation model for studying the mechanisms of human pancreas development and illustrates the potential of using human ES cells for the treatment of type I diabetes mellitus.

Key words: Human embryonic stem cells, Differentiation, Activin A, All-trans Retinoic Acid, Pancreas, Insulin-producing cells

1. Introduction

Islet transplantation has been suggested to be a promising treatment for type I diabetes. However, this therapy is not widely utilized because of the severe shortage of donor islets (1, 2). Human embryonic stem (ES) cells can be maintained in vitro for extended periods without loss of genetic stability, and are potential sources for generating a variety of specialized human cells needed for clinical applications. They hold the promise of serving as an unlimited source of insulin-producing donor cells for type I diabetes cell therapy (3, 4).

S. Ding (ed.), *Cellular Programming and Reprogramming: Methods and Protocols*, Methods in Molecular Biology, vol. 636, DOI 10.1007/978-1-60761-691-7_5, © Springer Science+Business Media, LLC 2010

Therefore, it is worthwhile to identify effective approaches to induce human ES cells to differentiate into functional insulin-producing cells in vitro.

It has been reported that human ES cells can spontaneously differentiate into insulin-producing cells in vitro (5, 6). A five-stage protocol similar to that described by Lumelsky et al. (7) has been used to induce human ES cell differentiation into insulin-producing cells (8). However, further investigation is still needed to confirm that human ES cells are indeed induced to become functional insulin-producing cells by these methods because of a possibility of exogenous insulin uptake (9). Therefore, it is necessary for us to explore more efficient induction approaches, which could mimic the pancreas development process in vivo and obtain the insulin-secreting cells with the characteristics of pancreatic beta cells. Activin A was the key induction factor for this step, which was reported to be indispensable for definitive endoderm differentiation from both mouse ES (10, 11) and human ES cells (12, 13). It was demonstrated that human ES cells were able to be differentiated into up to 80% definitive endoderm cells with Activin A treatment (12, 13). Currently, most human ES cells to pancreatic cell differentiation methods would choose Activin A to induce definitive endoderm specialization for the first step.

On the other hand, the quality and efficiency of this cell population would determine the final production of pancreatic islet cells from hES cell differentiation. RA is an important signaling molecule in the development of the early embryonic pancreas besides functions on induction of ectoderm and mesoderm development (14). During zebrafish development, increased RA signaling could induce remarkable anterior expansion of the pancreas and liver endoderm. Conversely, inhibition of RA signaling by BMS493 blocks early pancreas differentiation from embryonic endoderm (15). It was firstly showed that after Activin A and RA treatment in sequence, differentiated mouse ES cells could express pancreatic progenitor markers such as pdx1, hnf3ß, and $hnf4_{\alpha}$ (10). It was also further confirmed the role of RA in pancreatic specialization in nuclear-transfer mouse ES cells (16). Recently, more evidences (12, 17–19) showed that RA could induce pancreatic progenitor cell specification from human ES-derived definitive endoderm cells.

Here, we gave the detailed process of an effective approach in serum-free culture medium to induce human ES cells to differentiate into functional insulin-producing cells by combining Activin A, All-trans retinoic acid (RA) in chemically defined medium (CDM), and other maturation factors such as bFGF and Nicotinamide in islet maturation medium (IMM). Depending on this method, which consisted of an incubation with Activin A and RA in CDM followed by maturation in DMEM/F12 serum-free medium supplemented with bFGF and Nicotinamide, the human

ES cell-derived cells expressed islet specific genes such as pdx1, insulin, C-peptide, glut2, glucagon, and amylase. The C-peptide-positive cells were TUNEL-negative and the percentage of C-peptide-positive cells achieved was more than 15%. The secretion of insulin by these cells was responsive to variations in glucose levels. After transplantation into diabetic nude mice, 30% of the animals showed an obvious rescue of their hyperglycemia phenotype, and this condition was maintained for more than 6 weeks (12). Therefore, our approach offers a promising in vitro model for studying human pancreas development and also helps fulfill the urgent need of an ample supply of insulin-producing cells for cell transplantation therapy in type I diabetes.

2. Materials

2.1. Human ES Cell Line Expansion

1. The human ES cell lines H1 and H9 derived from WiCell Research Institute.
2. Mouse embryonic fibroblast cells (20).
3. Tissue culture plates from Falcon.
4. Mitomycin C Solution: 1 mg/mL mitomycin C (Roche) stock solution prepared in PBS (1:100 dilute before use). Store in the dark at 4°C, stable for 2 weeks.
5. Dispase Solution: Thaw the Dispase powder in PBS at the concentration at 0.5 mg/mL, store at −20°C, stable for 2–4 weeks.
6. Human ES cell culture medium: DMEM/F12 (Invitrogen) with 20% Knockout Serum Replacement (KSR) (Invitrogen) containing 8 ng/mL of bFGF (Invitrogen), Nonessential amino acids (1:100, Invitrogen), 4 mM l-Glutamine, 0.1 mM 2-Mercaptoethanol (Invitrogen), Penicillin/Streptomycin (1:100, Invitrogen); store at 4°C.

2.2. Human ES Cell Differentiation

1. Matrigel store at −80°C. Before differentiation, dilute Matrigel as 1:100 with 4°C DMEM/F12 medium. Coat tissue culture plates with diluted Matrigel and incubate at 37°C for 30 min to 1 h. Then the Matrigel-coated tissue culture plates are ready for use.
2. Human ES cell differentiation medium:
   (a) Chemical Defined Medium (CDM): 50% IMDM (Invitrogen) plus 50% F12 Nutrient Mixture (Invitrogen), supplemented with Insulin-Transferrin-Selenium-A (1:100, Invitrogen), 450 mM Monothioglycerol (Sigma), and 5 mg/mL Albumin Fraction V (Sigma).

(b) Islet Maturation Medium (IMM): DMEM/F12, Insulin-Transferrin-Selenium-A (1:100) and 2 mg/mL albumin fraction V (Sigma).

3. Factors on human ES cell to insulin-producing cell differentiation: Activin A (R&D System) 50–100 ng/mL (see Note 1), all-trans Retinoic Acid (Sigma) 10^{-6}M, bFGF (Invitrogen) 10 ng/mL, and Nicotinamide (Sigma) 10 mM.
4. Ultra Low Attachment culture dishes (Costar) for Suspension culture.

2.3. Human ES Cell Transplantation

1. 4–6-week-old BALB/c male nude mice or NOD/SCID mice.
2. Streptozotocin (STZ, Sigma).
3. GlucoTREND2 (Roche) for blood glucose test.
4. Mouse antihuman nuclei monoclonal antibody (1:30, CHEMICON) for in vivo human ES-derived cell detection.

3. Methods

The sketch of human ES cell to insulin-producing cell differentiation protocol is summarized in Fig. 1. All incubations are performed in a standard, humidified, cell culture incubator, at 37°C in 5% CO_2.

3.1. Day 0 Prepare hES Cells for Differentiation

1. hES cell lines are maintained following the typical protocol from WiCell Research Institute (WiCell's Proven Protocols).
2. Before induction, hES cells are splitted as 1:3 (60% confluent) and replated onto 1% Matrigel-coated tissue culture dishes. hES cells are incubated with hES culture medium overnight for attachment.

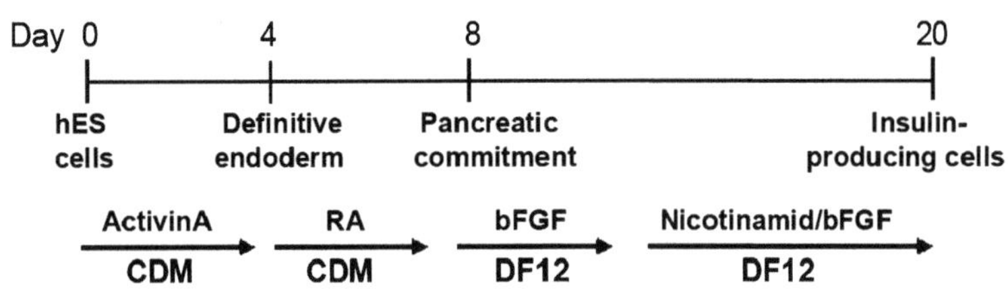

Fig. 1. The sketch of human ES cell differentiation protocol. Undifferentiated human ES cells were first cultured in CDM containing Activin A for 4 days. Then, the differentiated cells were further induced with RA in CDM for 4 days and transferred from CDM culture medium into DMEM/F12 islet maturation medium with bFGF added as a pancreatic cell maturation factor for 3 days. Finally, the differentiated cells were switched to DMEM/F12 islet maturation medium containing bFGF and nicotinamide for another 5 days

3.2. Days 1–4 Induce Human ES Cells to Differentiate into Definitive Endoderm Cells

1. Next day, aspirate the human ES cell culture medium from the dishes and wash with PBS.
2. Change the medium into CDM or DMEM/F12 medium with 50–100 ng/mL Activin A and replace with fresh medium every day.
3. After 4-day induction with CDM or DMEM/F12 medium plus Activin A, the 70–80% human ES cells would differentiate into definitive endoderm cells, which expressed definitive endoderm specific markers such as Sox17 and Cxcr4.

3.3. Days 5–8 Pancreatic Lineage Specification from Differentiated Human ES Cells

After 4-day differentiation with CDM or DMEM/F12 medium plus 50–100 ng/mL Activin A, human ES cells are further induced with 10^{-6} M RA in CDM or DMEM/F12 medium for another 4 days. At this stage, some Pdx1-positive cells appear which indicate the early pancreatic cell specification.

3.4. Days 9–20 Insulin-Producing Cell Matured from Differentiated Human ES Cells

1. Following the treatment with Activin A and RA, we transferred the differentiated cells from CDM or DMEM/F12 medium to the IMM containing 10 ng/mL bFGF as a pancreatic cell maturation factor for 3 days.
2. The differentiated cells were switched to IMM containing 10 mM Nicotinamide and 10 ng/mL bFGF for another 3–7 days for insulin-producing cell maturation (see Note 2).

3.5. Transplantation of the Differentiated Insulin-Producing Cells Derived from hES Cells In Vivo

1. Prior to transplantation, Streptozotocin (STZ) was injected i.p. at 40 mg/kg/day into 4–6-week-old BALB/c male nude mice or NOD/SCID mice for 5 days to induce experimental diabetes. The blood glucose of STZ-induced nude mice should be more than 15 mM.
2. When the STZ-treated mice had developed diabetes, about 1×10^6 differentiated cells in the final induction stage were transplanted into the left renal capsule. PBS or cells without Activin A and RA induction were used as control.
3. After more than 1 month, the cell-transplanted kidneys were removed from those mice whose blood glucose was rescued after induced cell transplantation. Blood glucose was measured by GlucoTREND2 from snipped tail. If the diabetes of STZ-treated mice were rescued, the blood glucose of the induced cell-transplanted mice should be maintained at normal levels (<13.9 mM) for nearly 6 weeks.
4. At least 1 month after transplantation, the Cryostat sections of the operated kidneys were prepared and C-peptide expression of the transplanted cells in the renal capsule was confirmed by immunohistochemistry. The hES-derived cells were detected by Mouse antihuman nuclei monoclonal antibody (1:30).

4. Notes

1. We identified that 50 ng/mL Activin A could be enough to obtain nearly 70% definitive endoderm cells from human ES cells. However, higher concentration of Activin A (100 ng/mL) would be helpful to increase the differentiation efficiency.
2. At the final induction stage, insulin-producing cells could be detected in the adhension culture. We discovered that if the differentiated cells were digested by 0.5 mg/mL dispase and transferred into Ultra Low Attachment culture dishes for suspension culture, insulin production efficiency of the differentiated cells would be increased.

References

1. Shapiro AM, Lakey JR, Ryan EA, et al. (2000) Islet transplantation in seven patients with type 1 diabetes mellitus using a glucocorticoid-free immunosuppressive regimen. *N Engl J Med.* **343,** 230–238.
2. Hussain MA, Theise ND. (2004) Stem-cell therapy for diabetes mellitus. *Lancet* **364,** 203–205.
3. Thomson JA, Itskovitz-Eldor J, Shapiro SS, et al. (1998) Embryonic stem cell lines derived from human blastocysts. *Science* **82,** 1145–1147.
4. Cowan CA, Klimanskaya I, McMahon J, et al. (2004) Derivation of embryonic stem-cell lines from human blastocysts. *N Engl J Med.* **350,** 1353–1356.
5. Assady S, Maor G, Amit M, et al. (2001) Insulin production by human embryonic stem cells. *Diabetes* **50,** 1691–1697.
6. Brolén G, Heins N, Edsbagge J, et al. (2005) Signals from the embryonic mouse pancreas induce differentiation of human embryonic stem cells into insulin-producing beta-cell-like cells. *Diabetes* **54,** 2867–2874.
7. Lumelsky N, Blondel O, Laeng P, et al. (2001) Differentiation of embryonic stem cells to insulin-secreting structures similar to pancreatic islets. *Science* **292,** 1389–1394.
8. Segev H, Fishman B, Ziskind A, et al. (2004) Differentiation of human embryonic stem cells into insulin-producing clusters. *Stem Cells* **22,** 265–274.
9. Hansson M, Tonning A, Frandsen U, et al. (2004) Artifactual insulin release from differentiated embryonic stem cells. *Diabetes* **53,** 2603–2609.
10. Shi Y, Hou L, Tang F, et al. (2005) Inducing embryonic stem cells to differentiate into pancreatic β cells by a novel three-step approach with activin A and all-trans retinoic acid. *Stem Cells* **23,** 656–662.
11. Yasunaga M, Tada S, Torikai-Nishikawa S, et al. (2005) Induction and monitoring of definitive and visceral endoderm differentiation of mouse ES cells. *Nat Biotechnol.* **23(12),** 1542–1550.
12. Jiang W, Shi Y, Zhao D, et al. (2007) In vitro derivation of functional insulin-producing cells from human embryonic stem cells. *Cell Res.* **17,** 333–344.
13. D'Amour KA, Agulnick AD, Eliazer S, et al. (2005) Efficient differentiation of human embryonic stem cells to definitive endoderm. *Nat Biotechnol.* **23,** 1534–1541.
14. Maden M. (2001) Role and distribution of retinoic acid during CNS development. *Int Rev Cytol.* **209,** 1–77.
15. Stafford D, Prince VE. (2002) Retinoic Acid signaling is required for a critical early step in Zebrafish pancreatic development. *Curr Biol.* **12,** 1215–1220.
16. Jiang W, Bai Z, Zhang D, et al. (2008) Differentiation of mouse nuclear transfer embryonic stem cells into functional pancreatic beta cells. *Diabetologia* **51,** 1671–1679.
17. D'Amour KA, Bang AG, Eliazer S, et al. (2006) Production of pancreatic hormone-expressing endocrine cells from human embryonic stem cells. *Nat Biotechnol.* **24,** 1392–1401.
18. Shim JH, Kim SE, Woo DH, et al. (2007) Directed differentiation of human embryonic

stem cells towards a pancreatic cell fate. *Diabetologia* **50(6)**, 1228–1238.

19. Zhang D, Jiang W, Liu M, et al. (2009) Highly efficient differentiation of human ES cells and iPS cells into mature pancreatic insulin-producing cells. *Cell Res.* **19(4)**, 429–438.
20. Robertson EJ. (1997) Derivation and maintenance of embryonic stem cell cultures. *Methods Mol Biol.* **75**, 173–184.

Chapter 6

Mesoderm Cell Development from ES Cells

Takumi Era

Abstract

Pluripotent stem cells such as embryonic stem (ES) and induced pluripotent stem (iPS) cells have attractive attention as a source of cells for use in therapeutic application. However, as the in vitro differentiation culture does not provide usefully positional information for cell type definition, this system definitely requires visible markers to identify and monitor the intermediates that present on the way of differentiation. We have been developing the cell surface markers against the various types of mesoderm in the ES cell culture. Using it, we have identified the intermediates of mesoderm and dissected their differentiation pathways in ES cell differentiation. The method described here could be useful for inducing and purifying mesoderm cells from iPS as well as ES cell cultures.

Key words: Embryonic stem cell (ES cell), Induced pluripotent stem cell (iPS cell), Mesoderm, Vascular endothelial growth factor receptor 2 (VEGFR2), Platelet-derived growth factor receptor α (PDGFRα)

1. Introduction

1.1. Mesoderm Development in the Embryos

In mouse embryo, mesoderm development starts at E6.5 and, for a short time, dramatically produces three major types of mesoderm: organizer, embryonic mesoderm, and extraembryonic mesoderm (1, 2). The most initial mesoderm appears at a proximal region in epiblasts of embryo as an early gastrula organizer (EGO) (3). EGO migrates into anterior part of embryo and becomes mid gastrula organizer (MGO) that contributes to axial mesoderm. While organizer migrates, the epiblast at posterior region subsequently begins to transform to second type of mesoderm, embryonic mesoderm, in primitive streak (4). Along with the elongation of primitive streak distally, embryonic mesoderm become to diversify region-specifically two types of mesoderms, paraxial and lateral mesoderm, which eventually forms a majority

S. Ding (ed.), *Cellular Programming and Reprogramming: Methods and Protocols*, Methods in Molecular Biology, vol. 636, DOI 10.1007/978-1-60761-691-7_6,

of mesoderm progenies such as bone and blood cells, respectively. The epiblasts at the proximal part of embryo also produce a third type of mesoderm; extraembryonic mesoderm. The precursors of this mesoderm move into the nascent streak and migrate to extra-embryonic part in which they mainly give rise to primitive hematopoietic cells (HPCs) and endothelial cells (ECs) (5, 6).

1.2. Flat Culture for In Vitro ES Cell Culture

The major aim of in vitro ES cell culture is to establish the culture condition that induces ES cell to the efficient differentiation for specific cell lineages (7). Embryoid-body (EB) formation method seemed to be suitable for attaining this. The method is based on the idea that ES cell differentiation requires the environments which are similar to those present in the actual embryo. However, EB exhibits a complex structure that disturbs the cells inside to meet the appropriate signals from outside. As a result, the culture conditions around EB are not able to exclude differentiation into unnecessary lineages. Previously, our study demonstrated that EB culture is less efficient in inducing mesendoderm cells expressing *Goosecid*, which is one of the markers for EGO, than the two-dimensional (2D) culture on collagen IV-coated dishes (8). This result indicates an inherent limitation of EB system in guiding ES cell differentiation, as uncontrollable complexity is inevitably associated with three-dimensional architecture in EB. To overcome the obstacle of EB formation, we prefer to use the flat culture system rather than EB formation method for an in vitro ES cell differentiation.

1.3. Differentiation of Mesoderm Cells in In Vitro ES Cell Culture

ES cells have the multiple potentials to give rise to a whole cell types in mouse body and to undergo unlimited symmetrical divisions with maintaining its pluritotency (9). The high ability for differentiation and unlimited growth capacity leads us to expect to utilize it as the source of cell therapies such as transplantation. Moreover, the forced differentiation system of ES cell in vitro has been expected to use as a good tool to find the developmental pathways into the specific cell lineage and to dissect them from others. However, as ES cell differentiation culture does not provide usefully positional information for cell type definition, this system definitely requires visible markers to identify and monitor the intermediates that present on the way of differentiation. In fact, availability of Vascular Endothelial Growth Factor Receptor 2 (VEGFR2, FLK1) that marks the subtypes of mesoderm cells with a potential to give rise to HPCs and ECs facilitates our understanding on the developmental pathways of these lineages (10–12). Another important surface marker involving in mesoderm development is Platelet-derived growth factor receptor alpha (PDGFRα) that is mainly expressed in paraxial mesoderm during mouse embryogenesis (13–15). We have exploited these markers for dissecting the differentiation course of ES cell-derived mesoderm cells. Our previous results obtained from in vitro ES cell culture shows that PDGFRα^{+}VEGFR2^{+} cell (DP) which initially appears at day 3.5 ES cell culture is a common precursor for PDGRFRα$^+$VEGFR2$^-$(PSP) and PDGFRα$^-$VEGFR2$^+$(VSP) cells (16). Based on the results of in vitro fate analysis, we found a new differentiation pathway in which the DP gives rise to both the PSP and the VSP that eventually differentiate into bone and cartilage cells, and HPCs and ECs, respectively (Fig. 1) (16). These indicate that PSP and VSP populations represent the paraxial and lateral mesoderm populations in actual mouse embryo, respectively. The analyses for gene expression in both populations also support the hypothesis that PSP and VSP correspond to paraxial and lateral mesoderms, respectively (Fig. 2).

2. Materials

2.1. ES Cell Lines

With numerous ES cell lines currently available, we recommend feeder-free ES cell lines such as CCE, EB3, EB5, and E14tg2a (17, 18). Before real experiments, each cell line should first be examined for its ability to generate VEGFR2$^+$ and PDGFRα$^+$ cells. CCE is usually analyzed in our laboratory.

2.2. ES Cell Maintenance

1. KNOCKOUT-Dulbecco's modified Eagle's medium (KO-DMEM, Invitrogen) is stored at 4°C.
2. Leukemia inhibitory factor (LIF) (Chemicon International) was purchased as 1×10^7 U/mL in a rubber-capped vial. Use a 1-mL syringe with a needle to push 1 mL of air into the bottle and pull out all the liquid. Aliquots of 100 μL each (1×10^6 U) are stored in sterilized cryotubes at –80°C.
3. Fetal bovine serum (FBS) pretested for ES cells (see Note 1).
4. 0.1% (w/v) Gelatin.

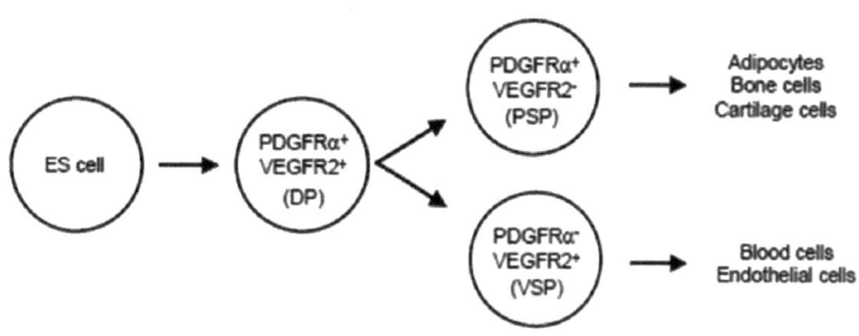

Fig. 1. Differentiation pathway of mesoderm in in vitro ES cell culture. The analyses of differentiated ES cells reveal the three types of mesoderm cells, PDGFRα$^+$VEGFR2$^+$ population (PDGFRα and VEGFR2 double positive population, DP), PDGFRα$^+$VEGFR2$^-$ population (PDGFRα single positive population, PSP) and PDGFRα$^-$VEGFR2$^+$ population (VEGFR2 single positive population, VSP). The DP is the most immature and can give rise to both the PSP and the VSP. Both the VSP and the PSP exhibit the specific properties of paraxial and lateral mesoderm respectively

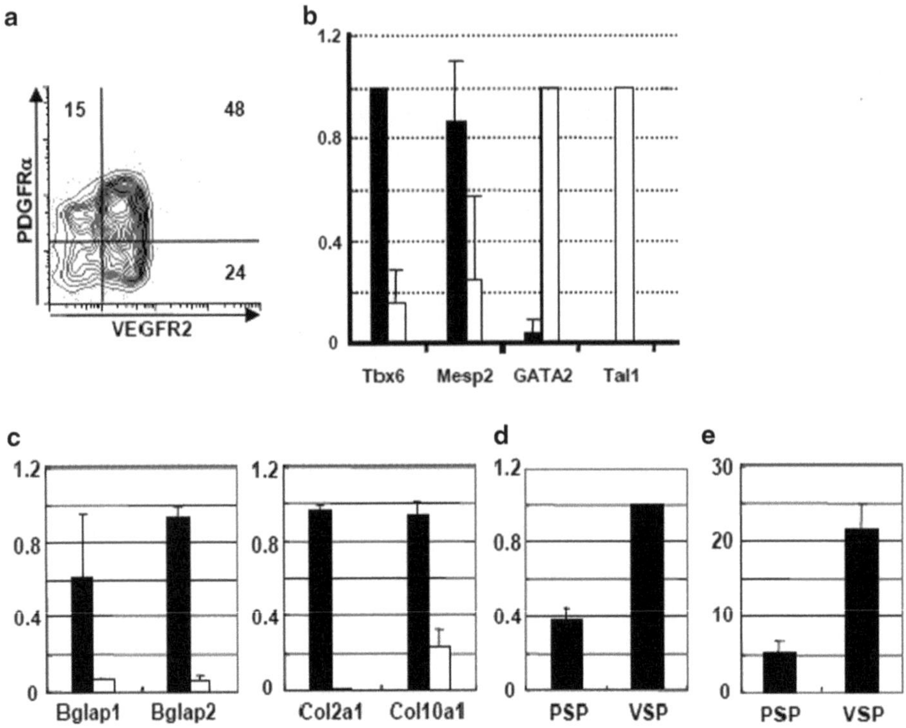

Fig. 2. Fate of the ES cell-derived mesoderm cells. (**a**) Day4 ES cell differentiation. CCE ES cells are cultured on type IV collagen-coated dishes with the differentiation medium in the absence of LIF. Four days after the induction, differentiated ES cells are harvested and the expression of PDGFRα and VEGFR2 are examined by FACS. Four populations (PDGFRα$^+$VEGFR2$^+$, DP; PDGFRα$^+$VEGFR2$^-$, PSP; PDGFRα$^-$VEGFR2$^+$, VSP; PDGFRα$^-$VEGFR2$^-$, DN) are observed in day 4 differentiated ES cells. (**b**) Marker expression by quantitative RT-PCR (qPCR). PSP expresses the markers specific for paraxial mesoderm such as Tbx6 and Mesp2. In contrast, VSP expresses the markers specific for lateral mesoderm such as GATA2 and Tal1, suggesting that it represents lateral mesoderm in actual embryo. *Black and white* squares indicate the expression in PSP and VSP, respectively. (**c**) Marker expression of bone and cartilage cells. The ES cell-derived mesoderm populations are cultured under distinct conditions that allow the differentiation of osteocytes or chondrocytes. After differentiation, RNA is purified and the expression levels of individual specific markers are measured by qPCR. Culture cells derived from the PSP expresses osteogenesis-(Bglap1 and Bglap2, *left panel*) and chondrogenesis-(col2a1 and col10a1, *right panel*) related genes at higher level than that of the VSP. *Black and white* squares indicate the expression in PSP and VSP, respectively. (**d**) The expression level of βHI in the cultures of PSP and VSP. The expression level of βHI is measured by qPCR and normalized by GAPDH expression level. Culture of VSP exhibits the higher expression of βHI than that of PSP. This suggests that VSP generates hematopoietic cells more efficiently than PSP. (**e**) The number of endothelial colonies derived from different mesoderm populations. 500 sorted cells are cultured on confluent OP9 cell layer for 3 days. Endothelial colonies are visualized by VE-cadherin immunostaining. The number of VE-cadherin$^+$ colonies is counted in each well of 24-well plates. (Error bars = SD). The frequency of endothelial progenitors in the PSP is a quarter of those of the VSP

Add 0.5 g Gelatin into 500 mL deionized water (culture-grade) and autoclaved. Store at room temperature.

5. Dulbecoco's Phosphate Buffered Saline without calcium and magnesium chloride (D-PBS).
6. 2-Mercaptoethanol (2-ME): stock solution: 1,000× (0.1 M). Add 70 μL 2-ME to 10 mL PBS and sterilized by 0.2-μm filter. Store up to 4 weeks at 4°C. Final concentration in medium: 10^{-4} M.

7. L-Glutamine (200 mM; 100×) and Penicillin-streptomycin (P/S; 100×) are stored in 15-mL centrifuge tubes as 5 mL aliquots at −20°C.
8. Non-Essential Amino Acids (NEAA; 100×) is stored at 4°C.
9. 0.25% (w/v) trypsin-EDTA is stored in 10 mL aliquots at −20°C.
10. 6 and 10 cm culture dishes (Becton Dickinson).
11. ES cell culture medium: KO-DMEM, 15%FBS, 10^{-4} 2-ME, 2 mM L-glutamine, 1× P/S, 0.1 mM NEAA, 1,000 U/mL LIF. Store up to 4 weeks at 4°C.

2.3. OP9 Stromal Cell Maintenance

1. Minimum essential medium α medium (αMEM) with ribonucleosides and deoxyribonucleosides.
2. FBS pretested for OP9 cells (see Note 1).
3. D-PBS.
4. L-Glutamine and Penicillin-streptomycin.
5. 0.05% (w/v) trypsin-EDTA is stored in 10 mL aliquots at −20°C.
6. OP9 culture medium: αMEM, 20%FBS, 2 mM L-glutamine, 1× P/S.

2.4. In Vitro ES Cell Differentiation Without OP9

1. αMEM.
2. FBS pretested for in vitro ES cell differentiation (see Note 2).
3. 2-ME: Final concentration in medium: 5×10^{-5} M.
4. L-Glutamine (200 mM 100×).
5. Penicillin-streptomycin (100×).
6. BIOCOAT Collagen IV-coated 10 cm dish (Becton Dickinson).
7. Differentiation culture medium: αMEM, 10%FBS, 5×10^{-5} M 2ME, 2 mM L-Glutamine, 1× P/S (see Note 3).

2.5. In Vitro ES Cell Differentiation with OP9

1. αMEM.
2. FBS pretested for in vitro ES cell differentiation (see Note 2).
3. L-Glutamine (200 mM 100×).
4. Penicillin-streptomycin (100×).
5. OP9 differentiation culture medium : αMEM, 20%FBS, 2 mM L-Glutamine, 1× P/S.

2.6. Purification of Mesoderm Cells

1. D-PBS.
2. Cell dissociation buffer.
3. 0.25% (w/v) Trypsin-EDTA.

4. Neutralization buffer for Cell dissociation buffer and Trypsin-EDTA: D-PBS, 10%FBS.
5. Normal mouse serum (NMS). NMS can be prepared from in-house or can be purchased (Chemicon international). Sterilized by 0.2-μm filter and aliquots of 500 μL each are stored in sterilized tubes at −20°C.
6. Hank's balanced Salt Solution (HBSS) (10×).
7. HBSS/BSA: 1× HBSS, 1% bovine serum albumin (BSA).
8. HBSS/BSA/PI: HBSS/BSA with 5 μg/ml propidium iodide.
9. Anti-VEGFR2 (AVAS12): Phycoerythrin-conjugated (eBioscience) and Allophycocyanin-conjugated (eBioscience).
10. Anti-PDGFRα (APA5)(eBioscience) Biotin-conjugated.
11. Allophycocyanin-conjugated Streptavidin (SAV-APC, eBioscience).

2.7. Bone Cell Differentiation from ES Cell-Derived Mesoderm Cells

1. Dulbecco's modified Eagle's medium (DMEM).
2. FBS pretested for in vitro ES cell differentiation (see Note 2).
3. L-Glutamine (200 mM 100×).
4. Penicillin-streptomycin (100×).
5. Dexamethasone.
6. Ascorbic acid 2-phosphate.
7. β-glycerophosphate.
8. Recombinant Human BMP4 (R&D systems).
9. 24-well culture plate (Becton Dickinson).
10. Bone cell differentiation medium: DMEM, 10%FBS, 2 mM L-Glutamine, 1× P/S, 0.1 μM Dexamethasone, 50 μM ascorbic acid 2-phosphate,10 mM β-glycerophosphate, 10 ng/mL BMP4.
11. 4% Paraformaldehyde solution (PFA).
12. Alizarin red S.
13. Ammonium hydroxide solution (28%).
14. Alizarin red staining solution:
    (a) Solution A: 1 g Alizarin red S, 100 mL distilled water.
    (b) Solution B: 0.1 mL Ammonium hydroxide solution,100 mL distilled water.
    (c) Mix solution A well. Adjust the pH 6.36–6.40 with solution B. The pH is critical, so make fresh or check pH if the solution is more than 1 month old. Keep at room temperature up to 6 months.

2.8. Cartilage Cell Differentiation from ES Cell-Derived Mesoderm Cells

1. αMEM.
2. FBS pretested for in vitro ES cell differentiation (see Note 2).
3. L-Glutamine (200 mM 100×).
4. Penicillin-streptomycin (100×).
5. Dexamethasone.
6. Ascorbic acid 2-phosphate.
7. Recombinant Human TGF-β3 (R&D systems).
8. Recombinant Human BMP2 (R&D systems).
9. 24-well culture plate (Becton Dickinson).
10. Cartilage cell differentiation medium: αMEM, 10%FBS, 2 mM L-Glutamine, 1× P/S, 0.1 μM Dexamethasone, 170 μM ascorbic acid 2-phosphate.
11. 4% PFA.
12. Alcian Blue.
13. Glacial acetic acid.
14. Alcian blue staining solution:
    (a) Solution 1: 3 mL glacial acetic acid, 97 mL distilled water.
    (b) Solution 2: 1 g Alucian blue, 100 mL Solution 1.
    (c) Mix Solution 2 well for 30 min. Then, filtrate it through filter paper. Store at 4°C up to 6 months.

2.9. Hematopoietic Cell Differentiation from ES Cell-Derived Mesoderm Cells

1. αMEM.
2. FBS pretested for in vitro ES cell differentiation (see Note 2).
3. L-Glutamine (200 mM 100×).
4. Penicillin-streptomycin (100×).
5. Cytokines: Recombinant human Erythropoietin (hEpo) (R&D systems), Recombinant human Interleukin-3 (hIL-3) (R&D systems), Recombinant human Stem cell factor (SCF) (R&D systems).
6. 6-well culture plate (Becton Dickinson).
7. HPC differentiation medium: αMEM, 20%FBS, 2 mM L-Glutamine, 1× P/S, 2 U/mL hEpo.

2.10. Endothelial Cell Differentiation from ES Cell-Derived Mesoderm Cells (ES Cell-Derived Endothelial Colony Assay)

1. αMEM.
2. FBS pretested for in vitro ES cell differentiation (see Note 2).
3. L-Glutamine (200 mM 100×).
4. Penicillin-streptomycin (100×).
5. 24-well culture plate (Becton Dickinson).
6. EC differentiation medium: αMEM, 20%FBS, 2 mM L-Glutamine, 1× P/S.

3. Methods

Feeder-independent ES cells such as CCE and E14tg2a are used for an in vitro ES cell differentiation because of the easy maintenance.

3.1. ES Cell Culture

All dishes, flasks, and plates should be gelatinized before use.

3.1.1. Gelatin Coating of Dishes

1. Add enough 0.1% gelatin solution to cover the plate surface. 6 cm dish – 3 mL, 10 cm dish – 7 mL, 1 well of 24-well plate – 0.5 mL.
2. Let the solution sit for at least 10 min at room temperature.
3. Aspirate the gelatin solution completely just before use.

3.1.2. Thawing of ES Cells

1. ES cells are removed from liquid nitrogen storage or deep-freezer (–150°C) and quickly thawed in a 37°C water bath.
2. Transfer ES cells into a 15-mL centrifuge tube containing 10 mL of 37°C prewarmed ES cell culture medium.
3. Spin down ES cells at low speed (190 × *g* for 5 min) at room temperature. Remove medium by suction and resuspend the cell pellet in 2 mL of ES cell culture medium by gently repeated pipetting.
4. More than 2×10^6 ES cells are transferred to a gelatin-coated 10 cm dish containing 8 mL of prewarmed ES cell culture medium and cultured in a tissue culture incubator (37°C, 5% CO_2).
5. Change entire medium daily until semiconfluent.

3.1.3. Passage of ES Cells

1. Once the ES cells grow to 70% confluence, passage them to new tissues culture dishes treated by gelatin. They should be passaged every 2–3 days as described below.
2. Aspirate medium and wash cells once with 37 prewarmed D-PBS. Volume of D-PBS needed for: 6-cm dish – 3 mL and 10-cm dish – 6 mL.
3. To remove ES cells from dish, add 0.5 mL of 37°C prewarmed 0.25% trypsin-EDTA into the 10 cm dish. Incubate in tissue culture incubator at 37°C for 5 min.
4. Add 5 mL of 37°C prewarmed ES cell culture medium and break up the cell aggregates by repeated pipetting 8–15 times (see Note 4).
5. Transfer ES cells into a 15-mL centrifuge tube and spin down them at 270 × *g* for 5 min at room temperature. Resuspend the cell pellets in 5 mL of 37°C prewarmed ES cell culture medium. Count cell number and seed ES cells at:

Seeding number	Days needed for confluence	Confluent cell number
1×10^6/10-cm dish	3 days	2–3×10^7/dish
2×10^6/10-cm dish	2 days	2–3×10^7/dish
8×10^6/10-cm dish	1 days	2–3×10^7/dish

6. Daily complete medium change is required until confluent.

3.1.4. Cell Freezing

1. Prepare 2× freezing solution: 20% dimethyl sulfoide (DMSO), 80% FBS. Keep on ice. Make fresh every time.
2. Remove cells from dish as in Subheading 3.1.3.
3. Resuspend 4×10^6 cells in 0.25 mL ice-cold FBS and keep on ice.
4. Add an equal amount of 2× freezing solution. Freeze the cells at −80°C overnight. The next day, transfer vials to a liquid nitrogen tank or ultra deep-freezer (−150°C).

3.2. Maintenance of OP9 Stromal Cell Line for In Vitro ES Cell Differentiation

For maintenance of OP9 stromal cell line, the over-confluent condition should be avoided because the cells that undergo an overgrowth will stop their growth. We recommend 90% confluence on passage (see Note 5).

3.2.1. Thawing of OP9 Cells

1. Thaw frozen vial in a 37°C water bath.
2. Transfer OP9 cells into a 15-mL centrifuge tube containing 10 mL of 37°C prewarmed OP9 culture medium.
3. Spin down OP9 cells at 190×*g* for 5 min at room temperature. Resuspend cells in 2 mL of OP9 culture medium by gently repeated pipetting.
4. Seed 5×10^5 OP9 cells/6-cm dish and cultured in a tissue culture incubator (37°C, 5% CO_2).

3.2.2. Passage of OP9

1. Aspirate medium and wash cells once with 37°C prewarmed D-PBS. Volume of D-PBS needed for: 6-cm dish – 3 mL and 10-cm dish – 6 mL.
2. To remove OP9 cells from dish, add 0.5 mL of 37°C prewarmed 0.05% trypsin-EDTA. Incubate in tissue culture incubator at 37°C for 5 min.
3. Add 5 mL of 37°C prewarmed OP9 culture medium and break up the cell aggregates by repeated pipetting 8–15 times.
4. Transfer OP9 cells into a 15-mL centrifuge tube and spin down them at 270×*g* for 5 min at room temperature. Resuspend the cell pellets in 5 mL of 37°C prewarmed ES cell

culture medium. We usually obtain: 7–8 × 10^5 cells/6-cm dish and 1.2–1.6 × 10^6/10-cm dish.

5. Seed 2–4 × 10^5 cells/10-cm dish.
6. OP9 cells should not be cultured for longer than 1 month after thawing. In addition, over 20-passage OP9 cells should not be used for in vitro ES cell differentiation (see Note 5).

3.2.3. Storing of OP9

1. Prepare 2× freezing solution. Keep on ice. Make fresh every time.
2. Remove cells from dish as in Subheading 3.2.2.
3. Resuspend 6–8 × 10^5 cells in 0.25 mL ice-cold 100% FBS and keep on ice.
4. Add an equal amount of 2× freezing solution. Freeze the cells at −80°C overnight. The next day, transfer vials to a liquid nitrogen tank.

3.3. In Vitro ES Cell Differentiation

Before induction, ES cells should be maintained for at least 1 week after thawing. For the differentiation into hematopoietic and ECs, we recommend the condition in the presence of OP9 cells.

3.3.1. Induction of Mesoderm Cells Without OP9 Cells

1. Aspirate medium and wash cells *twice* with 37°C prewarmed D-PBS. Volume of D-PBS needed for: 6-cm dish – 3 mL and 10-cm dish – 6 mL.
2. To remove ES cells from dish, add 0.5 mL of 37°C prewarmed 0.25% trypsin-EDTA into the 10 cm dish. Incubate in tissue culture incubator at 37°C for 5 min.
3. Add 5 mL of 37°C prewarmed *differentiation culture medium* and break up the cell aggregates by repeated pipetting 8–15 times.
4. Transfer ES cells into a 15-mL centrifuge tube and spin down them at 270 × *g* for 5 min at room temperature. Resuspend the cell pellets in 5 mL of 37°C prewarmed differentiation culture medium.
5. In the case of feeder-dependent ES cells, to remove feeder cells, ES cells harvested are incubated at 37°C with 5% CO_2 for 30 min. Collect the floating cells by pipetting gently.
6. Add 8 × 10^4 undifferentiated ES cells into 10-cm collagen IV-coated dish containing 15 mL of prewarmed differentiation culture medium.
7. Change a half of medium (~8 mL) on day 3.

3.3.2. Induction of Mesoderm Cells with OP9 Cells

1. Prepare 10-cm dish with 90% confluent OP9 cells. OP9 is splitted 3 days before in vitro ES cell differentiation.
2. Aspirate medium and wash cells *twice* with 37°C prewarmed D-PBS. Volume of D-PBS needed for: 6-cm dish – 3 mL and 10-cm dish – 6 mL.

3. To remove ES cells from dish, add 0.5 mL of 37°C prewarmed 0.25% trypsin-EDTA into the 10 cm dish. Incubate in tissue culture incubator at 37°C for 5 min.
4. Add 5 mL of 37°C prewarmed *OP9 differentiation culture medium* and break up the cell aggregates by repeated pipetting 8–15 times.
5. Transfer ES cells into a 15-mL centrifuge tube and spin down them at 270 ×*g* for 5 min at room temperature. Resuspend the cell pellets in 5 mL of 37°C prewarmed OP9 differentiation culture medium.
6. In the case of feeder-dependent ES cells, to remove feeder cells, ES cells harvested are incubated at 37°C with 5% CO_2 for 30 min. Collect the floating cells by pipetting gently.
7. Add 8×10^4 undifferentiated ES cells into 10-cm dish containing both 90% confluent OP9 and 10 mL of prewarmed OP9 differentiation culture medium.

3.3.3. Purification of Mesoderm Cells from the Culture Without OP9

On day4 culture day, the ES cells induced into mesoderm differentiation are harvested and analyzed by FACS to examine the expression pattern of mesoderm markers (Fig. 2a) (see Note 6).

1. Aspirate medium and wash cells *twice* with 37°C prewarmed D-PBS.
2. To remove the differentiated cells from dish, add 8 mL of 37°C prewarmed cell-dissociation buffer into the 10 cm dish. Incubate in tissue culture incubator (37°C, 5% CO_2) for 15 min.
3. Add 8 mL of 37°C prewarmed D-PBS with 10% FBS and break up the cell aggregates by repeated pipetting 10–15 times.
4. Transfer the cells into a 50-mL centrifuge tube and spin down them at 270 ×*g* for 5 min at 4°C. Resuspend the cell pellets in 8 mL of ice-cold D-PBS with 10% FBS. Count cell number. Transfer 1×10^7 cells into 15-mL centrifuge tube and spin down at 270 ×*g* for 5 min at 4°C.
5. Resuspend the cell pellet in 100 μL of ice-cold NMS and incubate the single-cell suspensions for 20 min on ice.
6. Add an appropriate concentration of PE-labeled anti-VEGFR2 and biotin-labeled anti-PDGFRα mAbs to cell suspension in NMS and incubate for 20 min on ice.
1. Add 10 mL of ice-cold HBSS/BSA into cell solution and spin down it at 270 ×*g* for 5 min at 4°C.
7. Resuspend cell pellet in 100 μL of ice-cold HBSS/BSA and add an appropriate concentration of SA-APC. Incubate it for 20 min on ice.

8. Wash the cells *twice* with 10 mL ice-cold HBSS/BSA. Resuspend the cells in 1 mL of ice-cold HBSS/BSA/PI for dead cell exclusion.
9. Analyze and sort VEGFR2$^+$PDGFRα$^+$, VEGFR2$^+$PDGFRα$^-$, and/or VEGFR2$^-$PDGFRα$^+$ according to your experiments (Fig. 2a, b) (see Note 7).

3.3.4. Purification of Mesoderm Cells from the Culture with OP9

On *day 5* culture day, the ES cells induced into mesoderm differentiation are harvested and analyzed by FACS to examine the expression pattern of mesoderm markers. We recommend you to investigate VEGFR2 expression but not PDGFRα as PDGFRα is also expressed in OP9 stromal cells.

1. Prepare the cells as in Subheading 3.3.3 (from steps 1 to 5).
2. Add an appropriate concentration of PE-labeled or APC-labeled anti-VEGFR2 to cell suspension in NMS and incubate for 20 min on ice.
3. To wash the cells, add 10 mL of ice-cold HBSS/BSA into the cell solution and spin down it at 270 × *g* for 5 min at 4°C.
4. Repeat above washing once and resuspend cell pellet in 100 μL of ice-cold HBSS/BSA/PI.
5. Analyze and sort VEGFR2$^+$ mesoderm cells (Fig. 2b) (see Note 7).

3.4. Differentiation into Descendants of the Mesoderm Cells

3.4.1. Induction of Bone Cells

1. Gelatinize the wells of 24-well plate as in Subheading 3.1.1.
2. Seed 1–3 × 10^3 ES-derived mesoderm cells purified by FACS into 1 well of gelatinized 24-well plate with 1 mL of bone cell differentiation medium.
3. Change a half of medium (~0.5 mL) every 3 days.
4. The calcium deposit can be observed around on day 28 (Fig. 2c) (see Note 8).

3.4.2. Alizarin Red Staining

To confirm the bone cell formation, specific staining is needed.

1. To fix the cells, add 1 mL of 4%PFA into the well. Keep at room temperature for 10 min.
2. Wash twice by D-PBS, 5 min, room temperature.
3. Add 1 mL of Alizarin red staining solution and keep for 5 min at room temperature.
4. Quickly wash 5–6 times by D-PBS.
5. Observation: Calcium deposit is stained to red color.

3.4.3. Induction of Cartilage Cells

1. Resuspend ES cell-derived mesoderm cells in cartilage cell differentiation medium at 8 × 10^6/mL concentration and put 10 μL of this solution on a well of 24-well plate.

2. Incubate in tissue culture incubator at 37°C for 30 min.
3. Add slowly 1 mL of prewarmed cartilage cell differentiation medium with 10 ng/mL *TGFβ3* into the well of plate.
4. One week later, change completely the medium by pre-warmed cartilage cell differentiation medium with 10 ng/mL *BMP2*. Do *not* add *TGFβ3*.
5. Change a half of cartilage cell differentiation medium with 10 ng/mL *BMP2* (~0.5 mL) every 3 days.
6. Analyze the cartilage generation on day 21 (Fig. 2c) (see Note 8).

3.4.4. Alcian Blue Staining

To confirm the cartilage cell formation, specific staining is needed.

1. To fix the cells, add 1 mL of 4%PFA into the well. Keep at room temperature for 10 min.
2. Wash twice by D-PBS, 5 min, room temperature.
3. Add 1 mL of 3% glacial acetic acid into the well and keep for 5 min at room temperature.
4. Discard the glacial acetic acid and add 1 mL of Alcian blue staining solution. Keep for 30 min at room temperature.
5. Quickly wash 5–6 times by 3% glacial acetic acid.
6. Observation: Muco-glycoprotin is stained to blue color.

3.4.5. Induction of Hematopoietic Cells

3.4.5.1. Generation of Primitive Erythrocytes

1. To prepare 6-well plate with confluent OP9 stromal cells, one confluent OP9 10-cm dish is spilt to four 6-well plates 3 days before the experiment. Prepare VEGFR2$^+$ ES cell-derived mesoderm cells as in Subheading 3.3.
2. Resuspend 1×10^4 VEGFR2$^+$ ES cell-derived mesoderm cells in 2 mL of HPC differentiation medium and seed to a well of 6-well plate with the 90% confluent OP9 stromal cell.
3. Analyze primitive erythrocytes that appeared after 3–4 days.

3.4.5.2. Generation of Definitive Hematopoietic Cells

1. Prepare 6-well plate with 90% OP9 stromal cells and VEGFR2$^+$ ES cell-derived mesoderm cells as in Subheading 3.4.5 (step 1).
2. Resuspend 1×10^4 VEGFR2$^+$ ES cell-derived mesoderm cells in 2 mL of HPC differentiation medium with 10 ng/mL hIL3 and 100 ng/mL SCF.
3. Seed it to a well of 6-well plate with the 90% confluent OP9 stromal cell.
4. Change the medium every 3–4 days. In general, TER119$^+$ definitive erythroid cells initially appear in culture after 3 days. Gr-1$^+$ mature myeloid cells appear after 5–7 days. To confirm

the definitive erythropoiesis, the expression of β-hemoglobin gene in the culture is examined (Table 1 and Fig. 2d).

3.4.6. Endothelial Cell Colony Assay

1. One confluent OP9 10-cm dish is spilt to four 24-well plates 3 days before the experiment. Prepare VEGFR2$^+$ ES cell-derived mesoderm cells as in Subheading 3.3.
2. Resuspend 5×10^2 to 1×10^3 VEGFR2$^+$ ES cell-derived mesoderm cells in 2 mL of EC differentiation medium.
3. Seed it to a well of 24-well plate with the 90% confluent OP9 stromal cell.
4. Sheets of ECs growing on OP9 stromal cells can be observed after 3 days (Fig. 2e) (see Note 9).

4. Notes

1. Test ten different sera for ES and OP9 stromal cells. Select a serum lot that supports a good growth rate of ES and OP9 cells. Refer to Methods Subheadings 3.1.3 and 3.2.2 for the growth rates of ES and OP9 cells, respectively. Cell and colony morphologies are also the key factors to select a good serum.
2. FBS is a critical factor to induce a high rate of PDGFRα$^+$ and VEGFR2$^+$ cell induction. Lot no. checks of sera are highly recommended for finding appropriate serum lot. They are usually examined by the induction rate of ES cell-derived mesoderm cells. In general, using 20 different sera lots, the frequency of PDGFR2$^+$ and VEGFR2$^+$ cells generated after 4 days under the condition without OP9 ranged from 30 to 60%.
3. The medium for differentiation is used less than 4 week as old medium affects the induction rate of mesoderm cells.
4. ES cells have to be plated as single cells, otherwise, ES cells will differentiate even in the presence of LIF. A long term culture (more than 2 weeks) induces ES cells to accumulate genetic mutation including chromosomal abnormality. Therefore, thaw new ES cells every 2–3 week.
5. The condition of OP9 stromal cells influences the generation rate of ES cell-derived mesoderm cells. OP9 cells should not be cultured for longer than 1 month after thawing. OP9 cells may lose the ability to support ES cell differentiation as they easily differentiate into adipocytes by the long time culture. In addition, High passage (>20 passages) easily induce ES cells to be transformed and may lose their ability to support the in vitro ES cell differentiation.

6. In the case of CCE ES cells, 50–60% of day 4 differentiated ES cells express PDGFRα and VEGFR2 (16).
7. To confirm the ES cell-derived mesoderm cells, the gene expression specific for paraxial and lateral mesoderm cells is examined by quantitative RT-PCR (qPCR). We use Tbx6 and Mesp2, and GATA2 and Tal1 for paraxial and lateral mesoderm markers, respectively (*see* Table 1 and Fig. 2b) (16). GAPDH is used as a control (*see* Table 1).
8. The qPCR method is useful for examining the presence of bone and cartilage cells. Several markers such as Bglap 1 and 2, and Col2a1 and Col10a1 are suitable for defining the bone

Table 1
Primers for quantitative RT-PCR

Gene		Sequence
GAPDH	Sense	5′-GGAGCGAGACCCCACTAACA-3′
	Antisense	5′-GCCTTCTCCATGGTGGTGAA-3′
Tbx6	Sense	5′-CCCAACTATGCAGCCAACACT-3′
	Antisense	5′-CTGTGTGATCCTAGGGTTCTGGTA-3′
Mesp2	Sense	5′-CTGAAAACCTTGGGAACAGGAT-3′
	Antisense	5′-GGCTCTTTCTAGGGACTGGTGTAA-3′
GATA2	Sense	5′-CGGCCTCTTCTTCTGCAGG-3′
	Antisense	5′-TGGTACTTGACGCCATCCTTG-3′
Tal1	Sense	5′-CCCACCAGACAAGAAACTAAGCA-3′
	Antisense	5′-GGCCAGGAAATTGATGTACTTCA-3′
Bglap1	Sense	5′-GAGGACCATCTTTCTGCTCACTCT-3′
	Antisense	5′-GACATGAAGGCTTTGTCAGACTCA-3′
Bglap2	Sense	5′-GCGCTACCTTGGAGCTTCAG-3′
	Antisense	5′-CATACTGGTTTGATAGCTCGTCACA-3′
Col2a1	Sense	5′-CCTTGGACGCCATGAAAGTT-3′
	Antisense	5′-CTTGCTGCTCCACCAGTTTTT-3′
Col10a1	Sense	5’-CCTGGTTCATGGGATGTTTTATG-3’
	Antisense	5′-TGGCGTATGGGATGAAGTATTG-3′
βH1	Sense	5′-TGTTTACCCATGGACTCAGAGATTC-3′
	Antisense	5′-CTTTCTTGCCATGGGCTCTAA-3′

and cartilage cell lineages, respectively (see Table 1 and Fig. 2c) (16, 19).

9. Immuno-cytostaining is the easiest method for clarifying the presence of endothelial colonies in the culture. We routinely examine the expression of either VE-cadherin or PCAM-1(CD31) as the markers for ECs (16, 18).

Acknowledgements

I thank Dr. Nishikawa SI and Dr. Sakurai H for helpful discussion and supporting of these studies.

References

1. Tam PP and Beddington RS. (1987) The formation of mesodermal tissues in the mouse embryo during gastrulation and early organogenesis. *Development* **99**, 109–126.
2. Kimelman D. (2006) Mesoderm induction: from caps to chips. *Nat Rev Genet* **7**, 360–372.
3. Kinder SJ, Tsang TE, Wakamiya M, Sasaki H, Behringer RR, Nagy A and Tam PP. (2001) The organizer of the mouse gastrula is composed of a dynamic population of progenitor cells for the axial mesoderm. *Development* **128**, 3623–3634.
4. Lawson KA, Meneses JJ and Pedersen RA. (1991) Clonal analysis of epiblast fate during germ layer formation in the mouse embryo. *Development* **113**, 891–911.
5. Kinder SJ, Tsang TE, Quinlan GA, Hadjantonakis AK, Nagy A and Tam PP (1999) The orderly allocation of mesodermal cells to the extraembryonic structures and the anteroposterior axis during gastrulation of the mouse embryo. *Development* **126**, 4691–4701.
6. Zon LI. (1995) Developmental biology of hematopoiesis. *Blood* **86**, 2876–2891.
7. Keller G. (2005) Embryonic stem cell differentiation: emergence of a new era in biology and medicine. *Genes Dev* **19**, 1129–1155.
8. Tada S, Era T, Furusawa C, Sakurai H, Nishikawa S, Kinoshita M, Nakao K, Chiba T and Nishikawa S-I. (2005) Characterization of mesendoderm: a diverging point of the definitive endoderm and mesoderm in embryonic stem cell differentiation culture. *Development* **132**, 4363–4374.
9. Niwa H. (2007) How is pluripotency determined and maintained? *Development* **134**, 635–646.
10. Kabrun N, Buhring HJ, Choi K, Ullrich A, Risau W and Keller G. (1997) Flk-1 expression defines a population of early embryonic hematopoietic precursors. *Development* **124**, 2039–2048.
11. Nishikawa SI, Nishikawa S, Hirashima M, Matsuyoshi N and Kodama H. (1998) Progressive lineage analysis by cell sorting and culture identifies FLK1+VE-cadherin+cells at a diverging point of endothelial and hemopoietic lineages. *Development* **125**,1747–1757.
12. Yamashita J, Itoh H, Hirashima M, Ogawa M, Nishikawa S, Yurugi T, Naito M, Nakao K and Nishikawa S. (2000) Flk1-positive cells derived from embryonic stem cells serve as vascular progenitors. *Nature* **408**, 92–96.
13. Soriano P. (1997) The PDGF alpha receptor is required for neural crest cell development and for normal patterning of the somites. *Development* **124**, 2691–2700.
14. Kataoka H, Takakura N, Nishikawa S, Tsuchida K, Kodama H, Kunisada T, Risau W, Kita T and Nishikawa SI. (1997) Expressions of PDGF receptor alpha, c-Kit and Flk1 genes clustering in mouse chromosome 5 define distinct subsets of nascent mesodermal cells. *Dev Growth Differ* **39**, 729–740.
15. Takebe A, Era T, Okada M, Jakt LM, Kuroda Y and Nishikawa S-I. (2006) Microarray analysis of PDGFRα+populations in ES cell differentiation culture identifies genes involved in differentiation of mesoderm and mesenchyme including ARID3b that is essential for development of embryonic mesenchymal cells. *Dev Biol* **293**, 25–37.
16. Sakurai H, Era T, Jakt LM, Okada M, Nakai S, Nishikawa S, Nishikawa S-I. (2006) In vitro modeling of paraxial and lateral mesoderm

differentiation reveals early reversibility. *Stem Cells* **24**, 575–586.

17. Era T and Witte ON. (2000) Regulated expression of P210 Bcr-Abl during embryonic stem cell differentiation stimulates multipotential progenitor expansion and myeloid cell fate. *Proc Natl Acad Sci U S A* **97**, 1737–1742.
18. Era T, Izumi N, Hayashi M, Tada S, Nishikawa S and Nishikawa S-I. (2008) Multiple mesoderm subsets give rise to endothelial cells, whereas hematopoietic cells are differentiationed only from a restricted subset in embyomic stemm cell differentiation culture. *Stem Cells* **26**, 401–411.
19. Takashima Y, Era T, Nakao K, Kondo S, Kasuga M, Smith AG and Nishikawa S-I. (2007) Neuroepithelial cells supply an initial transient wave of MSC differentiation. *Cell* **129**, 1377–1388.

Chapter 7

Directed Differentiation of Red Blood Cells from Human Embryonic Stem Cells

Shi-Jiang Lu, Qiang Feng, Jennifer S. Park, and Robert Lanza

Abstract

Human embryonic stem cells (hESC) represent a new source of stem cells that can be propagated and expanded in vitro indefinitely, providing a potentially inexhaustible and donorless source of cells for human therapy. The ability to create banks of hESC lines with matched or reduced incompatibility could potentially reduce or eliminate the need for immunosuppressive drugs and/or immunomodulatory protocols altogether, for example, O-type RhD$^-$ lines for generation of universal red blood cells (RBC). Hematopoietic differentiation of hESCs has been extensively investigated in vitro, and hematopoietic precursors as well as differentiated progeny representing erythroid, myeloid, macrophage, megakaryocytic, and lymphoid lineages have been identified in differentiating hESC cultures. Previous studies also generated primitive erythroid cells from hESCs by embryoid body (EB) formation and coculturing with stromal cells. However, the efficient and controlled differentiation of hESCs into homogeneous RBC populations with oxygen-carrying capacity has not been previously achieved. In this chapter, we describe a robust system that can efficiently generate large numbers of hemangioblasts from multiple hESC lines using well-defined conditions and produce functional homogeneous RBCs with oxygen-carrying capacity in large scale. The homogeneous erythroid cells can be used for further mechanism studies.

Key words: Human embryonic stem cells (hESC), Embryoid body (EB), Hemangioblasts, Erythroid cells, Red blood cells (RBC)

1. Introduction

Since their discovery in 1998 (1), human embryonic stem cells (hESC) have been considered a promising source of replacement cells for human therapy. However, successful clinical translation depends on the efficient and controlled differentiation of hESCs toward specific cell lineages and the generation of homogeneous transplantable cell populations. One potentially important application is the use of hESC-derived hematopoietic stem cells to generate

S. Ding (ed.), *Cellular Programming and Reprogramming: Methods and Protocols*, Methods in Molecular Biology, vol. 636,
DOI 10.1007/978-1-60761-691-7_7, © Springer Science+Business Media, LLC 2010

mature blood cells, such as erythrocytes, for transfusion and the treatment of leukemia and other blood diseases. Differentiation of hESCs into hematopoietic cells has been extensively investigated in vitro, and hematopoietic precursors as well as differentiated progeny representing erythroid, myeloid, macrophage, megakaryocytic, and lymphoid lineages have been identified in differentiating hESC culture systems (2–8).

Red blood cells (RBCs) – the oxygen-carrying component of the blood – are required in over half of all anemic patients admitted to intensive care units in the USA (9–11). Unfortunately, the supply of transfusable RBCs, especially "universal" donor type O Rh-negative, is often insufficient, especially in the battlefield environment due to the lack of blood type information and the limited timeline for transfusion. Although alternative sources of progenitors for the generation of transfusable RBCs have been investigated, including cord blood, bone marrow, and peripheral blood (12–14), it is clear that even after expansion and differentiation, these progenitors represent donor-limited sources of RBCs. Moreover, the low prevalence of O(−) type blood in the general population (<8% in Western countries and <0.3% in Asia) further intensifies the consequences of blood shortages for emergency situations where blood typing may not be possible. hESCs and induced pluripotent stem (iPS) cells (15, 16) represent a new source of stem cells that can be propagated and expanded in vitro indefinitely, providing a potentially inexhaustible and donorless source of RBCs for human therapy.

The generation of erythrocytes has been achieved from hESCs either by embryoid body (EB) formation (6) or coculturing with stromal cells followed by isolation of CD34+ cells and further expansion/differentiation (17–19). Chang et al. (6) generated erythroid cells from hESCs by isolating and expanding nonadherent cells of day-14 EBs in a span of 15–56 days of culture time. The definitive-like, but nucleated erythroid cells obtained from the above approach, however, coexpressed high levels of embryonic ε- and fetal γ-globins with little or no adult β-globin. Olivier et al. (17) and Qiu et al. (18) have developed a method for a relatively large scale ($0.5–5 \times 10^7$ cells) production of erythroid cells from hESCs. In their method, hESCs were cocultured for 14–35 days with human fetal liver stromal cells (FHB-hTERT) to produce CD34+ cells that were seeded in a four-step culture system. In steps 1 and 2, cocktails of cytokines were used to promote the proliferation and maturation of erythroid precursors. In steps 3 and 4, erythriod cells were transferred onto mouse bone marrow stromal cells (MS5) to facilitate terminal maturation. Similar to the results observed by Chang et al. (6), the erythroid cells expressed mainly the embryonic ε- and fetal γ-globins, only a trace amount of the adult β-globin gene was detected by real-time PCR (mRNA level), but not by HPLC (protein level) analysis.

Recently, Ma et al. (19) showed that, by using immunostaining with globin chain specific monoclonal antibodies, almost 100% of hESC-derived erythrocytes expressed the adult β-globin chain after cocultured with murine fetal liver-derived stromal cells (mFLC) in vitro, suggesting erythrocytes derived from hESCs are capable of switching on the expression of definitive adult β-globin chain. However, the majority of the erythrocytes obtained by coculturing with mFLCs were nucleated.

Mammalian erythropoiesis is a complex multistep process that encompasses the differentiation of hematopoietic stem cells to mature erythrocytes. Erythropoiesis in mammals consists of two waves: (1) primitive erythropoiesis initiated in the yolk sac with the generation of large *nucleated* erythroblasts, and (2) definitive erythropoiesis arising from the fetal liver with the development of smaller *enucleated* erythrocytes. This complex differentiation process involves many steps including the differentiation of early erytroid progenitors (burst-forming units-erythroid, BFU-E), via late erythroid progenitors (colony-forming units-erythroid, CFU-E), and finally morphologically recognizable erythroid precursors (20). Nuclear condensation is a key event in late stages of erythropoiesis and enucleation is the final step in the development of mature erythrocytes. Thus, the presence or absence of a nucleus has long been accepted as a key distinguishing feature for primitive and definitive erythroid cells. Although enucleation of erythroblasts was structurally studied by electron microscopy almost half a century ago (21), little is known about the underlying mechanism(s). It has been suggested that enucleation is the result of asymmetric cell division involving extrusion of a pycnotic nucleus enveloped by the plasma membrane (22), and that Rac GTPases and their effector mDia2 play important roles in the process (23). Studies also suggest that direct contact of erythroblasts with macrophages promotes nuclear extrusion, and that knock-out of erythroblast-macrophage-protein (Emp) results in the failure of enucleation (24–26). However, two groups recently demonstrated that although macrophages play a role in the maturation of erythroblasts, they are neither sufficient nor required for red cell enucleation in the mouse system (27, 28).

We recently developed a two-step strategy that can efficiently and reproducibly generate hemangioblasts from hESCs (29). This differentiation system uses a defined serum-free medium and eliminates the use of feeder cells. One characteristic of hESC-derived hemangioblasts is that they efficiently generate BFU-E and CFU-E colonies when cultured in serum-free hematopoietic colony forming cell (CFC) medium. This prompted us to investigate whether hemangioblasts can be used as an intermediate to generate erythrocytes on a clinically relevant scale. A procedure was developed to generate functional RBCs (blood types A, B, O, and both RhD+ and RhD−) on a large scale from multiple hESC lines using hemangioblasts as intermediates on a large scale under serum-free conditions suitable for scale-up and clinical translation (30). Three steps are critical for the efficient scale-up of RBCs: (1) Generation of hemangioblasts with high efficiency and high density, without disruption of their colony forming environment. (2) Expansion of hemangioblasts to erythroblasts in a high cell density. (3) Culture of erythroblasts in semisolid media containing methylcellulose to provide optimal conditions for maximum expansion and erythroid purity. We have used this approach to generate 10^{10} to 10^{11} pure erythroid cells from one 6-well plate hESCs ($\approx 1 \times 10^7$ cells), which is over a 1,000-fold more efficient than previously reported (17). We have also demonstrated that the oxygen equilibrium curves of the hESC-derived erythroid cells are comparable to normal transfusible RBCs and respond to changes in pH and 2,3-diphosphoglyerate. Importantly, the cells underwent multiple maturation events in vitro, including a progressive decrease in size and increase in glycophorin A expression, and chromatin and nuclear condensation. This process resulted in the extrusion of the pycnotic nucleus in up to 60% of the cells. The enucleated erythrocytes appeared morphologically identical to normal RBCs with a diameter of approximately 6–8 μm. These cells also possess the capacity to express the adult definitive β-globin chain upon further maturation in vitro. Globin chain specific-PCR and -immunofluorescent analysis show that the cells increase expression of β-globin from 0 to 15% after in vitro culture. The results show that it is feasible to differentiate and mature hESCs into functional oxygen-carrying erythrocytes on a clinically applicable scale. The identification of a hESC line and generation of an iPS cell line with a O(−) genotype would permit the production of ABO and RhD compatible (and pathogen-free) "universal donor" RBCs.

2. Materials

2.1. hESC Culture

1. bFGF stock solution: Add 1.25 ml of protein-containing medium (use hES-BM plus 20% Serum Replacement) to a vial containing 10 μg of bFGF. This makes 8 μg/ml stock solution. Make 240 μl aliquots and freeze at −20°C.
2. PMEF growth medium: To a 500 ml bottle of high glucose DMEM add 6 ml Penicillin/Streptomycin (100× solution), 6 ml Glutamax-1 (100× solution), and 50 ml Fetal Bovine Serum (FBS), sterilize by 0.22 μm filtration, and store at 4°C.
3. hESC Basal Medium (hESC-BM): To a 500 ml bottle of KO-DMEM add 6 ml Penicillin/Streptomycin, 6 ml

Glutamax-1, 6 ml Nonessential amino acids (NEAA), (100× solution), and 0.6 ml β-mercaptoethanol (1,000× solution), and store at 4°C.

4. hESC growth medium (hESC-GM): To 200 ml of hES-BM add 40 ml Knockout Serum Replacement, 240 μl of human LIF for 10 ng/ml, and 240 μl of bFGF for 8 ng/ml. Sterilize by 0.22 μm filtration, and store at 4°C.
5. Gelatin (0.1%): Dissolve 0.5 g of gelatin (from porcine skin, Sigma) in 500 ml of warm (50–60°C) Milli-Q water. Cool down to room temperature, sterilize by 0.22 μm filtration.
6. Mitomycin C: Add 2 ml of sterile Milli-Q water to a vial (2 mg) of lyophilized Mitomycin C to make 1 mg/ml stock solution. The solution is light-sensitive and is good for 1 week at 4°C (see Note 1).

2.2. Embryoid Body (EB) Formation

1. VEGF solution: Add 1 ml of PBS (Ca^{++}, Mg^{++}-free) with 1% BSA to a vial containing 50 μg of human recombinant $VEGF_{165}$ (R & D Systems). Make 100 μl aliquots and freeze at −20°C. This makes a 50 μg/ml stock solution.
2. BMP-4 solution: Add 1 ml of PBS (Ca^{++}, Mg^{++}-free) with 1% BSA to a vial containing 10 μg of human recombinant BMP-4 (R & D Systems). Make 100 μl aliquots and freeze at −20°C. This makes 10 μg/ml stock solution.
3. Embryoid Body (EB) formation medium-1: Transfer 20 ml of StemLine II hematopoietic stem cell expansion medium (Sigma) into a 50 ml tube, and add 2 ml of Penicillin/Streptomycin, 20 μl of VEGF, and 100 μl of BMP-4. Sterilize by 0.22 μm filtration, and store at 4°C up to a month.
4. EB formation medium-2: Transfer 20 ml of StemLine II hematopoietic stem cell expansion medium into a 50 ml tube, and add 2 ml of Penicillin/Streptomycin, 20 μl of VEGF, 100 μl of BMP-4, and 80 μl of StemSpan Cytokine Cocktail (Stem Cell Technologies). Sterilize by 0.22 μm filtration, and store at 4°C up to a month.

2.3. Blast Cell Growth and Expansion

1. Flt3-ligand solution: Add 1 ml of PBS (Ca^{++}, Mg^{++}-free) with 1% BSA to a vial containing 25 μg of human recombinant Flt3 ligand (R & D Systems). Make 100 μl aliquots and freeze at −20°C. This makes a 25 μg/ml stock solution.
2. TPO solution: Add 1 ml of PBS (Ca^{++}, Mg^{++}-free) with 1% BSA to a vial containing 25 μg of human recombinant TPO (R & D Systems). Make 100 μl aliquots and freeze at −20°C. This makes a 25 μg/ml stock solution.
3. Blast cell growth medium (BGM): To a 100 ml bottle of serum-free hematopoietic CFC medium (Stem Cell Technologies)

add 1 ml of penicillin/streptomycin, 1 ml of Ex-Cyte growth enhancement media supplement (Millipore), 100 μl of VEGF, 200 μl of Flt3 ligand, 200 μl of TPO, 250 μl of bFGF, and 2 μg/ml tPTD-HoxB4 fusion protein (see Note 2). Mix well by shaking. Stand for 5–10 min, then aliquot 2.5–3 ml/tube by using a 10 ml-syringe with a 16 or 18 gauge needle. Store at −20°C.

2.4. Erythroid Enucleation

Basal enucleation medium: StemLine II hematopoietic stem cell expansion medium supplemented with 40 μg/ml *myo*-inositol (Sigma), 10 μg/ml folic acid, 160 μM monothioglycerol (Sigma), 120 μg/ml holo-human transferring (Sigma), 10 μg/ml insulin, 90 ng/ml iron (III) nitrate, 900 ng/ml iron (II) sulfate, 10 mg/ml BSA (Stem Cell Technologies), 4 mM L-glutamine, and 1% penicillin–streptomycin. Sterilize by 0.22 μm filtration.

3. Methods

3.1. Preparation of PMEF Feeder

Prepare primary mouse embryo fibroblasts (PMEF) from12.5 dpc CD-1 mouse embryos with the heads on as previously described (31).

1. Plate and grow early passage PMEFs (<P5) in 150 mm tissue culture plate to confluency
2. Add 10 μg/ml mitomycin-C to the media and incubate at 37°C for 3 h
3. Rinse mitomycin-C treated PMEFs 3 times with PBS
4. Add 4 ml of 0.05% trypsin/0.53 mM EDTA and incubate at 37°C for 2–4 min, then add 10 ml of PMEF medium to inactivate trypsin.
5. Collect inactivated PMEFs by centrifugation at 1,000 rpm (210g) for 5 min.
6. Count and plate inactivated PMEFs onto 0.1% gelatin coated 6-well plates at a density of 7.5×10^5 cells/well (in a 6-well plate) in PMEF medium. PMEF feeders should be prepared at least 1 day before culturing hES cells and remain suitable up to 5 days.

3.2. Culture of Undifferentiated hESCs

Culture undifferentiated hESCs as previously described (32). Add 2 ml of hESC-GM to PMEF plates and equilibrate in the CO_2 incubator for 30 min or longer before plating hESCs for helpful recovery of hESCs (see Note 3).

3.2.1. Thawing Frozen hESCs

1. Take a vial of frozen hESCs out from liquid nitrogen, immediately put into in a 37°C water bath, constantly agitating the

vial while ensuring that the neck of the vial is above the water level.

2. When last sliver of ice in vial remains (after about a minute in 37°C). Spray the vials with 70% isopropanol, using a 1 ml pipetman, add warm hESC-GM medium to the contents of the vial drop-wise with gentle agitation.
3. Transfer the contents immediately into a blue polypropylene 15 ml conical tube with 10–15 ml warm hESC-GM medium and centrifuge at 1,000 rpm (210g) for 4 min.
4. Aspirate the supernatant, add 1 ml hESC-GM and gently resuspend the cells using a 1 ml pipetman with 2–4 repetitions.
5. Transfer the cells to the prepared PMEF plates with equilibrated hESC-GM medium. Spread the cells evenly throughout the well by moving the plate several times in two directions, at 90° to each other, and avoid swirling.
6. Check the cells the next day. If there are many dead cells or the medium has changed color, change 2/3 of the medium. Otherwise, do not change it for another day. On the second day after passage, change half (1.5 ml) of the medium every 24 h until the cells reach 70–80% confluence (see Note 4).

3.2.2. Splitting hESCs

1. Rinse the plate of hESCs with 2 ml Ca^{2+}, Mg^{2+}-free PBS 2–3 times, and add 1 ml 0.05% trypsin-0.53 mM EDTA to 1 well of 6-well plate (see Note 5).
2. Incubate at RT for 2–3 min, then pipette with a P1000 pipetman to produce smaller cell clumps (2–5 cells, see Note 6).
3. Collect cells by adding 2 ml of PMEF medium, and spin down at 1,000 rpm (210g) for 4 min.
4. Resuspend cells in 3 ml of hESC-GM and replate in 3 wells of 6-well plate with preformed PMEF feeder equilibrated with 2 ml hESC-GM.
5. Change half (1.5 ml) of the medium every 24 h until the cells reach 70–80% confluence.

3.3. Induction of hESC Differentiation (EB Formation) and Expansion of Blast Cells

The method for the generation of hemangioblasts (blast cells) from hESCs has been described previously (29, 30, 33).

1. Collect undifferentiated hESCs by trypsinization. Usually 1 well of 80% confluent, high quality undifferentiated hESCs (Fig. 1, see Note 7) will generate approximately 2 million cells.
2. Plate cells in EB formation medium-I at a density of 2.5–5.0×10^5 cells/ml using Costar Ultra-low 24-well plate and incubate for 48 h. EBs form during the first 24 h.

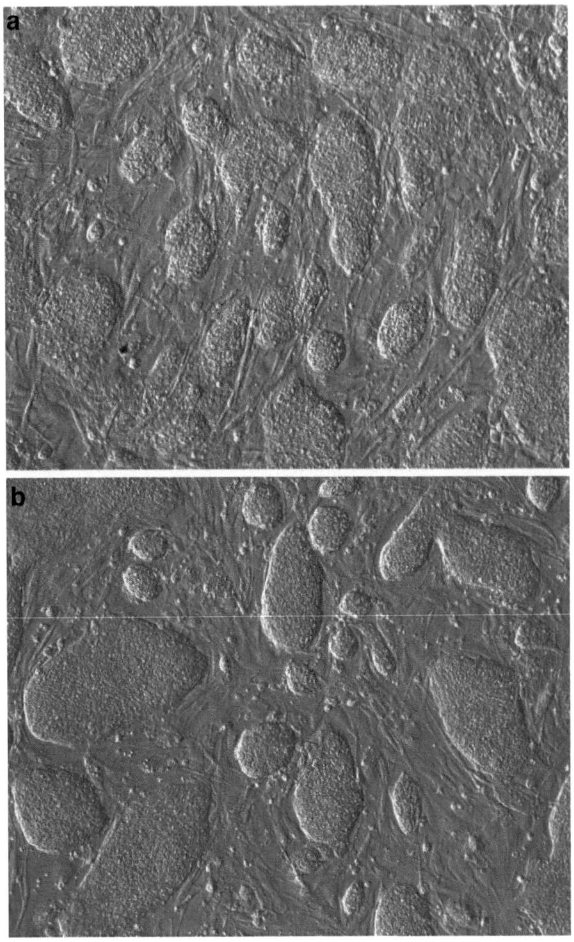

Fig. 1. High quality undifferentiated hESC colonies grown on top of PMEF feeders. (**a**) H1 (WA01) cell line and (**b**) MA01 cell line

3. Remove half (0.5 ml) of the medium with a P1000 pipetman after 48 h, and add 0.5 ml of EBs formation medium-II without disturbing the EBs. Continue incubation for another 36 h.
4. Transfer EBs after 80–84 h of culturing (total EB formation time) into a 15 ml conical tube. Let stand for 1 min, and aspirate medium gently.
5. Add 0.5 ml trypsin/EDTA and mix gently. Incubate 37°C for 2–5 min.
6. Pipette vigorously with a P1000 pipetman to dissociate EBs. If visible clumps still remain, incubate another 1–2 min and repeat pipetting as above until no visible clumps can be seen.

7. Add 2 ml serum-containing medium such as PMEF medium and pass through a 22G needle three times followed by a 40 μm strainer.
8. Count cells; usually 1 well of high quality hES cells should generate 1.5–2 million EB cells. If yield is low, efficiency may not be good.
9. Spin down cells at 1,000 rpm (210g) for 4–5 min, and resuspend cells in Stemline II hematopietic stem cell expansion medium at a density of $2–5 \times 10^6$ cells/ml.
10. Mix $1.0–1.5 \times 10^5$ cells (<0.1 ml) with 2.5–3.0 ml of BGM. Votex for 10 s and let stand for 5 min.
11. Transfer the BGM-cell mixture to 1 well of 6-well Costar Ultra low plate by using a 3 ml syringe attached to a 16G needle, and incubate at 37°C with 5% CO_2.
12. Check blast colony growth after 4 days. Usually blast colonies are visible at 3 days, and after 4–6 days, grape-like blast colonies can be easily identified under microscopy (Fig. 2). After 6–7 days, large, grape-like blast colonies can be picked up using a mouth-glass capillary tube or a P10 pipette tip for in vitro lineage differentiation studies and for in vivo functional studies.

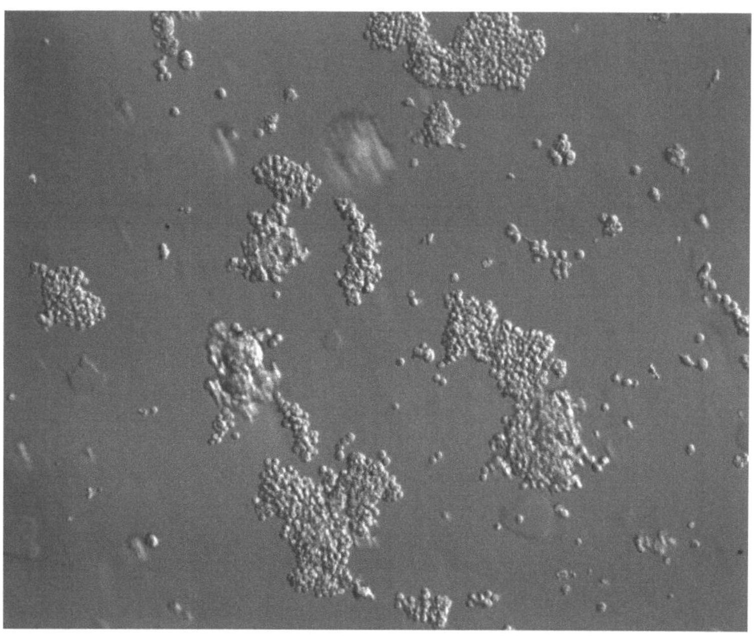

Fig. 2. Blast colonies derived from hESCs. Individual cells of day 3.5 EBs were plated in blast cell growth medium and incubated for 6 days (×100)

13. For erythroid cell differentiation, blast colonies are left on the plate for further incubation with the addition of erythroid cell expansion medium.

3.4. Erythroid Cell Differentiation and Expansion

The method for erythroid cell differentiation and maturation from hESCs has been described previously (30).

1. Add 2 ml of BGM on top of the original BGM without disturbing the blast colonies after 6–7 days and incubate for an additional 3–4 days. Add BGM to keep the density of blast cells at $1–2 \times 10^6$ cells/ml (Fig. 3).
2. At this point, the cell density is often very high ($\geq 2 \times 10^6$/ml). Add equal volumes of BGM, containing an additional 3 units/ml of Epo (total Epo is 6 units/ml, R & D Systems) without HoxB4, incubate for an additional 5 days.
3. Transfer the cell-BGM mixtures into 100 mm Costar Ultra Low dish and mix with equal volume of Stemline II hematopoietic stem cell expansion medium containing SCF (100 ng/ml) and Epo (3 unit/ml). Incubate for an additional 3 days.
4. Add 1/2 volume of Stemline II hematopoietic stem cell expansion medium containing SCF (100 ng/ml), Epo

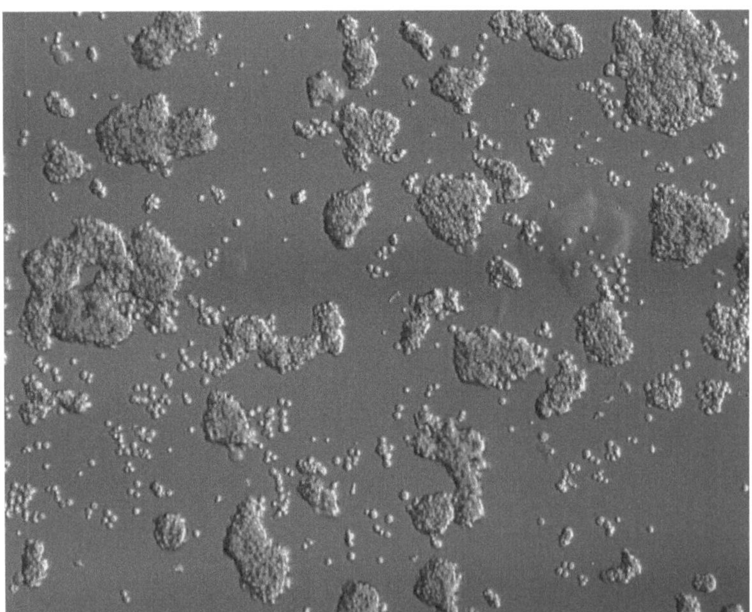

Fig. 3. Erythroid cell colonies derived from hESCs: Blast colonies were grown in blast cell growth medium (BGM) for 6 days after plating individual cells of day 3.5 EBs, then added with more BGM with 3 U/ml Epo and incubated for 3–5 days. Majority of blast colonies differentiated into erythroid cell colonies after longer (9–11 days) growth time with extra (6 U/ml) Epo

Fig. 4. Erythroid cells from hESCs. After prolonged growth (13–15 days), erythroid cells were spread out and colonies were merged, indicating that cells need to be spitted

(3 unit/ml), and 0.5% methylcellulose (see Note 8) every 2–3 days (Fig. 4). (When the cells reach confluence, it is very important to split the cells at a ratio of 1:3 to allow maximum expansion for an additional 7 days [cell density should keep at $\leq 2–4 \times 10^6$/ml].)

5. To enrich erythroid cells, dilute the cells in five volumes of IMDM medium plus 0.5% BSA, collect cells by centrifugation at 1,000 rpm (210g) for 5–10 min (see Note 9). Wash the cell pellets twice with IMDM medium containing 0.5% BSA, and plate in tissue culture flasks overnight to allow nonerythroid cells (usually the larger cells) to attach. The nonadherent erythroid cells are then collected by brief centrifugation (Fig. 5).
6. For further maturation, plate erythroid cells in StemPro-34 SCF medium containing SCF (100 ng/ml) and Epo (3 unit/ml) at a density of 2×10^6 cells/ml. Culture the cells for 6 days with media changes every 2 days.
7. Collect the cells and plate in StemPro-34 containing only Epo (3 unit/ml) for 4–5 more days. After this maturation step, about 15% of these cells express the adult β-globin chain gene.

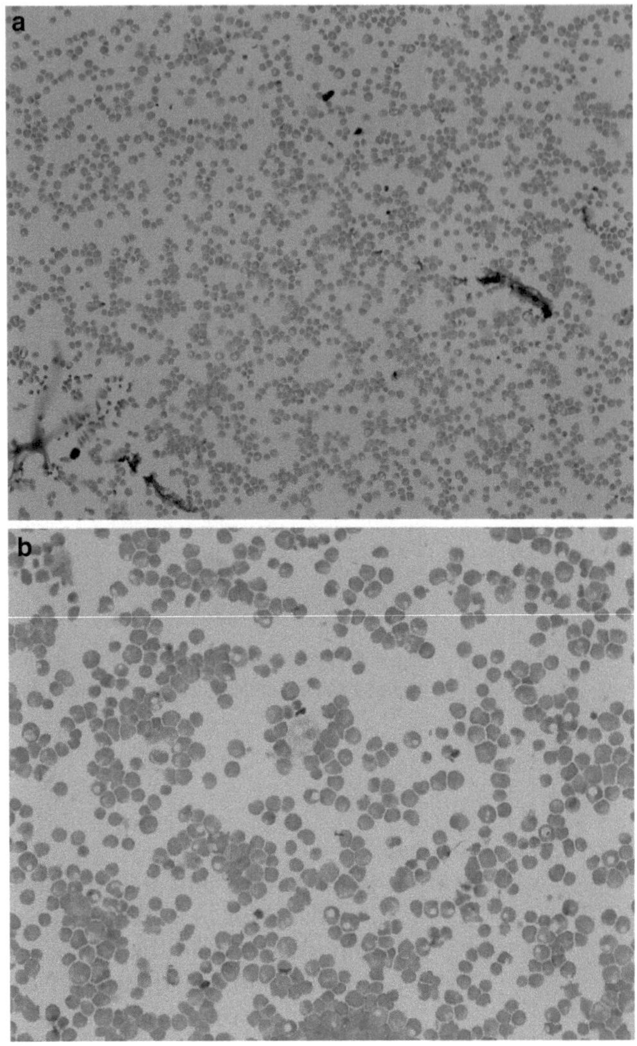

Fig. 5. Enriched erythroid cells: Erythroid cells were plated in tissue culture flasks and incubated over night, and nonadherent were then collected. Majority of erythroid cells are nucleated (**a**, ×100; **b**, ×200)

3.5. Enucleation of Erythroid Cells In Vitro

Culture blast cells as described above up until day 7.

1. Add five volumes of IMDM on day 7 blast cell dishes, pass through a 40-μm strainer, and collect the cells by centrifugation at 1,000 rpm (210g) for 10 min.
2. Plate cells in basal enucleation medium with the addition of 1 μM hydrocortisone, 100 ng/ml SCF, 5 ng/ml IL3, and 3 U/ml Epo at a density of 1×10^6 cells/ml in Costar Ultra low 100 mm dishes. Incubate for 7 day with the addition of more medium every 2 days to maintain the cell density at 1×10^6 cells/ml.

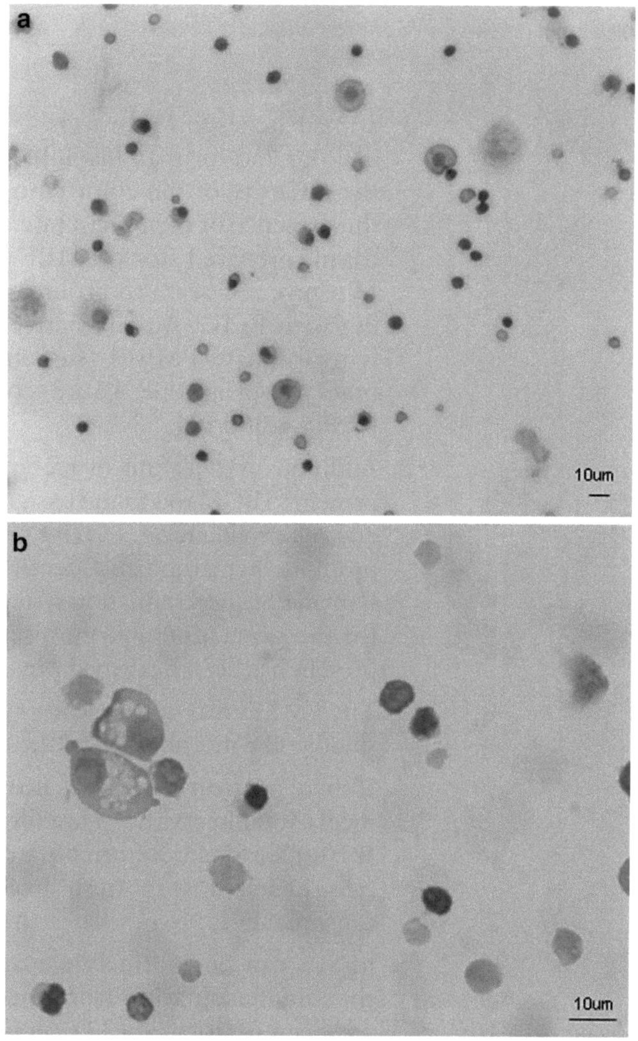

Fig. 6. Enucleation of erythroid cells from hESCs: Erythroid cells from hESCs were cultured in vitro for 4 weeks in Stemline II media with supplements and cocultured with OP9 stromal cells on day 36. On day 42, cells were cytospun and stained with Wright-Giemsa dye (**a**, ×200; **b**, ×1,000). We have noticed that integrity of some enucleated cells was damaged when microscope oil was applied on the slides

3. Remove the original culture medium and refresh with basal enucleation medium containing Epo (3 U/ml). Incubate for another 21 days. Change and refresh medium every 2 days to maintain the cell density of 2×10^6 cells/ml.
4. Transfer cells onto OP9 mouse stromal cells (2×10^6 cells/ml, see Note 10) in basal enucleation medium with 3 U/ml of Epo. Incubate up to 14 day with medium change every 3 days. Cytospin cell samples every 2–3 days and stain with Wright-Giemsa dye to check for nuclear condensation and eventually enucleation using microscope (Fig. 6).

4. Notes

1. Some batches of Mitomycin C appear to become very light in color and form an insoluble precipitate, so always check the intensity of the color (should be deep purple) and for the presence of the precipitate. Do not use if different from freshly prepared stock. PMEF treated with such discolored Mitomycin C seem to proliferate and show intensive labeling with BrdU. An alternative to Mitomycin-C treatment is to irradiate PMEFs (suspended in 5-ml medium) in a blue polypropylene 15-ml conical tube with a dose of 3,500–4,000 rad.
2. Addition of 2 μg/ml of recombinant tPTD-HoxB4 fusion protein to BGM was found to significantly enhance hematopoietic cell proliferation. HoxB4 protein has also been shown to promote hematopoietic development in both mouse and human ESC differentiation systems (34–39). Detailed method for the generation and purification of tPTD-HoxB4 fusion protein has been described previously (39).
3. The PMEFs may appear less confluent in hES-GM due to the spindle-like morphology of the cells, which is expected.
4. If there are many dead cells floating in the culture, collect all media and filter with 0.4 μm filter or centrifuge at 1,000 rpm (210g) for 5 min. Return filtrate or supernatant to well. Keep cells hydrated with fresh hESC-GM during filtration or centrifugation.
5. hESCs can be routinely passaged by trypsin/EDTA after the initial adaptation from mechanically passaged cultures has been performed (32). In our experience trypsin works better than widely used collagenase IV because it produces smaller cell clumps (2–5 cells) and single cells that form more uniformly distributed and similarly sized colonies, while collagenase passaging results in larger colonies that show more extensive differentiation and have to be passed either at a lower splitting ratio or before the desired density of the culture is reached. Overall, trypsin/EDTA passaging allows scaling up the culture 3–4 times faster than collagenase.
6. hESCs grown in Knockout Serum Replacement sometime form thicker and hard-to-digest monolayers after long time (5–6 days) in culture. If no clear gaps are visible after 3 min digestion (or no rounding up of the colonies under the microscope), such plates need to be incubated longer at 37°C for a few minutes. If pipetting is done too soon, a

large clot of undigested PMEFs may form, with all the colonies trapped inside, and retrieving the colonies from such detached aggregates is difficult or even impossible. We get better results when we use a P1000 automatic pipette instead of a cell culture pipette for suspending the cells in both trypsin and medium. After the pellet is resuspended in 1 ml of medium, more medium can be added with a cell culture pipette as desired.

7. We found that the quality of hESCs is one of the most important factors for high-efficient generation of blast cells and red blood cells (RBC). High quality hESCs usually generate a high number of EB cells (e.g., 2×10^6 high quality hESCs will generate $\approx 2\text{–}3 \times 10^6$ EB cells after 3.5 days). High quality hESCs are defined by colonies with tight borders and minimal signs of differentiation as seen under the microscope. Cells are about 80% confluent but not touching each other and grown at a moderate rate. Cells split 1:3 become confluent in 3–5 days and stain almost 100% positive with markers of pluripotency. Cells also form uniform EBs 24 h after replating.
8. We noted that the presence of 0.2–0.5% methylcellulose in the differentiation and expansion medium prevents cells from aggregating, resulting in enhanced expansion. To prepare 0.5% methylcellulose, add 170 ml of Stemline II hematopoietic stem cell expansion medium to one bottle (40 ml) of 2.6% methylcellulose. Mix vigorously by shaking the bottle and store at 4°C.
9. Due to the presence of methylcellulose in the culture medium, you may need to increase the speed (up to 15,000 rpm) and time (15 min) to spin down all cells. We noted that a minimum dilution of five times with warm medium (37°C) aids in collection of these cells.
10. We have observed that blast cells cultured in this condition without stromal layers resulted in 10–30% enucleation, while culturing on human MSC stromal cells resulted in approximately 30% enucleation, and on mouse OP9 stromal cell layers further enhanced the enucleation process. Approximately 30–65% of erythroid cells were enucleated when these cells were transferred to OP9 stromal layers from nonstromal 5-week cultures and cocultured for 7 days. Erythroblasts should be transferred to new OP9 cultures every 3–4 days. To prepare OP9 stromal cell feeder, OP9 cells were expanded in α-MEM medium with 20% FBS, then cultured to 80–100% confluence.

References

1. Thomson JA, Itskovitz-Eldor J, Shapiro SS et al. (1998) Embryonic stem cell lines derived from human blastocysts. Science;282: 1145–1147.
2. Kaufman DS, Hanson ET, Lewis RL et al. (2001) Hematopoietic colony-forming cells derived from human embryonic stem cells. Proc Natl Acad Sci U S A;98:10716–10721.
3. Lu S-J, Li F, Vida L et al. (2004) CD34+CD38- hematopoietic precursors derived from human embryonic stem cells exhibit an embryonic gene expression pattern. Blood;103: 4134–4141.
4. Vodyanik MA, Bork JA, Thomson JA et al. (2005) Human embryonic stem cell-derived CD34+ cells: efficient production in the coculture with OP9 stromal cells and analysis of lymphohematopoietic potential. Blood;105: 617–626.
5. Wang L, Menendez P, Shojaei F et al. (2005) Generation of hematopoietic repopulating cells from human embryonic stem cells independent of ectopic HOXB4 expression. J Exp Med;201:1603–1614.
6. Chang KH, Nelson AM, Cao H et al. (2006) Definitive-like erythroid cells derived from human embryonic stem cells coexpress high levels of embryonic and fetal globins with little or no adult globin. Blood;108: 1515–1523.
7. Qiu C, Hanson E, Olivier E et al. (2005) Differentiation of human embryonic stem cells into hematopoietic cells by coculture with human fetal liver cells recapitulates the globin switch that occurs early in development. Exp Hematol;33:1450–1458.
8. Zhan X, Dravid G, Ye Z et al. (2004) Functional antigen-presenting leucocytes derived from human embryonic stem cells in vitro. Lancet;364:163–171.
9. Corwin HL, Parsonnet KC, Gettinger A. (1995) RBC transfusion in the ICU. Is there a reason? Chest;108:767–771.
10. Littenberg B, Corwin H, Gettinger A et al. (1995) A practice guideline and decision aid for blood transfusion. Immunohematology;11: 88–94.
11. Corwin HL, Gettinger A, Pearl RG et al. (2004) The CRIT Study: anemia and blood transfusion in the critically ill – current clinical practice in the United States. Crit Care Med;32:39–52.
12. Leberbauer C, Boulme F, Unfried G et al. (2005) Different steroids co-regulate long-term expansion versus terminal differentiation in primary human erythroid progenitors. Blood;105:85–94.
13. Giarratana MC, Kobari L, Lapillonne H et al. (2005) Ex vivo generation of fully mature human red blood cells from hematopoietic stem cells. Nat Biotechnol;23:69–74.
14. Miharada K, Hiroyama T, Sudo K et al. (2006) Efficient enucleation of erythroblasts differentiated in vitro from hematopoietic stem and progenitor cells. Nat Biotechnol;24: 1255–1256.
15. Yu J, Vodyanik MA, Smuga-Otto K et al. (2007) Induced pluripotent stem cell lines derived from human somatic cells. Science;318: 1917–1920.
16. Takahashi K, Tanabe K, Ohnuki M et al. (2007) Induction of pluripotent stem cells from adult human fibroblasts by defined factors. Cell;131:861–872.
17. Olivier EN, Qiu C, Velho M et al. (2006) Large-scale production of embryonic red blood cells from human embryonic stem cells. Exp Hematol;34:1635–1642.
18. Qiu C, Olivier EN, Velho M et al. (2008) Globin switches in yolk-sac-like primitive and fetal-like definitive red blood cells produced from human embryonic stem cells. Blood;111: 2400–2408.
19. Ma F, Ebihara Y, Umeda K et al. (2008) Generation of functional erythrocytes from human embryonic stem cell-derived definitive hematopoiesis. Proc Natl Acad Sci U S A;105:13087–13092.
20. Ney PA. (2006) Gene expression during terminal erythroid differentiation. Curr Opin Hematol;13:203–208.
21. Skutelsky E, Danon D. (1967) An electron microscopic study of nuclear elimination from the late erythroblast. J Cell Biol;33: 625–635.
22. Chasis JA, Prenant M, Leung A et al. (1989) Membrane assembly and remodeling during reticulocyte maturation. Blood;74: 1112–1120.
23. Ji P, Jayapal SR, Lodish HF. (2008) Enucleation of cultured mouse fetal erythroblasts requires Rac GTPases and mDia2. Nat Cell Biol;10:314–321.
24. Hanspal M, Hanspal JS. (1994) The association of erythroblasts with macrophages promotes erythroid proliferation and maturation: a 30-kD heparin-binding protein is involved in this contact. Blood;84:3494–3504.

25. Soni S, Bala S, Gwynn B et al. (2006) Absence of erythroblast macrophage protein (Emp) leads to failure of erythroblast nuclear extrusion. J Biol Chem;281:20181–20189.
26. Hanspal M, Smockova Y, Uong Q. (1998) Molecular identification and functional characterization of a novel protein that mediates the attachment of erythroblasts to macrophages. Blood;92:2940–2950.
27. Spike BT, Dibling BC, Macleod KF. (2007) Hypoxic stress underlies defects in erythroblast islands in the Rb-null mouse. Blood;110: 2173–2181.
28. Isern J, Fraser ST, He Z et al. (2008) The fetal liver is a niche for maturation of primitive erythroid cells. Proc Natl Acad Sci U S A;105:6662–6667.
29. Lu SJ, Feng Q, Caballero S et al. (2007) Generation of functional hemangioblasts from human embryonic stem cells. Nat Methods;4: 501–509.
30. Lu SJ, Feng Q, Park JS et al. (2008) Biologic properties and enucleation of red blood cells from human embryonic stem cells. Blood;112:4475–4484.
31. Klimanskaya I, Chung Y, Becker S et al. (2007) Derivation of human embryonic stem cells from single blastomeres. Nat Protoc;2: 1963–1972.
32. Klimanskaya I, McMahon J. (2004) Approaches of derivation and maintenance of human ES cells: detailed procedures and alternatives. In: Lanza Rea, ed. Handbook of Stem Cells. Volume 1: Embryonic Stem Cells. New York, USA: Elsevier/Academic Press,: 437–449.
33. Lu SJ, Luo C, Holton K et al. (2008) Robust generation of hemangioblastic progenitors from human embryonic stem cells. Regen Med;3:693–704.
34. Helgason CD, Sauvageau G, Lawrence HJ et al. (1996) Overexpression of HOXB4 enhances the hematopoietic potential of embryonic stem cells differentiated in vitro. Blood;87:2740–2749.
35. Kyba M, Perlingeiro RC, Daley GQ. (2002) HoxB4 confers definitive lymphoid-myeloid engraftment potential on embryonic stem cell and yolk sac hematopoietic progenitors. Cell;109:29–37.
36. Wang Y, Yates F, Naveiras O et al. (2005) Embryonic stem cell-derived hematopoietic stem cells. Proc Natl Acad Sci U S A;102:19081–19086.
37. Bowles KM, Vallier L, Smith JR et al. (2006) HOXB4 overexpression promotes hematopoietic development by human embryonic stem cells. Stem Cells;24:1359–1369.
38. Pilat S, Carotta S, Schiedlmeier B et al. (2005) HOXB4 enforces equivalent fates of ES-cell-derived and adult hematopoietic cells. Proc Natl Acad Sci U S A;102: 12101–12106.
39. Lu SJ, Feng Q, Ivanova Y et al. (2007) Recombinant HoxB4 fusion proteins enhance hematopoietic differentiation of human embryonic stem cells. Stem Cells Dev;16: 547–560.

Chapter 8

Directed Differentiation of Neural-stem cells and Subtype-Specific Neurons from hESCs

Bao-Yang Hu and Su-Chun Zhang

Abstract

We describe a chemically defined protocol for efficient differentiation of human embryonic stem cells (hESCs) to neural epithelial cells and then to functional spinal motor neurons. This protocol comprises four major steps. Human ESCs are differentiated without morphogens into neuroepithelial cells that form neural tube-like rosettes in the first 2 weeks. The neuroepithelial cells are then specified to OLIG2-expressing motoneuron progenitors in the presence of retinoic acid (RA) and sonic hedgehog (SHH) in the following 2 weeks. These OLIG2 progenitors generate postmitotic, HB9 expressing motoneurons at the fifth week and mature to functional motor neurons thereafter. The protein factor SHH can be replaced by a small molecule purmorphamine in the entire process, which may facilitate potential clinical applications. This protocol has been shown equally effective in generating motor neurons from human induced pluropotent stem (iPS) cells.

Key words: Stem cells, Motor neuron, Spinal cord, Neural differentiation, Motor neuron disease, Neuromuscular junction

1. Introduction

Directing human embryonic stem cells (hESCs) to specific lineages is prerequisite for using hESCs to model early human development and for applying the hESCs-derived lineages in clinic. In the past decade, various protocols have been presented to differentiate hESCs to neuroectodermal cells (1, 2) including the spinal motor neurons (3–5). These differentiation protocols vary considerably in the starting hESCs, feeder cells, unknown factors (e.g., sera and conditioned media), efficiency, and cell purity (6). We developed a series of neural differentiation protocols, including the one described here, for two objectives: modeling the early

S. Ding (ed.), *Cellular Programming and Reprogramming: Methods and Protocols*, Methods in Molecular Biology, vol. 636, DOI 10.1007/978-1-60761-691-7_8, © Springer Science+Business Media, LLC 2010

human brain development; and producing enriched/pure populations of functional neural cells for therapeutics.

The protocol was devised based on the developmental principle underlying motoneuron development. Spinal motor neurons are differentiated from neuroepithelial (NE) cells in a very narrow band of the ventral neural tube called the pMN domain, where the progenitors express the helix-loop-helix transcription factor Olig2. These Olig2-expressing progenitors are specified in the presence of a particular amount of sonic hedgehog (Shh) that is released from the notochord and subsequently the floor plate (7, 8). Through interaction of Olig2 and neurogenic transcription factors including Ngn2 and Pax6, the Olig2-expressing progenitors differentiate to postmitotic motor neurons during the neurogenesis phase and express motoneuron-specific transcription factors such as HB9 and Isl1 while downregulating Olig2 (9–12). Thereafter, HB9-expressing motoneuorns mature and express choline acetyltransferase (ChAT), an enzyme that catalizes the synthesis of the transmitter acetylcholine for transmitting signals through the neuromuscular junctions.

Generation of spinal motoneurons from hESCs follows the same basic steps of neuroectoderm induction, motoneuron progenitor specification, differentiation, and maturation of postmitotic motoneurons (Fig. 1). hESCs are removed from the mouse embryonic fibroblast (MEF) feeder to initiate differentiation. In the serum-free culture condition, these hESCs differentiate to the neuroectoderm fate in 2 weeks (13). During the neural induction

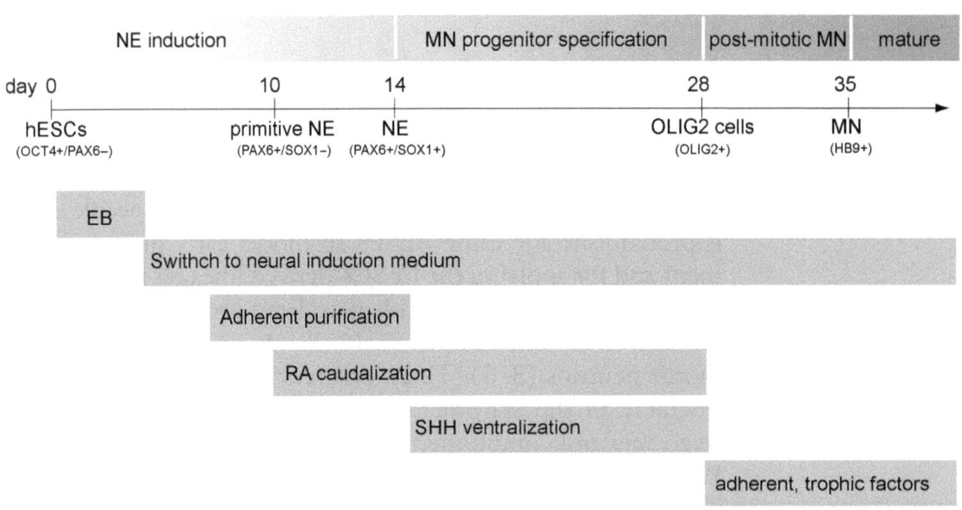

Fig. 1. Scheme of differentiation of spinal cord motoneurons from hESCs. The hESCs are directed to neuroepithelial cells in the first 2 weeks. These neureopithelial cells are patterned to OLIG2-expressing motoneuron progenitors in the subsequent 2 weeks in response to RA and SHH (or purmorphamine). Finally, the progenitors differentiate to post-mitotic motoneurons in the presence of neurotrophic factors. The process employs a simple serum-free neural differentiation medium for motoneuron differentiation. The adherent culture during the neural induction phase is uniquely designed for directly observing neuroepithelial differentiation and for purifying the neuroepithelial cells by removing the non-neural colonies

phase, the hESC aggregates are reseeded from day 7 onto a culture surface free of feeder to form individual monolayer colonies, allowing an even exposure to morphogens and a synchronized differentiation of the neuroepithelia. By the end of the second week (day 14–17), NE cells, in the readily identifiable neural tube-like rosettes (1), develop. They express a panel of neuroectoderm transcription factors including PAX6 and SOX1.

NE cells generated in this way bear an anterior phenotype by expressing OTX2 (13). Hence, it is necessary to caudalize and ventralize the NE cells to generate spinal motor neurons. We found that early NE cells at day 10, also referred to as primitive NE cells (13), present higher competence to respond to morphogens including RA (3, 14). Therefore, the anterior NE cells are patterned with retinoic acid (RA) and SHH for the subsequent 2 weeks. This treatment results in the induction of OLIG2-expressing ventral spinal progenitors in the fourth week. These OLIG2 cells become postmitotic in the fifth week and express MN transcription factors like HB9 and ISL1. The MNs, when growing on substrate, extend substantial projections and express distinctive ChAT, indicating gradual maturation. When cocultured with myoblasts, these hESC-derived MNs form characteristic neuromuscular junctions. The 5-week in vitro differentiation process coincides with the appearance of motor neurons in the ventral horn of the developing human spinal cord at the fifth to sixth week.

The protocol is the modification of our previous reports (3, 14). Major modifications include streamlined procedure, simplified media, the use of more potent recombinant SHH (resulting from a mutation at the N-terminus), and application of small molecules capable of activating SHH signaling in human cells (15). The optimized protocol typically generates about 50% of HB9 expressing motoneruons among the total hESC progenies. This protocol has recently been tested effective for differentiating human iPS cells to spinal motor neurons (16).

2. Materials

2.1. Stock Solutions

1. ACCUTASE (Innovative Cell Technology, San Diego, CA, cat. no. AT104): ready to use.
2. Ascorbic acid (200 μg/mL): Dissolve 2 mg ascorbic acid in 10 mL PBS. Aliquot and store at −80°C.
3. B27 supplement without vitamin A: 50× (Invitrogen, cat. no. 12587-010).
4. BDNF, GDNF, IGF1 (100 μg/mL): Dissolve 100 μg of growth factor in 1 mL sterilize distilled water, aliquot and store at −80°C.

5. beta-Mercaptoethanol (14.3 M): ready to use.
6. Boric acid buffer (pH 8.4): In 100 mL distilled water add 0.927 g H_3BO_3 and 0.6 g NaOH. Adjust pH to 8.4 by adding HCl.
7. Bovine serum albumin (BSA): Dissolve 50 μg BSA powder in 50 mL PBS, filter and store at −80°C.
8. Cyclic AMP (1 mM): Dissolve 4.914 mg cyclic AMP in 10 mL sterilized water. Aliquot and store at −80 C.
9. Dispase (1 U/mL): Dissolve 50 U Dispase (Invitrogen; cat. no. 17105-041) in 50 mL F12/DMEM. Warm at 37°C for 15 min. Filter with a 50 mL-Steri-flip. Be aware that the right amount of 50 U dispase varies among lots.
10. Dulbecco's modified eagle medium: Nutrient mixture F-12 1: 1 (DMEM/F12, Invitrogen, cat. no. 11330). Ready to use.
11. FGF2 (100 μg/mL): Dissolve 100 μg bFGF in 1 mL sterilized PBS with 0.1% bovine serum albumin (BSA).
12. Heparin (1 mg/mL): Dissolve 10 mg heparin in 10 mL DMEM medium, aliquot and store at −80°C.
13. Knockout serum replacer (Invitrogen; cat. no. 10828). Store stock in −80°C. Make aliquots of 50 mL and store at −20°C if it cannot be used up in a week after thaw.
14. Laminin from human placenta: Ready to use.
15. L-Glutamine solution (200 mM). Ready to use. Make aliquots of 5 mL and store at −20°C.
16. MEM nonessential amino acids solution 100× (Invitrogen, cat. no. 11140): ready to use.
17. N2 supplement 100× (Invitrogen, cat. no. 17502-048).
18. Poly-L-Ornithine 10× (1 mg/mL): Add 0.1 g poly-L-ornithine to 100 mL pH 8.4 boric acid buffer. Filter through a 0.22 μm teflon filter.
19. Purmorphamine (10 mM): Dissolve 5 mg purmorphamine in 480 μL ethanol and 480 μL DMSO, aliquot and store at −20°C. The working concentration range of purmorphamine is very narrow. Prepare the stock solution as accurately as possible. When adding stock solution into the culture medium, use the smallest tip and a well-calibrated pipetteman.
20. Retinoic Acid (RA, 100 mM): Dissolve 50 mg RA in 1.67 mL DMSO. Aliquot 50 μL into brown microtubes and store at −80°C. RA is extremely sensitive to UV light, air, and oxidizing agents, especially in solution. It is recommended to use all the powder immediately after opening the ample. Dilute each aliquot with 4.95 mL ethanol and store at −20°C as working stock solution. Try not to use working stock solution older than 2 weeks.

21. SHH (100 μg/mL): Dissolve 100 μg of SHH in 1 mL sterilized PBS with 0.1% BSA. Aliquot 100 μL into sterilized tubes and store at −80°C.
22. Trypsin inhibitor (1 mg/mL): Dissolve 50 mg trypsin inhibitor in 50 mL DMEM/F12 and filter through 50 mL Steriflip.

2.2. Media

1. Human ESC growth medium (500 mL): Sterilely combine 392.5 mL DMEM-F12, 100 mL Knockout serum replacer, 5 mL MEM nonessential amino acids solution, 2.5 mL of 200 mM L-glutamine solution (final concentration of 1 mM), and 3.5 μL 14.3 M 2-Mercaptoethanol (final concentration of 0.1 mM). The medium can be stored at 4°C for up to 7–10 day.
2. Neural differentiation medium (DMEM/F12/N2, 500 mL): Sterilely combine 489 mL of DMEM/F12, 5 mL N2 supplement, 5 mL MEM nonessential amino acids solution, and 1 mL of 1 mg/mL Heparin. The medium can be stored at 4°C for up to 2 weeks. For neuronal differentiation, add cAMP (1:10,000), ascorbic acid (1:1,000), BDNF (1:10,000), GDNF (1:10,000), and IGF-1 (1:10,000) before use.

2.3. Antibodies

2.3.1. For Neuroepithelial Cell Identity

1. Pax6 (monoclonal, Developmental Studies Hybridoma Bank-DSHB): use at 1:5,000 for immunostaining on cultured cells.
2. Sox2 (monoclonal, R&D systems MAB2018): use at 1:1,000.
3. Sox1 (goat IgG, R&D AF3366): use at 1:1,000.

2.3.2. For Regional Identity

1. Otx2 (goat IgG, R&D AF1979): use at 1:2,000.
2. Bf1 (FoxG1, Rabbit IgG) : use at 1:1,500.
3. HoxB4 (rat IgG, DSHB 112): use at 1:50.

2.3.3. For Neurons and Progenitors

1. βIII-tubulin (Rabbit IgG, Covance PRB-435P): use at 1:5,000.
2. Synapsin (Rabbit IgG. CALBIOCHEM 574777): use at 1:250.

2.3.4. For Motor Neurons and Progenitors

1. Olig2 (goat IgG, Santa Cruz SC-19969): use at 1:500.
2. MNR2 (HB9, monoclonal antibody, DSHB 81.5C10): use 1:50.
3. ChAT (goat IgG, Chemicon AB144P): use at 1:500.

2.4. Culture Substrate Preparation

1. Poly-L-ornithine coated coverslips: In a sterile hood, put one sterilized coverslip in each well of a 24-well plate. Add 75 μL of 0.1 mg/mL Poly-ornithine onto each coverslip. Incubate plates at 37°C overnight. The next day, aspirate Poly-ornithine

off and let the coverslips dry for approximately 30 min. Wash three times with 1 mL sterile water for each well. Leave the plate open in the hood until completely dry. Cover plates, wrap in foil, label with date, and store at −20°C.

2. Laminin coated 6-well plate: Dilute laminin with fresh neural differentiation medium at final concentration of 20 μg/mL. Put 300 μL of laminin solution into each well of a 6-well plate. Let the medium hold as a big drop and spread within the central area of the well. Do not let the medium drain to the edge. Incubate the plate at 37°C for 1 h. Laminin is very easy to be absorbed by plastic and tends to form aggregates in room temperature. Store laminin at −80°C and thaw at 4°C before using. Try not aliquot laminin to plastic tubes from the original glass vial.

3. Methods

The undifferentiated state of the starting hESCs is a prerequisite for efficient differentiation of NE cells and subsequent functional motor neurons. Presence of partially differentiated hESCs or contamination of differentiated hESC colonies will result in unsynchronized neural differentiation and reduce the differentiation efficiency.

In the multiple-step process, we use adherent culture mode except the suspension culture steps in the initial separation of hESCs from MEF and in the purification of NE cells. The adherent culture allows direct visualization of neural differentiation, including the neural tube-like rosettes during NE induction and neural progenitor migration and neurite outgrowth in the neuronal differentiation phase.

In the NE induction phase, we employ a colony culture. Almost all the colonies possess neural tube-like rosettes or at least 90% of the total differentiated cells represent NE cells that express PAX6 and SOX1. The colony culture permits readily removal of nonneural colonies. Once nonneural colonies are scraped from the culture, 95–99% among the total population should be PAX6+ cells. This ensures subsequent neural differentiation efficiently.

Motoneuron progenitor population reaches a peak in the fourth week. If purmorphamine replaces SHH in the protocol, it increases the proportion of OLIG2-expressing cells from 50 to 60–80% of the total cells.

Differentiation of OLIG2-expressing motoneuron progenitors to HB9-expressing postmitotic motor neurons takes another week. By the end of the fifth week, the HB9-expressing cells account for half of the total cells. The HB9-expressing cells rarely migrate away from the cluster; rather, they stay in the cluster or immediate periphery of the cluster and extend extremely long processes (axons) that often travel throughout the entire 11-mm diameter coverslip. Dissociating the OLIG2-expressing progenitor spheres often results in a significant motor neuron loss, thus we use small clusters of MNs for final differentiation.

After the fifth week, the motor neurons can be further cultured for several weeks or months depending on the applications. Other mature motoneuron markers, e.g., ChAT and VaChAT, will appear over time.

3.1. Induction of Neuroepithelial Cells

3.1.1. Lift hESC Colonies from MEF Feeder Cells

1. Culture hESCs in a 6-well plate with MEF feeder cells and feed the cultures daily for 5–7 days. These hESCs grow as colonies and express OCT4 uniformly (Fig. 2a).
2. Remove the old medium and rinse each well of hESCs with 2 mL warm DMEM/F12 for 2 min. Remove DMEM/F12.
3. Add 1 mL freshly made dispase (1 U/mL) to each well of a 6-well plate, incubate the cultures at 37°C for 3–5 min. Carefully observe the cells under a microscope every 3 min (see Note 1). When the edge of hESC colonies starts to curl, aspirate the Dispase off.
4. Gently rinse the well with 2 mL DMEM/F12. The hESC colonies are now loosely attached and very easy to dislodge. Remove the medium carefully without disturbing the colonies.

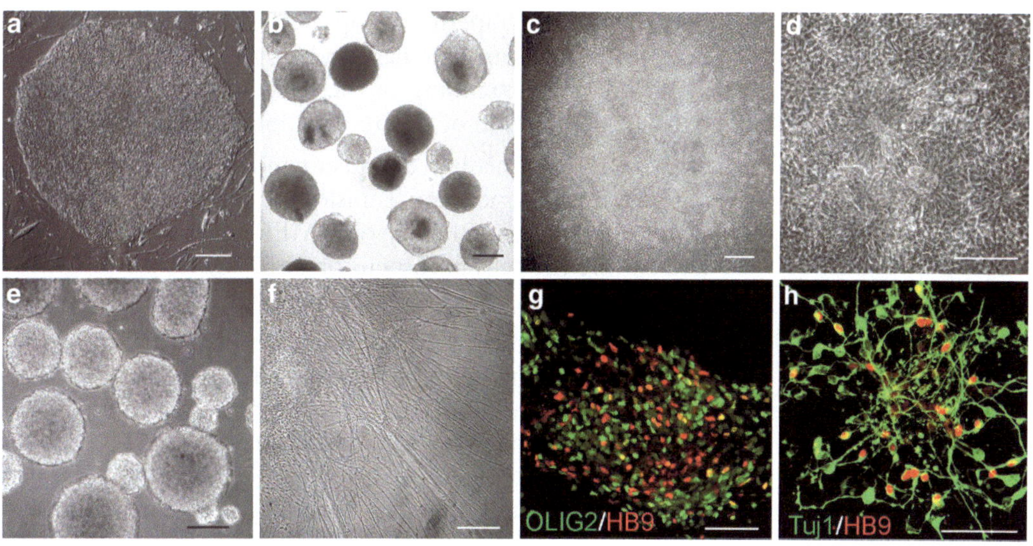

Fig. 2. Differentiation of motor neurons from hESCs. (**a**) hESCs growing on MEF feeder as a uniform colonies. (**b**) After lifting the hESCs from the MEF and growing in suspension, the hESCs aggregate to spheres. (**c**) From day 10, columnar epithelial cells appear within the hESC colonies. The columnar cells are starting to organize into rosettes in the colony. (**d**) At day 15, neural tube-like rosettes are obvious. (**e**) The motoneruon progenitors are cultured in suspension. (**f**) Extensive axonal projections come out from the clusters a week after adherent culture. (**g**) At the fourth week, HB9-expressing motoneurons present among the OLIG2-expressing progenitors. (**h**) HB9 motorneurons are also positively stained for neuronal marker Tuj1+ at the fifth week

5. Add 2 mL of fresh hESC medium to each well. Lift the colonies off by gently swirling the plate and/or pipetting. This usually leaves the MEF attaching to the well. If there are substantial numbers of hESC colonies remain attached, blow off the colonies by pipetting. Pipette the hESC colonies to the size of 50–100 μm. Limit the times of pipetting to no more than 5.
6. Gently collect the colonies with a 5-mL serological pipette or a 1,000 μL-pipette tip into a 15-mL conical tube. Spin down the cells at 50 × *g* for 1 min; alternatively, let the hESC colonies sink by standing the tube for 3 min.
7. Aspirate the medium without disturbing the pellet or colonies. Resuspend the hESC colonies with fresh hESC medium, and wash the cells. Remove the medium carefully.
8. Resuspend the hESC colonies with 5 mL hESC medium without FGF2.
9. Transfer the cells to a T25 or T75 culture flask. Cultures from one 6-well plate go to one T75-flask with 40 mL medium or three T25-flasks in 12 mL medium (see Note 2). Record the date when hESC colonies are lifted off from feeder cells as day 0 of differentiation.

3.1.2. Formation of hESC Aggregates (or Embryoid Bodies)

1. Next day (day 1), the lifted hESCs colonies generally round up as individual spheres with some dead cells and cell debris floating in the medium. Briefly pipette the clusters with a 5-mL serological pipette to strip the attached debris off from the cell cluster. Stand the flask and let the aggregates sink for 5 min. A simple standing of the flask but no centrifugation will allow separation of the hESC colonies from debris.
2. Remove the old medium and resuspend the hESC aggregates with fresh hESC medium. Transfer the culture to a new flask, if there are substantial carry-over MEF adhering to the flask, and culture the cells in suspension at 37°C.
3. In the next few days, hESC aggregates (also termed embryoid bodies) become brighter overtime (Fig. 2b). They should be floating in the medium without attaching to the flask (see Note 3). Observe and feed the cells daily with fresh hESC medium using the procedure described above.

3.1.3. Differentiation of Primitive Neuroepithelial Cells

1. From day 4, switch the culture medium to the neural differentiation medium and feed the cells in the same way every other day.
2. On day 7, collect the hESC aggregates to a 15-mL conical tube, centrifuge at 50 × *g* for 2 min. An additional wash with DMEM/F12 is optional to remove the dead cells and facilitate the attachment of the clusters to the culture surface.

3. Aspirate off the medium and resuspend the hESC aggregates with 5 mL neural differentiation medium. Transfer the cells to a 60-mm Petri-dish.
4. Seed 20–25 clusters evenly to each well of a laminin coated 6-well plate in 300 μL medium (see Note 4).
5. Clusters usually attach to the culture surface within 12 h (see Note 5). Add 2 mL of fresh neural differentiation medium to each well. Continue culturing by feeding the cells with 2 mL neural differentiation medium every other day.
6. 3–4 days after attachment (around 10 days from hESC differentiation), carefully examine the morphology of the attached clusters. Columnar epithelial cells appear in the colony center and radially line up (Fig. 2c). This time point represents a critical step toward the specified fates upon presence of appropriate morphogen (see Note 6).
7. On day 14, check the morphology again under a microscope. The columnar epithelial cells proliferate quickly and form multiple layers, forming neural tube-like rosettes (Fig. 8.2d) (see Note 7).

3.2. Specification of Olig2-Expressing Motoneuron Progenitors

1. From day 10 of differentiation, feed the primitive neural epithelial cells with fresh neural differentiation medium supplemented with RA at the final concentration of 0.1 μM (1:10,000 of the stock solution). RA patterns the cells to the upper spinal cord phenotype which expresses HOXB4. Feed the culture every other day for another 5 days.
2. On day 15, carefully observe the cells. Evaluate the quality of differentiation based on the rosette formation (see Note 7).
3. Remove the old medium, add 2 mL of fresh neural differentiation medium to each well of the 6-well plate.
4. Gently blow the clusters with a 1-mL pipette to detach the neural tube-like rosettes in the colonies. Keep the pipette tip within the medium to avoid bubbles when pipetting. The rosettes detach easily while the flat cells at the peripheral part of the colony should remain attached.
5. Collect the rosette clumps into a 15-mL conical tube, briefly triturate the clumps with a 5- or 10-mL serological pipette up and down twice. It is not necessary to break up the clumps too much.
6. Centrifuge at 50 × *g* for 2 min at room temperature. Remove the medium, resuspend the clusters in 5 mL fresh neural differentiation medium. Transfer the culture to a T25 or T75 flask (cells from three wells may be added to one T25 flask). Culture the cells at 37°C.

7. Add SHH at 100 ng/mL and RA at 0.1 μM. In the optimized protocol, we use a small molecule purmorphamine as a replacement of SHH. Purmorphamine at 1 μM ventralize the neural epithelial cells similarly as SHH at 100 ng/mL.
8. Feed the cultures by replacing 2/3 of the medium every other day using the same medium. Simply stand the flask for 2–3 min and aspirate the supernate, and then add the fresh medium. Clusters of neuroepithelia tend to become spherical within 2 days, typically 100–200 μm in diameter (Fig. 2e; see Note 8).
9. If the spheres grow bigger than 300 μm, break them with a fire polished Pasteur pipette. Alternatively, incubate the big clusters with accustase at 37°C for 3 min followed by gentle pipetting.
10. On day 23, OLIG2-expressing progenitors should be detected. The highest population of OLIG2-expressing motoneuron progenitors appears at the end of the fourth week (day 28). This can be done by either plating the cells onto a coverslip for immunostaining or FACS analysis after immunostaining on dissociated cells.

3.3. Generation of Spinal Motor Neurons

1. From the fifth week (day 29), the motoneuron progenitors are differentiating to postmitotic motoneurons. Plate the progenitor spheres onto glass coverslips that are coated with polyornithine and laminin (2–4 clusters/coverslip in a 24-well plate) in the presence of 50 μL of the neural differentiation medium. The medium is supplemented with BDNF, GDNF, IGF1, cAMP (1 μM), ascorbic acid (AA, 200 ng/mL) whereas RA and SHH are reduced to 50 nM and 50 ng/mL, respectively. Incubate at 37°C for at least 2 h or overnight until attachment.
2. The next day, feed the attached spheres with 0.5 mL of fresh neural differentiation medium with the above supplements. Feed the cells every other day for long-term differentiation (see Note 9).
3. 2 days after plating, the spheres flatten and some cells migrate out of the spheres. Long neurites start to extend from the sphere.
4. By the end of the fifth week, extensive neurites grow out of the sphere (Fig. 2f). Immunostaining will reveal HB9-expressing cells in the sphere whereas Olig2-expressing progenitors decrease (Fig. 2g). The HB9+ neurons also express tubulin (Fig. 2h). With the optimized protocol above half of the total cells are positive for HB9 staining (see Note 10).

3.4. Maturation of Spinal Motor Neurons

1. For longer cultures of mature motoneurons, continue feeding the cultures with the same neural differentiation medium every other day.
2. Mature, ChAT positive motor neurons begin to appear at around the sixth week (day 40–42) and increase overtime (see Note 11). Meanwhile, HB9-expressing cells decrease. During the sixth and seventh week, some cells coexpress HB9 in the nucleus and ChAT in the cytoplasm and neurites.
3. The motor neurons survive for several weeks on coverslips. They usually die in an environment without target cells. When cocultured with myocytes, the motoneuron axons induce clustering of acetylcholine receptors, which can be visualized by bungarotoxin staining.

3.5. Passaging Neuroepithelial Spheres

The cell clusters in suspension grow big over time. When the spheres are larger than 300 μm in diameter, they should get dissociated to smaller ones for continued growth and expansion. We usually split the big clusters using two simple procedures.

3.5.1. Passage Cells Using Polished Pasteur Pipette

1. Feed the cells with fresh neural differentiation medium containing SHH and RA the day before splitting.
2. Prepare the pipette for dissociating the cells before taking the cells from incubator for dissociation. Choose the cotton plugged 9″-Pasteur pipettes. In the hood, fire polishes the Pasteur pipette tip and adjusts the diameter of the inner lumen to around 200 μm. Carefully check the tip and the lumen every 3–5 s.
3. Heat the pipette at 2 cm from the tip. Gently bend the pipette to 145–150°. The curve helps shearing the spheres when they pass through the pipette.
4. Cool down the pipette to room temperature. Prepare extra pipettes of different lumen sizes. Keep the pipette sterile in hood.
5. Take out the cells from incubator, lean the flask at 45° to let the clusters sink to a corner.
6. Rinse the pipette with the supernate three times. Cells will stick on glass pipette if not well rinsed before using.
7. Take all the clusters into the glass pipette using the pipette aid. If clusters are too big to be sucked into the pipette, change to another one with a bigger lumen size. Blow out the spheres into the medium in flask. The shearing force breaks the clusters to smaller pieces without completely dissociating to single cells.
8. If necessary, triturate the remaining large spheres one more time but do not triturate the spheres for more than twice. If the spheres are not broken, it indicates that the Pasteur pipette is not appropriately narrowed and bended.

9. Culture the cells at 37°C.
10. Feed the cells with fresh medium containing the supplements needed from the third day. Since the spheres are small, in the first two feedings simply add fresh medium.

3.5.2. Splitting Big Neuroepithelial Spheres Using ACCUTASE

Alternatively, the bigger clusters are dissociated with ACCUTASE. The enzymatic effect of ACCUTASE is not as powerful as trypsin and no enzyme inhibitor is needed after dissociation. Store the ACCUTASE in aliquots of 5 mL or 10 mL at –20°C. Thaw the frozen aliquot in a refrigerator overnight before using.

1. Collect the big clusters into a 15- or 50-mL conical tube.
2. Centrifuge at 50 g for 2 min to pellet the cells.
3. Remove the medium from the tube.
4. Add 1 mL of ACCUTASE to each tube. Resuspend the pellet by gentle shaking or tapping.
5. Incubate the clusters in ACCUTASE at 37°C for 3–5 min. Inspect and gently shake the tubes every 2 min.
6. When the clusters look loose and/or the solution looks foggy, add 9 mL medium to the tube and centrifuge at $50 \times g$ for 2 min.
7. Remove the medium containing ACCUTASE as clean as possible without disturbing the palette.
8. Add 800 μL medium back to the tube. Pipette the clusters with a 1-mL tip up and down gently for less than five times.
9. Let the cells stand for 2 min. Transfer the medium containing single cells and small clusters to the flask prefilled with the fresh medium. Leave the big clusters in the tube.
10. Repeat steps 8 and 9 to further break the rest large clusters.
11. The cells aggregate to small clusters within hours. Feed the cells with fresh medium the next day.

3.6. Plate Cells on Coverslips that Are Coated with Polyornithine and Laminin

1. Leave the frozen 24-well plate with precoated coverslips in room temperature for 20 min.
2. Dilute the laminin with neural differentiation medium at a final concentration of 20 μg/mL.
3. Add 50 μL of medium containing laminin and spread evenly on top of the coverslip precoated with poly-ornithine. Leave the plate in incubator for an hour.
4. Remove the medium.
5. Transfer the motoneuron progenitor clusters to a Petri-dish, pick up 3–5 small clusters and seed them in 50 μL medium onto the precoated coverslips.
6. Incubate at 37°C for 2 h. Once the cells have attached, add 500 μL of medium to each well.

7. The cells on coverslips can be fixed for staining once they have attached. For long-term culture, feed the cells every other day.

4. Notes

1. Do not leave hESCs in dispase longer than 15 min. Longer incubation in dispase may result in poor survival of the lifted hESCs.
2. The cell density significantly affects the neural differentiation. High density significantly compromises the efficiency of neural specification.
3. hESC aggregates (embryoid bodies) free of feeder cells usually float and do not attach to the culture surface. Feeder fibroblasts around the hESCs may reform feeder, which results in EB attachment. Gently tapping the bottom of the flask will release the loosely attached EBs. Briefly pipette the EBs to remove the dead cells and feeder cells, and then transfer the EBs to a new flask.
4. Do not seed the colonies in a high density. The ideal density is that after 7 days of growth attached clusters remain as individual colonies without merging to each other. Incubate the culture at 37°C overnight.
5. After cultured in suspension for a week well differentiated hESC aggregates tend to attach to plastic surface. Dead cells around the clusters may interfere with the attachment of the aggregates. Wash the aggregates with neural differentiation medium and plate them again onto a new plate coated with laminin (20 μg/mL). Alternatively, addition of 10% FBS into the culture overnight will promote the attachment of the aggregates. The FBS should be removed right after the aggregate attachment. Presence of FBS will inhibit neural differentiation.
6. We refer to these columnar epithelial cells as early neural rosettes. Surrounding the neural epithelial rosettes are flat round cells which are likely of the neural crest lineage. The neural epithelial cells at this state express Pax6 and many other neural transcription factors but not Sox1. We refer to cells at this stage as primitive neural epithelial cells. These primitive neural epithelial cells are responsive to morphogens like RA and SHH for regional patterning. Therefore, we will start the process of motoneuron specification at day 10.
7. Within the clusters, multiple neural tube-like rosettes appear. Immunostaining will indicate that these cells express both PAX6 and SOX1. The cells in the form of neural tube-like rosettes

attach to the substrate loosely whereas the flat cells in the surrounding attach more tightly. We refer to these cells as definitive neural epithelial cells. Thus, it takes about 2 weeks for hESCs to differentiate to neural epithelial cells. The readily identifiable rosettes formation is a valuable parameter to judge the quality of differentiation. Partially differentiated hESCs, overly damaged EBs, or early RA-treated culture may result in poor rosette formation. If there are colonies that do not possess rosettes, these are usually nonneural colonies. Scrap those colonies with a pipette tip after marking them using an objective marker that is mounted in a phase contrast scope. This step will minimize, if not eliminate, the contamination of nonneural cells.

8. Clusters of nonneural lineage may be present in the culture if the nonneural colonies are not scraped before lifting. Instead of forming bright round spheres, those clusters are usually grey or dark with irregular shapes. Should there be any nonneural cell contamination in culture, the partially differentiated hESCs are inevitably the source. These partially differentiated hESCs usually generate "bad colonies" which can be easily recognized by direct observation. Mark the "bad colonies" and manually remove them in the step of "rosettes" formation.
9. Motor neuron progenitors represent a vulnerable population in culture. Enzymatic disaggregation of neuroepithelial spheres can damage the population thus resulting in very few motor neurons. Mild dissociation of the progenitor clusters with accutase (for 3–5 min) can facilitate monolayer formation after attachment. Plating cells at a higher density (30,000 cells/11-mm coverslip), or seeding small clusters (100–200 μm) will help cell survival. Addition of B27 and low concentration of SHH/RA in culture will also help minimize cell death.
10. We have noticed that the HB9 antibodies from different sources vary significantly in terms of specificity. The MNR2 (HB9, monoclonal antibody, DSHB 81.5C10) is a reliable antibody for staining motoneurons from various species including human. The Chemicon Inc is releasing a new polyclonal anti-HB9 that is developed in goat to replace its previous less-specific rabbit HB9 antibody.
11. When using antibodies against ChAT to label mature motor neurons, the available ChAT antibody may present strong background in cultured cells (though it stains ChAT-expressing cells in vivo very well). This is usually because of an inappropriate fixation of the enzyme. Using picric acid buffer for fixation and diluting the antibody will reduce the background. Try to use this antibody against ChAT at 1:500 dilution (goat IgG, Chemicon AB144P).

References

1. Zhang, S. C., Wernig, M., Duncan, I. D., Brustle, O., and Thomson, J. A. (2001) In vitro differentiation of transplantable neural precursors from human embryonic stem cells. *Nat Biotechnol.* **19**, 1129–33.
2. Reubinoff, B. E., Itsykson, P., Turetsky, T., Pera, M. F., Reinhartz, E., Itzik, A., and Ben-Hur, T. (2001) Neural progenitors from human embryonic stem cells. *Nat Biotechnol.* **19**, 1134–40.
3. Li, X. J., Du, Z. W., Zarnowska, E. D., Pankratz, M., Hansen, L. O., Pearce, R. A., and Zhang, S. C. (2005) Specification of motoneurons from human embryonic stem cells. *Nat Biotechnol.* **23**, 215–21.
4. Singh Roy, N., Nakano, T., Xuing, L., Kang, J., Nedergaard, M., and Goldman, S. A. (2005) Enhancer-specified GFP-based FACS purification of human spinal motor neurons from embryonic stem cells. *Exp Neurol.* **196,** 224–34.
5. Lee, H., Shamy, G. A., Elkabetz, Y., Schofield, C. M., Harrsion, N. L., Panagiotakos, G., Socci, N. D., Tabar, V., and Studer, L. (2007) Directed differentiation and transplantation of human embryonic stem cell-derived motoneurons. *Stem Cells.* **25**, 1931–9.
6. Zhang, S. C. (2006) Neural subtype specification from embryonic stem cells. *Brain Pathol.* **16**, 132–42.
7. Lu, Q. R., Yuk, D., Alberta, J. A., Zhu, Z., Pawlitzky, I., Chan, J., McMahon, A. P., Stiles, C. D., and Rowitch, D. H. (2000) Sonic hedgehog – regulated oligodendrocyte lineage genes encoding bHLH proteins in the mammalian central nervous system. *Neuron.* **25**, 317–29.
8. Zhou, Q., Wang, S., and Anderson, D. J. (2000) Identification of a novel family of oligodendrocyte lineage-specific basic helix-loop-helix transcription factors. *Neuron.* **25,** 331–43.
9. Ericson, J., Rashbass, P., Schedl, A., Brenner-Morton, S., Kawakami, A., van Heyningen, V., Jessell, T. M., and Briscoe, J. (1997) Pax6 controls progenitor cell identity and neuronal fate in response to graded Shh signaling. *Cell.* **90**, 169–80.
10. Mizuguchi, R., Sugimori, M., Takebayashi, H., Kosako, H., Nagao, M., Yoshida, S., Nabeshima, Y., Shimamura, K., and Nakafuku, M. (2001) Combinatorial roles of olig2 and neurogenin2 in the coordinated induction of pan-neuronal and subtype-specific properties of motoneurons. *Neuron.* **31**, 757–71.
11. Wichterle, H., Lieberam, I., Porter, J. A., and Jessell, T. M. (2002) Directed differentiation of embryonic stem cells into motor neurons. *Cell.* **110**, 385–97.
12. Lee, S. K., Lee, B., Ruiz, E. C., and Pfaff, S. L. (2005) Olig2 and Ngn2 function in opposition to modulate gene expression in motor neuron progenitor cells. *Genes Dev.* **19**, 282–94.
13. Pankratz, M. T., Li, X. J., Lavaute, T. M., Lyons, E. A., Chen, X., and Zhang, S. C. (2007) Directed neural differentiation of human embryonic stem cells via an obligated primitive anterior stage. *Stem Cells.* **25**, 1511–20.
14. Li, X. J., Hu, B. Y., Jones, S. A., Zhang, Y. S., Lavaute, T., Du, Z. W., and Zhang, S. C. (2008) Directed differentiation of ventral spinal progenitors and motor neurons from human embryonic stem cells by small molecules. *Stem Cells.* **26**, 886–93.
15. Hu, B. Y. and Zhang, S. C. (2009) Differentiation of spinal motor neurons from pluripotent human stem cells. *Nat Protoc* **4**, 1295–04.
16. Karumbayaram, S., Novitch, B. G., Patterson, M., Umbach, J. A., Richter, L., Lindgren, A., Conway, A. E., Clark, A. T., Goldman, S. A., Plath, K., Wiedau-Pazos, M., Kornblum, H. I., Lowry, W. E. (2009) Directed differentiation of human-induced pluripotent stem cells generates active motor neurons. *Stem Cells.* **27**, 806–11.

Chapter 9

Directing Human Embryonic Stem Cells to a Retinal Fate

Thomas A. Reh, Deepak Lamba, and Julianne Gust

Abstract

Substantial progress has been made in the development of methods to direct embryonic stem cells to differentiate into various regions of the central nervous system (CNS). We have used the current model of eye specification to develop a protocol for directing human embryonic stem cells to generate retinal progenitors and various types of retinal neurons. Our method uses a multistep protocol in which embryoid bodies are treated with inhibitors of both BMP signaling and canonical Wnt signaling to promote expression of key eye field transcription factors (EFTFs), as assayed by RT-PCR and immunofluorescence microscopy. The retinal progenitor cells spontaneously undergo differentiation into various types of retinal neurons, including photoreceptors, and this can be promoted by treatment with small molecule inhibitors of the Notch pathway.

Key words: Eye field, Retinal progenitor, Notch, Photoreceptor, hES cells, DAPT

1. Introduction

The neural retina is subject to a large number of conditions that lead to neuronal degeneration and visual impairment. Some of these disorders are linked to specific mutations in retinal-specific genes; for example, mutations in rhodopsin underlie many forms of Retinitis Pigmentosa. While gene therapy will likely be important in the treatment of these inherited degenerations, the diversity of mutations will pose a challenge; moreover, in many people, the retina is already severely degenerated by the third decade and they face a lifetime of blindness. In addition, retinal degeneration can be a complication of diabetes and uncontrolled glaucoma, and for these individuals, cell replacement therapy may be the only option. Finding an appropriate source for new retinal cells is an important challenge to the development of a cell-based therapy for retinal degenerations (1).

S. Ding (ed.), *Cellular Programming and Reprogramming: Methods and Protocols*, Methods in Molecular Biology, vol. 636,
DOI 10.1007/978-1-60761-691-7_9, © Springer Science+Business Media, LLC 2010

The advances in developmental biology over the past 15 years have provided the background for developing methods to direct embryonic stem cells toward various cell types and tissues. In the nervous system, protocols that were initially based on differentiation of cell lines have been replaced by methods based on manipulation of critical developmental signaling pathways (2, 3). Protocols for deriving various types of central nervous system neurons from human embryonic stem cells have now been published, including motor-neurons, dopaminergic midbrain neurons, and cerebral cortical neurons (4–6). All of these more recently published protocols rely on manipulation of developmental signaling pathways that are normally critical for regionalization of the CNS.

The eyes develop from the anterior neural plate, from a region of the developing CNS called the diencephalon. Two critical developmental signals regulate the formation of the anterior–posterior axis of the nervous system (as well as the rest of the animal): BMP and Wnt (7). Inhibition of the BMP pathway promotes neural plate development medially, while higher levels of BMP laterally promote epidermal development (Fig. 1). The overall level of canonical Wnt signaling is higher in posterior regions of the embryo, promoting hindbrain and spinal cord development. Endogenous inhibitors of both BMP and Wnt signaling are therefore important developmental regulators of the anterior–posterior axis; targeted deletion of

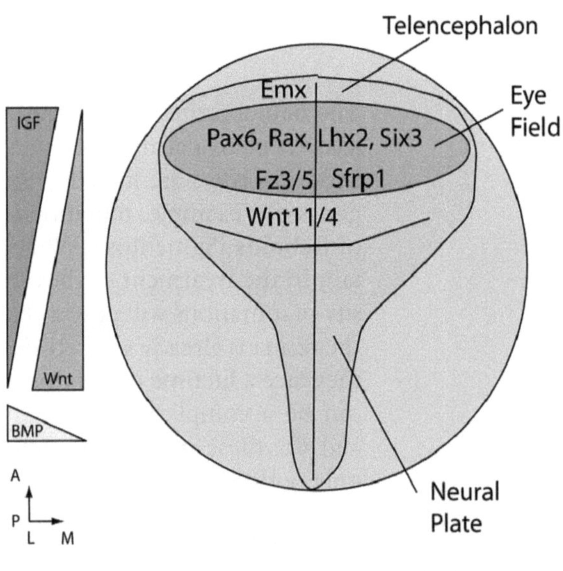

Fig. 1. Schematic of Eye field induction showing the early eye field characterized by the expression of EFTFs: Pax6, Rx, Lhx2, Six3, Fz3/5, and Sfrp1. The eye field is surrounded anteriorly by Emx expressing telencephalon and posteriorly by Wnt11 and Wnt4. Additionally, during early embryonic development, there is an antero–posterior gradient of IGF and Wnt with high IGF anteriorly and high Wnt posteriorly. Also, BMPs have a similar medio-lateral gradient with high BMP levels laterally

the genes coding for Dkk1, a potent Wnt inhibitor, along with noggin, a BMP antagonist, leads to almost the complete absence of the head in mice (8). These results, along with many others, have led to the current model of neural specification, in which anterior neural structures develop when both BMP and Wnt signaling are blocked.

Prior to overt eye formation, the presumptive eyes develop from the eye field, a specific region of the neural plate that can be identified prior to neural tube closure. Classic experiments demonstrated that the eye field region of the neural plate of amphibian embryos can be transplanted to the flank of another embryo and will give rise to an ectopic eye. More recently, a group of transcription factors that are both necessary and in some cases sufficient for eye development have been localized to this region (9), and named the eye field transcription factors (EFTFs). During embryonic development, the expression of the EFTFs is defined by a specific set of signals (10, 11), which are particularly effective at inhibiting the canonical Wnt signaling pathway (Fig. 1). At least three mechanisms inhibit Wnt signaling in the developing eye field:

1. The Wnt inhibitor, Sfrp1, is expressed within the eye field.
2. A signal from Wnt11 or Wnt4 activates the noncannonical Wnt pathway via Frizzled 5 or Frizzled, which also inhibits the canonical Wnt signal.
3. IGF activates the Akt pathway and this also blocks canonical Wnt signaling.

Based on these previous studies, we developed a protocol to promote eye field development from human embryonic stem cells (12, 13). We first generate embryoid bodies from undifferentiated human ES cells, treat with inhibitors of BMP and Wnt signaling, and then assay for the expression of EFTFs in the resulting cells. The methods described in this chapter outline the basic protocol for generating retinal progenitors, confirming commitment to this state with RT-PCR and immunofluorescence, and additional information for further directing and analyzing their subsequent differentiation into retinal neurons using immunofluorescence and intraocular transplantation.

2. Materials

2.1. Culture of Undifferentiated Human Embryonic Stem Cells

1. Cells: H-1 (WA-01) human embryonic stem cell line from Wicell. Mouse embryonic fibroblasts are primary lines derived from CF-1 mouse embryos.
2. Human ES cell medium contains 400 ml of Dulbecco's Modified Eagle's Medium (DMEM) with F12 from

Invitrogen, 100 ml of Knockout Serum Replacer from Invitrogen, 5 ml of nonessential amino acids, 5 ml of l-Glutamine, 5 ml of sodium pyruvate, and 3.5 μL of β-mercaptoethanol.

3. Medium is supplemented with 4 ng/ml of basic fibroblast growth factor (bFGF)

2.2. Generation of Retinal Cells from Undifferentiated Human ES Cells

1. Collagenase/Dispase Mix contains 10 mg/ml of Dispase and 10 mg/ml of Collagenase type IV.
2. Retinal Induction (RI) medium DMEM/F12, 10% Knockout serum replacer, N2 supplement, B27 supplement, 1× sodium pyruvate, 1× nonessential amino acids, 1 ng/ml mouse noggin (R&D Systems), 1 ng/ml human recombinant Dkk-1 (R&D Systems), and 1 ng/ml human recombinant insulin-like growth factor-1 (IGF-1) (R&D Systems).
3. EBS are cultured in Ultra-low attachment six-well plates (Costar, VWR).
4. Retinal differentiation (RD) medium contains DMEM/F12, N2 supplement, B27 supplement, 1× sodium pyruvate, 1× nonessential amino acids, 10 ng/ml mouse noggin (R&D Systems), 10 ng/ml human recombinant Dkk-1 (R&D Systems), 10 ng/ml human recombinant insulin-like growth factor-1 (IGF-1) (R&D Systems), and 5 ng/ml human recombinant bFGF.
5. Penicillin and streptomycin may be added to all medium solutions to prevent contamination.
6. All growth factors were dissolved in DMEM containing 0.1% of bovine serum albumin (BSA) and maintained as stocks at a concentration of 100 ng/μL in –80°C freezer.

2.3. Coating for Adherent Culture of Cells

1. Poly-D-lysine Hydrobromide MW 30–70K, lyophilized powder, cell culture tested. The poly-D-lysine is dissolved in sterile water at a concentration of 0.5 mg/mL and 1-mL aliquots are stored in 15-mL conical tubes at –20°C.
2. Coverslips (12 mm circular) are sterilized by autoclaving.
3. Tissue Culture plates.
4. Matrigel (Collaborative Research). Matrigel is supplied by the manufacturer as a frozen solution. Thaw the bottle slowly on ice (for several hours) to prevent gel formation. Make small (200 μL) aliquots using precooled tubes (15 mL) on ice and a prechilled pipette. If the Matrigel warms during the aliquotting, it will gel and not be effective for the cell cultures. Store aliquots at –20°C for up to 6 months.
5. DMEM medium.

2.4. Manipulation of Notch Pathway in Human ES Cell-Derived Retinal Cells

1. N-(N-(3,5-difluorophenacetyl)-l-alanyl)-S-phenylglycine t-butyl ester (DAPT, EMD Biosciences).
2. Dimethyl sulfoxide – molecular biology grade (DMSO).

2.5. Subretinal Transplantation of Retinal Cells in Adult Mice

1. Trypsin 2.5%.
2. Fetal bovine serum (FBS).
3. Stereotaxic apparatus with mouse and neonatal rat adaptor (Stoelting).
4. 1–5 μL graduated micropipets with plunger (Wiretrol II, Drummond).
5. Glass pipet puller (Sutter Instruments).
6. #11 scalpel blades (sterile).
7. Suture (5–0, silk braided, 3/8 curve 13-mm reverse cutting needle).
8. Ketamine (100 mg/ml).
9. Xylazine (20 mg/ml).
10. Bacitracin ophthalmic ointment (500 units/g).
11. 0.5% Proparacaine hydrochloride ophthalmic solution.
12. Petrolatum ophthalmic ointment.
13. Cyclosporine A for injection (5 mg/ml, Bedford Laboratories).

2.6. Quantitative PCR

1. Trizol.
2. Pestle motor and RNAse-free pestles (Kimble-Kontes).
3. Chloroform ACS-grade.
4. Isopropanol.
5. Ethanol 200-proof.
6. RQ1 RNAse-free DNAse.
7. Ribolock RNAse-Inhibitor.
8. RNAeasy mini kit.
9. Oligo-dT primers.
10. SuperScript II Reverse Transcriptase.
11. SYBR Green PCR Master Mix (Applera Corp).

2.7. Immunocyto chemistry

1. Block solution: Normal goat serum or Fetal bovine serum, Triton X-100, Phosphate-buffered saline.
2. Alexa Fluorophore conjugated secondary antibodies.

3. Methods

3.1. Culture of Undifferentiated Human Embryonic Stem Cells

H-1 cells are cultured on mouse embryonic fibroblasts in the human ES medium in a 37°C incubator in 5% CO_2

1. Thaw a cryogenic straw containing undifferentiated human ES cells by quickly transferring from liquid nitrogen tank to room temperature water.
2. After 30 s, rinse the straw with 70% ethanol, cut the edges of the straw and transfer the cells to a 15-ml tube. Add human ES medium drop-wise slowly.
3. Centrifuge the tube at 800 × *g* for 3 min.
4. Resuspend the cells in fresh human ES medium and transfer cells to a plate previously seeded with irradiated mouse embryonic fibroblasts. Put in incubator at 37°C and 5% CO_2
5. Change medium next day and thereafter every other day till 60–80% confluency.

3.2. Substrate for Adherent Culture of Cells

Adherent cultures of the retinal progenitors can be carried out either on glass coverslips coated with poly-D-lysine and Matrigel or tissue culture plates coated with Matrigel.

1. Prior to use, thaw poly-D-lysine and add sterile water to a final concentration of 50 μg/mL.
2. Place 25–30 coverslips in a large Petri dish for coating. Add poly-D-lysine solution to the dish and ensure all coverslips have sunk in the solution. Incubate the coverslips at 37°C for 15–30 min.
3. Remove all poly-D-lysine solution and wash the coverslips in sterile water three times for a minimum 5 min each. Wash the coverslips very well since poly-D-lysine in solution can be toxic to cells.
4. The coverslips can be dried and stored for up to 2 weeks in the Petri dish at 4°C, or used immediately.
5. When ready to use, put one coverslip in each well of a 24-well plate using flamed forceps. Then proceed to coat them with Matrigel.
6. To coat the coverslips, remove one aliquot of Matrigel from the −20°C freezer and place on ice for 30 min to thaw (200 μL is used for one 24-well plate).
7. Add 10 mL of ice cold HBSS+ to the 15-ml tube containing 200 μL of thawed Matrigel. Mix gently.
8. Immediately, put 0.5 mL of the dilute Matrigel solution into each well of a 24-well plate in which you have already placed poly-D-lysine-coated coverslips and place the plate in the incubator for 30 min at 37°C.
9. Remove the plate from the incubator and, under the sterile hood, remove nearly all of the liquid from the wells.

10. Let the plate dry in the hood uncovered for 15–30 min. The Matrigel will dry into a thin coating. Plate cells onto the Matrigel.
11. The above protocol can be directly applied to coat wells of a tissue culture plate.

3.3. Generation of Retinal Cells from Undifferentiated Human ES Cells (Fig. 9.2a)

1. Remove medium from a confluent plate of human ES cells and treat the plate with a combination of collagenase IV and dispase solution.
2. Replace the plate back to incubator and wait for 3–5 min until the edges of the human ES cell colonies just start to lift.

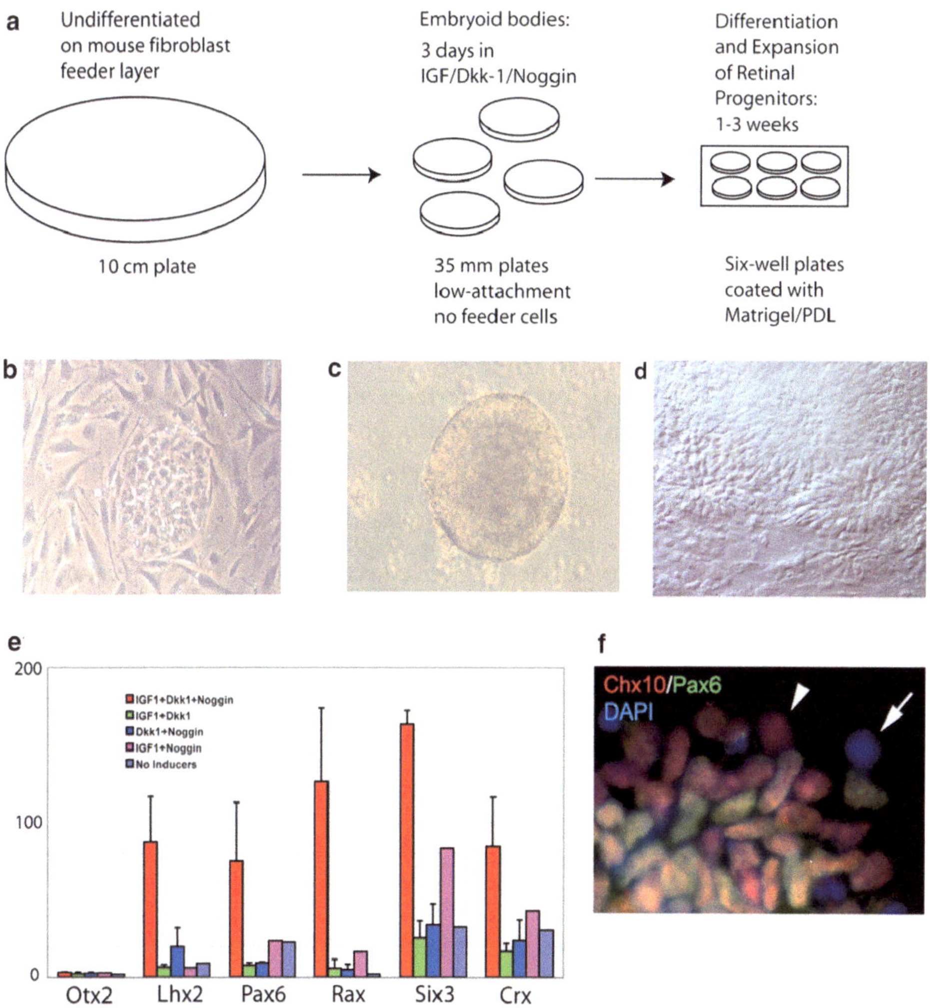

Fig. 2. (**a**) shows the schematic of the protocol for generation of retinal cells from human ES cells. (**b**–**d**) show representative images of undifferentiated human ES cells, embryoid bodies and plated rosettes of retinal progenitors respectively. (**e**) is a graph showing the 80–160-fold increase in expression of EFTFs and Crx, a photoreceptor marker following QPCR analysis. The graph also shows that all three factors, IGF, dkk1, and noggin are required for efficient retinal induction. In (**f**), the image shows the co-expression of Pax6 (*green*) and Chx10 (*red*) by the human ES cell-derived retinal progenitors. *Arrow* points to a cell that expresses neither protein, whereas the *arrowhead* points to an occasional cell expressing Chx10 but not Pax6

3. Remove the collagenase/dispase solution and gently and thoroughly rinse the cells twice with PBS.
4. Add RI medium to the plate and gently scrape the cells off the plate using the tip of a 5-ml pipet.
5. Triturate the colonies such that the final size of the colonies is approximately 250–400 μm.
6. Transfer the cell aggregates to an ultra-low attachment plate. Overnight, the colonies round up and form circular three-dimensional masses of cells called embryoid bodies (EBs) (Fig. 2c).
7. 48 h later, change the medium by placing the plate at an angle so as to allow the cells to settle down at the edge of the plate and carefully remove most of the medium.
8. 72 h later, transfer the EBs to a plate previously coated with Matrigel. Evenly distribute the cells by moving the plate front-to-back and side-to-side a few times.
9. After overnight incubation, all the EBs should stick down and start spreading out on Matrigel. Change medium with fresh RD medium.
10. Change medium every 2–3 days for up to 3 weeks. As the cells grow out they form rosette-like structures throughout the plate (Fig. 2d). Do not passage the cells for at least 2 weeks as this interferes with cell-to-cell interactions and in turn retinal progenitor proliferation.
11. After 3 weeks, the cells can be cultured in medium without any growth factors.

3.4. Analysis of Retinal Determination Using Quantitative Real-Time PCR

1. Harvest the ES-derived retinal cells from the plate using the Collagenase/Dispase mix and collect the cells as a pellet by centrifugation at 800 × *g* for 5 min.
2. Resuspend the pellet into 500 μL of TRIzol in a 1.5-mL RNase-free tube.
3. Homogenize tissue thoroughly with a Pellet Pestle Motor and fresh, RNase-free Pellet Pestles: samples can be frozen (−80°C) at this point indefinitely.
4. Add 200 μL RNase-free chloroform, vortex, centrifuge to separate layers, and transfer the top, aqueous layer to a fresh tube.
5. Re-extract a second time with chloroform to clear any remaining impurities.
6. Add an equal volume of 100% isopropanol, mix, and centrifuge at 15,000 × *g* for 10 min.
7. Wash pellet with 70% ethanol (RNase-free), decant, and air-dry. Resuspend in 40 μL of RNase-free H_2O. Samples can be stored at −80°C at this point.

8. Digest genomic DNA by adding RiboLock RNase-inhibitor, 10× RQ1 DNAse buffer, and RQ1 RNase-free DNase.
9. Incubate at 37°C for 30–60 min.
10. Remove genomic DNA by using the RNAeasy-cleanup procedure, part of the RNAeasy mini kit, according to manufacturer's instructions.
11. Elute in 20 μL of RNase-free water.
12. Synthesize cDNA using standard oligo-dT primed cDNA synthesis reaction with SuperScript II Reverse Transcriptase. Standard positive RT reaction mix:
    1 μg of Total RNA from above dissolved in 10 μL RNAse-free water
    1 μL oligo-dT primer (0.5 μg/μL)
    1 μL dNTPs (10 mM)
    Denature at 65°C, 5 min, place on ice, and then add the following:
    4 μL 5× SSII First Strand Buffer
    2 μL DTT (100 mM)
    1 μL RiboLock RNase-inhibitor
    1 μL SuperScript II RT
13. Incubate for 50–75 min at 42°C, heat kill RT at 70°C for 15 min (always include a no-RT control). We typically use half the amount for the no-RT reaction. Dilute reactions 1:3 or 1:4 with H_2O to prepare them for normalization via QPCR (store at −20°C).
14. The reaction mix for QPCR reaction is as follows:
    1 μL cDNA
    1 μL Forward primer (20mM)
    1 μL Reverse primer (20mM)
    7 μL H_2O
    10 μL SYBR Green PCR Master Mix
    20 μL total volume
15. In order to compare samples, normalize by assaying levels of control gene, i.e., β-actin. Adjust the sample concentrations according to the ratio of the cycle numbers. Set the threshold at the level that the fluorescence increase has reached a maximal slope. The cycle number difference in transcript levels measured between the experimental and control cDNA samples is used to calculate the fold difference. This fold difference is used in conjunction with the original sample volumes to dilute the more concentrated sample to that of the less concentrated sample. After sample concentrations are adjusted, retest β-actin levels to determine how well they were normalized.

16. Each positive control sample should be run in duplicate or triplicate and one negative RT reaction to check genomic DNA contamination.
17. Further QPCR analysis with additional primer pairs for retinal candidate genes including Pax6, Six3, Rx, Lhx2 (see Table 1). QPCR primer sets should be designed to amplify 50–200 bp amplicons, and should always be checked for specificity. Additionally, always include β-actin in each run to allow for more precise normalization of sample concentrations and accurate values in test primer sets.
18. Upon comparing expression levels of the eye field transcription factors, Pax6, Rx, Lhx2, and Six3, there should be an 80–160-fold upregulation of all of these genes at 1 week (Fig. 2e). We also found that IGF-1, dkk1, and noggin were each required for this retinal determination of the undifferentiated human ES cells.

Table 1
Primers for quantitative PCR

Gene	Primer pairs
β-actin	F-actcttccagccttccttc R-atctccttctgcatcctgtc
Pax-6	F-tctaatcgaagggccaaatg R-tgtgagggctgtgtctgttc
Lhx-2	F-tagcatctactgcaaggaagac R-gtgataaaccaagtcccgag
Six-3	F-ggaatgtgatgtatgatagcc R-tgatttcggtttgttctgg
Rx	F-gaatctcgaaatctcagccc R-cttcactaatttgctcaggac
Crx	F-atgatggcgtatatgaaccc R-tcttgaaccaaacctgaacc
Rhodopsin	F-tcatcatggtcatcgctttc R-catgaagatgggaccgaagt
S-Opsin	F-gatgaatccgacacatgcag R-ctgttgcaaacaggccaata
Recoverin	F-ccagagcatctacgccaagt R-cacgtcgtagagggagaagg
Engrailed-1	F-ccgcaccaccaactttttcat R-tggacagggtctctacctgc
Otx-2	F-gcagaggtcctatcccatga R-ctgggtggaaagagaagctg
Emx-1	F-aggtgaaggtgtggttccag R-agtcattggaggtgacatcg

Table 2
Primary antibodies for immunocytochemistry

Antibody	Company	Dilution
Mouse anti-Pax6	DHSB	1:250
Rabbit anti-Pax6	Covance research products	1:400
Mouse anti-tuj-1	Covance research products	1:1,000
Mouse anti-Rhosopsin	Gift from Dr. Molday	1:750
Mouse anti-Hu C/D	Invitrogen	1:200
Mouse anti-Nestin	Gift from Dr. E. Major	1:200
Rabbit anti-Nrl	Gift from Dr. A. Swaroop	1:500
Rabbit anti-Neurofilament	Chemicon	1:500
Rabbit anti-Crx	Gift from Dr. C. Gregory-Evans	1:2,000
Rabbit anti-Chx10	Gift from Dr. T. M. Jessel	1:1,000

3.5. Analysis of Retinal Determination Using Fluorescent Immunohisto chemistry

1. Fix the cells with 4% paraformaldehyde at 4°C for 30–60 min.
2. Rinse cells twice with PBS.
3. Incubate the cells in block solution (PBS with 5% serum and 0.5% triton X) for 30 min at room temperature.
4. Add primary antibody solution. The antibody is diluted to the suggested working concentration (Table 2) in the block solution.
5. Incubate overnight at 4°C.
6. Wash three times 5 min each in PBS.
7. Add secondary antibody solution (usually 1:500 Alexa-fluor conjugated at 1:500 diluted in block solution) and incubate in humidified chamber for about 1 h in the dark at room temperature.
8. Wash three times 5 min each in PBS.
9. Embed in Fluoromount-G.
10. After 3 weeks of the retinal determination protocol, the majority of the cells should show coexpression of Pax6 and Chx10 (approximately 70–80% cells (Fig. 2f). Also, many of the cells should be labeled with neuronal markers like Hu C/D, Neurofilament, Tuj1, as well as photoreceptor markers like Crx and Nrl.

3.6. Forced Differentiation of Human ESc-Derived Retinal Progenitors Using Inhibitors of Notch Pathway

The effects of manipulations in Notch signaling in the retina can be analyzed by using a γ-secretase inhibitor that blocks the presenilin/γ-secretase complex from releasing the Notch intracellular domain (NICD). This prevents its nuclear translocation and activation of effectors like the Hes family of genes. DAPT is a commercially available small molecule inhibitor of γ-secretase. Since Notch plays a key role in maintenance of stem cell identity of the progenitors, its inhibition forces them to differentiate (14, 15).

1. Culture human ES cell-derived retinal progenitors on Matrigel coated plates at high density.
2. Add DAPT or DMSO vehicle to the wells of the dissociated cells to desired final concentration. We found that concentrations between 10 and 50 μM strongly inhibit all Notch signaling.
3. Replace fresh medium with DAPT every day for 5–7 days.
4. Assay for effects of notch inhibition by qPCR and Immunohistochemistry. On quantitative PCR, there should be a 30–50-fold reduction in Notch effector Hes5. Upon immunostaining for marker of differentiated retinal neurons, one should observe an increase in number of Hu C/D, Tuj-1, and rhodopsin-expressing cells and a reduction in nestin-expressing cells (Fig. 3).

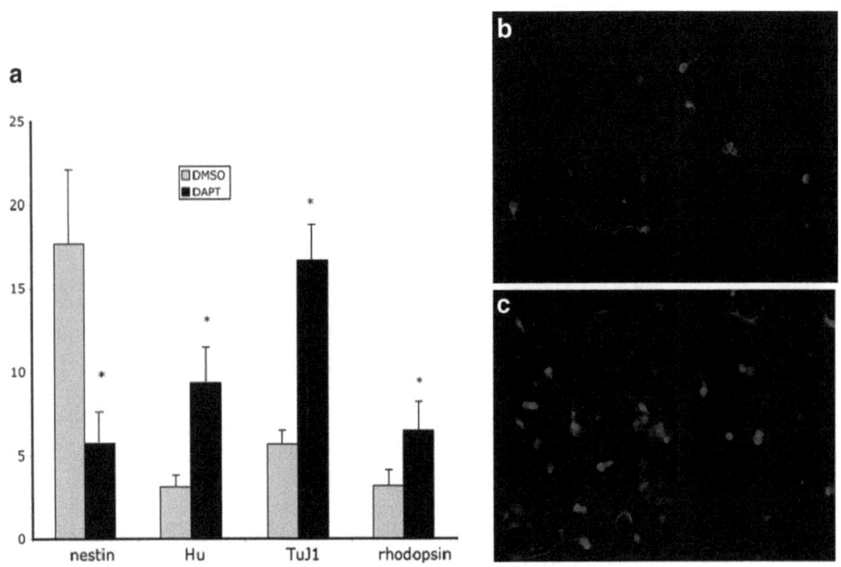

Fig. 3. Effect of DAPT on differentiation of human ES cell-derived retinal progenitors. (**a**) shows the percentage of cells expressing Nestin, HuC/D, Tuj1 and Rhodopsin following treatment with either DAPT or its vehicle DMSO. DAPT results in a doubling in the number of cells expressing differentiated markers and a reduction in nestin-expressing cells. (**b**, **c**) show representative micrographs of rhodopsin staining after 7 days of DMSO or DAPT, respectively

3.7. Subretinal Injection of Dissociated Cells Into Adult Mouse Recipients

One day prior to transplantation, the mice are started on Cyclosporine A injections (10 mg/kg/day) daily until euthanasia.

1. Prepare dissociated cell suspension by rinsing the cells with PBS followed by incubation in 0.2% trypsin for 5 min. As the cells lift off the plate, block the trypsin activity with 10% FBS in DMEM. Centrifuge the cells at $800 \times g$ and resuspend in medium at a concentration of 80,000–100,000 cells/μL.
2. One eye per animal is injected with cells, while the other eye serves as a control. We always inject the left eye to avoid confusion.
3. Prepare pipets by pulling them to a long (10 mm), gentle taper. The tip is then broken with forceps to a ~120 μm opening, which will be small enough to minimize injury, but large enough to let cells pass through easily. Just before starting the injections, mount the pipet into the holder, insert the plunger, and draw up 2–3μL of cell suspension.
4. Anesthetize the transplant recipient mouse with ketamine (130 mg/kg) and xylazine (8.8 mg/kg). Any other reliable method of anesthesia can be substituted. Use a toe pinch to assure the animal is fully anesthetized. The deep anesthesia needs to last at least 20 min, although with practice the injection only takes 3–5 min. Apply petrolatum ophthalmic ointment to the control eye to keep it moist, and apply a topical anesthetic (0.5% proparacaine hydrochloride ophthalmic solution) to the eye to be injected.
5. Position the stereotaxic apparatus under a dissection microscope. Mount the animal's head in the head holder. First insert the upper teeth into the notch of the tooth bar, and tighten the nose clamp just enough to keep the head level. Then insert the ear bars into the bony ear canal on each side, and tighten gently. Make sure the skull is firmly in place before proceeding.
6. From here on, work viewing through the microscope with good bright lighting. Gently lift the upper eyelid using forceps, then pass the suture through the upper eyelid. Fasten both ends of the suture material to the screws of the stereotax to pull the eyelid open. Often, this is enough to pop the eye out of its socket. Otherwise, put gentle pressure on the periorbital area with the blunt end of your forceps to bring the eyeball out of the socket.
7. Using a sterile #11 scalpel blade very carefully and gently cut away the sclera on the dorsal surface of the eye. The blade should be repeatedly passed across the sclera to gradually thin the tissue in a small area approximately 0.5 mm in diameter. Do not apply too much pressure as you may puncture the eye. If using a pigmented mouse, the dark coloration of

the pigment epithelium should be increasingly visible. Make sure the surface of the eye does not dry out. Continue cutting until a tiny area of retina (approximately the diameter of the pipet tip) is exposed at the center of the thinned area. Due to the high intraocular pressure, the retina will bleb out of the opening in the sclera, so it is very important to keep the cut as small as possible. Use a tissue to wick up any blood.

8. Advance the pipet tip toward the opening in the sclera. Keep it at a shallow angle (10–20°) relative to the surface to the eye to avoid puncturing the retina. Position the tip directly over the opening, then slip the tip just under the sclera with a tiny down-and-forward motion. Now apply very gentle outward traction on the eye by pulling the pipet back up. This helps to open up a space under the retina. Advance the pipet tip about 250 μm into the subretinal space.
9. Pressing on the plunger very slowly and lightly, inject 1 μL of cells into the subretinal space. Expect some of the injected cells to squirt back out around the needle tip, because the subretinal space is a potential space and does not admit much volume. Let the pipet rest in place for about 30 s to allow the injected volume to disperse in the subretinal space, then gently withdraw the pipet.
10. Remove the suture, and push the eyeball back into the socket. Remove the mouse from the holder, apply bacitracin ophthalmic ointment to the injected eye, and let the animal recover from the surgery. Using this method, we have not observed any intraocular or periocular infections. The injection site will heal within a few days, and the mice tolerate the procedure very well.
11. After survival periods of 1–4 weeks, the animals are sacrificed by a protocol that has been previously approved by the institutional IACUC (or similar animal care committee) and the eyes removed for analysis of transplanted cells by fixation (2 h to overnight 2–4% paraformaldehyde in PBS at 4°C), standard cryostat sectioning, and immunofluorescent labeling as described above.

References

1. Lamba, D., Karl, M., and Reh, T. (2008). Neural regeneration and cell replacement: a view from the eye. Cell Stem Cell 2, 538–549.
2. Sanchez-Pernaute, R., Studer, L., Bankiewicz, K.S., Major, E.O., and McKay, R.D. (2001). In vitro generation and transplantation of precursor-derived human dopamine neurons. J Neurosci Res *65*, 284–288.
3. Wichterle, H., Lieberam, I., Porter, J.A., and Jessell, T.M. (2002). Directed differentiation of embryonic stem cells into motor neurons. Cell *110*, 385–397.
4. Cho, M.S., Hwang, D.Y., and Kim, D.W. (2008). Efficient derivation of functional dopaminergic neurons from human embryonic stem cells on a large scale. Nat Protoc *3*, 1888–1894.

5. Dimos, J.T., Rodolfa, K.T., Niakan, K.K., Weisenthal, L.M., Mitsumoto, H., Chung, W., Croft, G.F., Saphier, G., Leibel, R., Goland, R., *et al.* (2008). Induced pluripotent stem cells generated from patients with ALS can be differentiated into motor neurons. Science *321*, 1218–1221.
6. Park, C.H., Minn, Y.K., Lee, J.Y., Choi, D.H., Chang, M.Y., Shim, J.W., Ko, J.Y., Koh, H.C., Kang, M.J., Kang, J.S., *et al.* (2005). In vitro and in vivo analyses of human embryonic stem cell-derived dopamine neurons. J Neurochem *92*, 1265–1276.
7. Niehrs, C. (2001). Developmental biology. Solving a sticky problem. Nature *413*, 787–788.
8. del Barco Barrantes, I., Davidson, G., Grone, H.J., Westphal, H., and Niehrs, C. (2003). Dkk1 and noggin cooperate in mammalian head induction. Genes Dev *17*, 2239–2244.
9. Zuber, M.E., Gestri, G., Viczian, A.S., Barsacchi, G., and Harris, W.A. (2003). Specification of the vertebrate eye by a network of eye field transcription factors. Development *130*, 5155–5167.
10. Cavodeassi, F., Carreira-Barbosa, F., Young, R.M., Concha, M.L., Allende, M.L., Houart, C., Tada, M., and Wilson, S.W. (2005). Early stages of zebrafish eye formation require the coordinated activity of Wnt11, Fz5, and the Wnt/beta-catenin pathway. Neuron *47*, 43–56.
11. Esteve, P., and Bovolenta, P. (2006). Secreted inducers in vertebrate eye development: more functions for old morphogens. Curr Opin Neurobiol *16*, 13–19.
12. Lamba, D.A., Gust, J., and Reh, T.A. (2009). Transplantation of Human Embryonic Stem Cell-Derived Photoreceptors Restores Some Visual Function in Crx-Deficient Mice. Cell Stem Cell *4*, 1–7.
13. Lamba, D.A., Karl, M.O., Ware, C.B., and Reh, T.A. (2006). Efficient generation of retinal progenitor cells from human embryonic stem cells. Proc Natl Acad Sci U S A *103*, 12769–12774.
14. Nelson, B.R., Hartman, B.H., Georgi, S.A., Lan, M.S., and Reh, T.A. (2007). Transient inactivation of Notch signaling synchronizes differentiation of neural progenitor cells. Dev Biol *304*, 479–498.
15. Nelson, B.R., and Reh, T.A. (2008). Relationship between Delta-like and proneural bHLH genes during chick retinal development. Dev Dyn *237*, 1565–1580.

Chapter 10

Bovine Somatic Cell Nuclear Transfer

Pablo J. Ross and Jose B. Cibelli

Abstract

Somatic cell nuclear transfer (SCNT) is a technique by which the nucleus of a differentiated cell is introduced into an oocyte from which its genetic material has been removed by a process called enucleation. In mammals, the reconstructed embryo is artificially induced to initiate embryonic development (activation). The oocyte turns the somatic cell nucleus into an embryonic nucleus. This process is called nuclear reprogramming and involves an important change of cell fate, by which the somatic cell nucleus becomes capable of generating all the cell types required for the formation of a new individual, including extraembryonic tissues. Therefore, after transfer of a cloned embryo to a surrogate mother, an offspring genetically identical to the animal from which the somatic cells where isolated, is born. Cloning by nuclear transfer has potential applications in agriculture and biomedicine, but is limited by low efficiency. Cattle were the second mammalian species to be cloned after Dolly the sheep, and it is probably the most widely used species for SCNT experiments. This is, in part due to the high availability of bovine oocytes and the relatively higher efficiency levels usually obtained in cattle. Given the wide utilization of this species for cloning, several alternatives to this basic protocol can be found in the literature. Here we describe a basic protocol for bovine SCNT currently being used in our laboratory, which is amenable for the use of the nuclear transplantation technique for research or commercial purposes.

Key words: Cloning, SCNT, Reprogramming, Oocyte enucleation, Nuclear transfer, Cell fusion, Oocyte activation, Micromanipulation, Bovine, Cattle

1. Introduction

Somatic cell nuclear transfer (SCNT) is a technique by which the nucleus of a somatic cell is introduced into an enucleated oocyte. As a result, the somatic nucleus is modified by the recipient oocyte's cytoplasm, allowing the development of the reconstructed embryo into a whole individual. The result is a genetic clone of the animal from which the donor cell was derived. The original idea of generating an animal from a somatic cell was

S. Ding (ed.), *Cellular Programming and Reprogramming: Methods and Protocols*, Methods in Molecular Biology, vol. 636, DOI 10.1007/978-1-60761-691-7_10, © Springer Science+Business Media, LLC 2010

proposed by Spemann as a way to test the developmental potential of a cell nucleus (1). However, the required technology to perform Spemann's proposed experiment was not available until the 1950s when Briggs and King developed nuclear transfer techniques in frogs, obtaining adult animals when injecting blastula cells into enucleated oocytes (2). Later, Gurdon produced feeding tadpoles from frog somatic cells (3). In mammals, nuclear transfer technology was developed several decades later. The transfer of a blastocyst cell (blastomere) nucleus into an enucleated mouse zygote was reported by Illmensee and Hoppe in 1981, with the development of adult animals (4). However, controversy surrounded these results, as other groups were unable to repeat the experiment. McGrath and Solter developed a more efficient technique by which the donor cell was fused, instead of the nucleus injected, with an enucleated zygote; however, they were not able to produce offspring when two-cell embryos and older were used as cell donors. (5). In 1986, Willadsen obtained offspring after fusing sheep MII oocytes with 8- or 16-cell embryonic blastomeres (6). Later, animal cloning from embryonic cells was successfully repeated in several species including cattle (7), rabbits (8), pigs (9), mice (10), and monkeys (11). Dolly the sheep was the first mammal to be cloned from a somatic cell nucleus (12) in 1997 and several other species followed, including cow (13), mouse (14), goat (15), pig (16), gaur (17), mouflon (18), rabbit (19), cat (20), rat (21), mule (22), horse (23), African Wildcat (24), dog (25), ferret (26), wolf (27), buffalo (28), and red deer (29). In each of these species, the efficiency remains very low, with less than 1% of nuclear transfers from adult cells developing into normal offspring. Nevertheless, the success of SCNT-cloning in several species underscores the totipotent potential of the somatic cell nucleus and the reprogramming ability of the MII oocyte, and presents nuclear transplantation as a powerful methodology to study the molecular mechanisms that regulate cell fate commitment, differentiation, and pluripotency.

In more applied areas, cloning by nuclear transfer has the potential to contribute substantially to animal agriculture, biotechnology, biomedicine, and preservation of endangered species. The success of adult SCNT with almost all agriculturally important species (12, 13, 15, 19, 30) confirms its usefulness for the clonal expansion of animals with superior genotypes. Moreover, SCNT makes possible germline genetic modifications in domestic species (13). Traits which have been considered for genetic modification include feed utilization, resistance to disease (thus reducing drug/antibiotic use), reduction of animal waste, and diversification of agricultural products, i.e., providing new economic opportunities in rural areas, and generation of new consumer products (31). SCNT can also be used for gene targeting, making additions or deletions in the genome feasible. Using this approach, cattle that lack the prion gene responsible for bovine spongiform encephalopathy were recently produced (32). Targeted modifications have also been successfully achieved in sheep (33) and pigs (34). Farm animals carrying genetic modifications have great potential in biotechnology. Engineered animals are being used as bioreactors for the production of pharmaceuticals and as potential organ donors for the human population. Further, SCNT offers an alternative means to preserve endangered, and even to recover extinct species. Wells et al. reported the use of SCNT to clone the last surviving animal of the Enderby Island cattle breed (35), and Lanza et al., using interspecies nuclear transfer, were able to clone an endangered species (*Bos gaurus*) (36). The same approach was used to clone Mouflons (an endangered breed of sheep) with tissue collected from dead animals (18). Although all of the above-described applications for SCNT are far-reaching, its broad implementation is hindered by low efficiency.

Advances in micromanipulation techniques have allowed an improvement in preimplantation development of reconstructed embryos; however, the full-term developmental potential of embryos produced by SCNT remains low. High rates of early pregnancy loss are commonly observed and a higher incidence of late-term abortion is often reported for SCNT embryos compared to embryos produced by fertilization. Also, higher mortality rates of offspring born from NT embryos are often reported.

Here we describe a basic protocol for bovine SCNT currently being used in our laboratory, which is amenable for the use of the nuclear transplantation technique for research or commercial purposes.

2. Materials

2.1. Equipment

1. Inverted fluorescent microscope with 4× and 20× Hoffman modulation contrast optics. Thermoplate/sheet (38.5°C) is recommended. Fluorescence illumination is required to visualize the HOECHST 33342 stained DNA at the time of enucleation. A pedal-controlled shutter that blocks UV light from the path of light is important to minimize exposure of the oocytes to UV irradiation. Also, a condenser that limits UV light to the center of the field of view will help minimize oocyte exposure to UV light.
2. Micromanipulation equipment attached to microscope. For holding pipette (usually mounted on the left side), a coarse manipulator is sufficient, since after setting up the position of this pipette it is not necessary to make continuous adjustments.

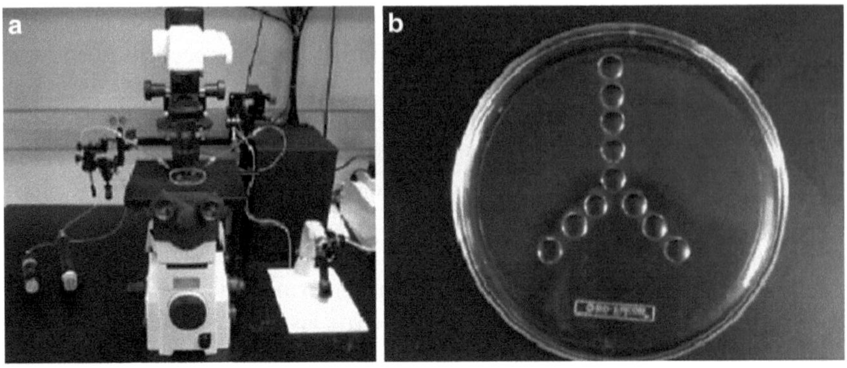

Fig. 1. Micromanipulation setup. (**a**) Microscope and micromanipulation equipment. (**b**) Layout of micromanipulation chamber

For the enucleation/transfer manipulator (usually mounted on the right side) a hydraulic controlled manipulator is required (Fig. 1).

3. Microinjectors: An air microinjector can be used for holding the oocyte. For enucleation/cell transfer, an oil-filled injector is preferred to achieve greater flux control.
4. Electrofusion generator: A square DC pulse generator capable of voltage and pulse duration adjustments.
5. Fusion chamber with 0.5 mm gap between electrodes.
6. CO_2 incubator.
7. Microdispensers (Drummond Scientific Co., Broomall, PA): Used for handling of oocytes/embryos. Alternatively, mouth-pipettes or tomcat catheters can be used.
8. Pipette puller.
9. Micro forge: Used to cut and model glass pipettes.
10. Micro grinder: Used to produce a bevel in the pipette tip.

2.2. Oocyte Collection and Maturation

1. Saline solution: 8.5 mg/mL NaCl.
2. Hepes-Buffered Hamster Embryo Culture Medium (HH) (37): 114 mM NaCl, 3.2 mM KCl, 2 mM $CaCl_2$, 0.5 mM $MgCl_2$, 0.1 mM Na pyruvate, 2 mM $NaHCO_3$, 10 mM HEPES, 17 mM Na lactate, 1× MEM nonessential amino acids, 100 IU/mL penicillin G, 100 μg/mL streptomycin, 3 mg/mL BSA. pH: 7.3–7.4; Osmolarity: 275 ± 10 mOsm/kg; Filter sterilize and store at 4°C for up to 30 days.
3. Medium 199 (Sigma).
4. Pyruvate stock solution: 11 mg of sodium pyruvate in 5 mL of M199, store at 4°C for up to 30 days.
5. LH stock: 3 mg Luteinizing Hormone (Sioux Biochem), 10 mg fatty acid free BSA, 1 mL saline solution. Aliquot and store at −20°C for up to 6 months.

6. FSH stock: 3 mg Follicular-Stimulating Hormone (Sioux Biochem), 10 mg fatty acid free BSA, 1 mL saline solution. Aliquot and store at −20°C for up to 6 months.
7. Estradiol stock: 1 mg 17 beta-estradiol in 1 mL absolute ethanol. Store at −20°C.
8. Gentamicin: 10 mg/mL (Gibco).
9. Fetal bovine serum (FBS).

2.3. Micropipette Preparation

1. Borosilicate glass capillaries 1 mm OD (outside diameter) × 0.75 mm ID (inside diameter).
2. Borosilicate glass capillaries 1 mm OD × 0.58 mm ID.
3. Chromerge glass cleaner.
4. 70% Ethanol.

2.4. Somatic Cell Nuclear Transfer

1. Hyaluronidase solution: 1 mg/mL hyaluronidase in HH medium. Divide in 1 mL single use aliquots and store at −20°C.
2. CB stock: 1 mg of Cytochalasin B in 200 μL DMSO. Aliquot and store at −20°C.
3. HOECHST stock: 1 mg Bisbenzimide (Hoechst 33342) in 1 mL of PBS. Store at −20°C protected from light.
4. Mineral oil.
5. Fluorinert FC-40 (Sigma).
6. Pronase: 10 IU/mL Pronase in HH medium, store at 4°C.
7. Fusion medium: 250 mM D-Sorbitol, 0.5 mM MgOAc, 1 mg/mL BSA, pH: 7.2.
8. Osmolarity: 255 ± 10 mOsm/kg. Filter, sterilize, and store at 4°C for 20 days.
9. Ionomycin stock (5 mM): Dissolve 1 mg Ionomycin in 267.6 μL DMSO. Aliquot and store at −20°C for up to 12 months.
10. DMAP stock (200 mM): Dissolve 163 mg of DMAP in 5 mL ddH2O in a 90°C water bath. Aliquot and store at −20°C.
11. KSOM embryo culture medium: Available commercially (Millipore) supplemented with 3 mg/mL BSA.

3. Methods

3.1. Oocyte Collection and Maturation

Oocytes for bovine SCNT are typically harvested from slaughterhouse-derived ovaries and matured in vitro. Alternatively, oocytes can be collected from animals by ultrasound-guided oocyte aspiration at mature or immature stages. We describe the protocol for collecting oocyte from slaughterhouse-derived ovaries only.

1. Prepare oocyte maturation medium by adding the following to 9 mL of Medium 199: 1 mL of FBS, 10 μL of FSH stock, 10 μL of LH stock, 100 μL of pyruvate stock, and 25 μL of gentamicin. Filter sterilize and then add 10 μL of estradiol stock (do not take the estradiol out of the −20°C freezer to avoid ethanol evaporation and estradiol concentration). Equilibrate in the incubator for at least 4 h.
2. Ovaries are transported from slaughterhouse to lab in a thermal container at room temperature.
3. Place ovaries in a colander and rinse them thoroughly using warm tap water.
4. Transfer oocytes to a beaker and add warm saline solution.
5. Keep the beaker with oocytes in a 30°C waterbath.
6. Aspirate follicles using an 18-G hypodermic needle. The needle is connected to a vacuum source that can be a 10-mL syringe or a vacuum pump. The use of a vacuum pump allows for faster oocyte collection. A 50-mL tube acting as a vacuum trap is used to collect the follicular fluid containing the oocytes (Fig. 2).

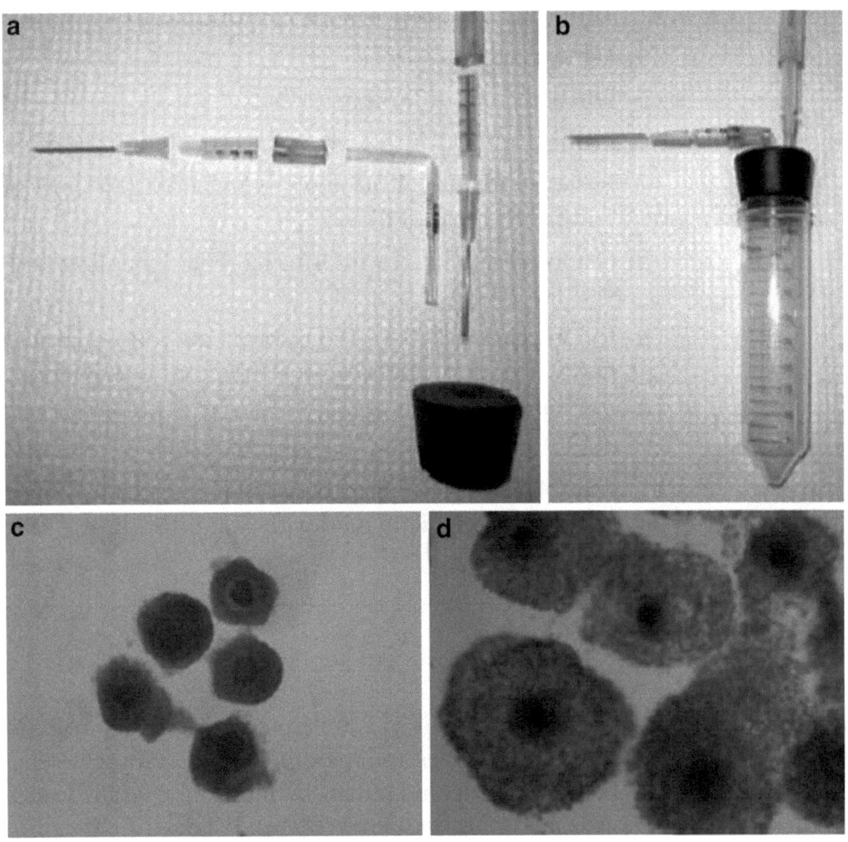

Fig. 2. Oocyte collection. (**a**) Aspiration assembly. (**b**) Dissembled aspiration assembly. (**c**) Immature oocytes. (**d**) Matured oocytes

An aspiration assembly is constructed with a rubber stopper and a 1-mL glass pipette. Bend the glass pipette at a 90° angle and cut the narrow end. Drill a hole in the center of the rubber stopper. Insert the pipette in the stopper through the hole. Cut a 1-mL plastic syringe in the middle and assemble the end that connects to the needle to the glass pipette using a small piece of tygon tubing. Ensure that the connection is air tight. Cut another 1-mL syringe and connect it to an 18-G needle, then insert the needle through the rubber stopper. Connect the vacuum pump to this piece of syringe with tygon tubing. For oocyte aspiration attach an 18-G needle to the aspiration assembly with the opening of the needle facing down and mount the assembly on a 50-mL tube. After using, rinse the assembly thoroughly with distilled water and spray with 70% ethanol. Let dry in a clean container.

7. Remove groups of 10–20 ovaries from the beaker and dry their surfaces with paper towels.
8. Aspirate follicles 2–8 mm in diameter. To aspirate the follicle content first penetrate the ovarian parenchyma and then the follicle. This will prevent a potential follicle rupture and loss of oocyte. Also, several follicles can be aspirated through the same hole by advancing the needle through the oocyte cortex.
9. Let the oocytes sediment in the follicular fluid and collect the sediment using a disposable plastic Pasteur pipette.
10. Add 2–3 mL of HH medium to a gridded 100-mm petri dish (the grid can be drawn with a marker on the external surface of the dish).
11. Disperse the liquid in the dish but without touching the edges.
12. Add the oocytes to the dish and allow 1 min for them to sediment (*see* Note 1).
13. Collect and transfer the oocytes to a 1-mL drop of HH medium.
14. Select good quality oocytes (homogeneous oocyte cytoplasm and at least three layers of cumulus cells; Fig. 10.2) and transfer them in groups of 50 into 100 μL drops of HH medium. Immediately after releasing the oocytes in clean HH drops, aspirate loose cells; this will help clean the oocytes in fewer washes therefore reduce handling of the oocytes.
15. Wash through another drop of HH and then transfer to a four-well dish containing 500 μL of preequilibrated maturation medium.
16. Incubate at 38.5°C, humidity to saturation, and 5% CO_2 in air.

3.2. Micropipette Preparation

Preparing good manipulation tools accounts for a great part of success in nuclear transfer technique. The micropipettes required to do nuclear transfer consist of a holding pipette, enucleation pipette, and cell transfer pipette. These pipettes differ in size and shape and are fashioned from glass capillaries. Making these pipettes will require some practice and trial and error at first, but proficiency in making micromanipulation tools is generally gained in a short period of time (few weeks).

3.2.1. Holding Pipette

The holding pipette is used to position the oocyte for enucleation and cell transfer. The external diameter of this pipette can range from 50 to 90% of that of the oocyte. To manipulate bovine oocytes, we typically prepare holding pipettes with an external diameter of 150 μM. The opening of the pipette is set at 20–30% the oocyte diameter, in our case approximately 30 μM. To produce holding pipettes we use 1 mm OD × 0.58 mm ID glass capillaries. The pipette is pulled using a pipette puller to achieve a lightly tapered and long tip. Then, the tip of the pipette is cut at the desired width using a diamond-tip pen. This can be easily performed by placing the pipette in the microforge where the desired diameter can be measured with the micrometer scale in the eyepiece (Fig. 3).

1. Pull the glass.
2. Place the pipette on the microforge in horizontal position and locate the desired width to be broken.
3. Pass a diamond-tip pen across the top surface of the pipette two or three times.
4. Apply pressure to the tip to break it. This should result in an even cut. If the cut is not even, discard the pipette and start over. Alternatively, the pipette can be cut using the technique described below for enucleation and transfer pipettes, although because of the larger size of holding pipette, this could be cumbersome.
5. Place the pipette in vertical position on top of the glass bead present in the microforge.
6. Set the heat to 90–100% of total power and lower the pipette close to the glass bead, but without touching it. The tip of the pipette will melt resulting in widening of the borders and shrinking of the inside diameter.
7. When the desired inside diameter is reached release the heat.
8. Finally, the pipette is bent by heating on the gas burner at a distance of approximately 1–2 cm from the tip.
9. Set the gas burner to produce the smallest flame possible. Hold the pipette horizontally in front of you and move it on top of the flame at the desired place to be bent. As the glass

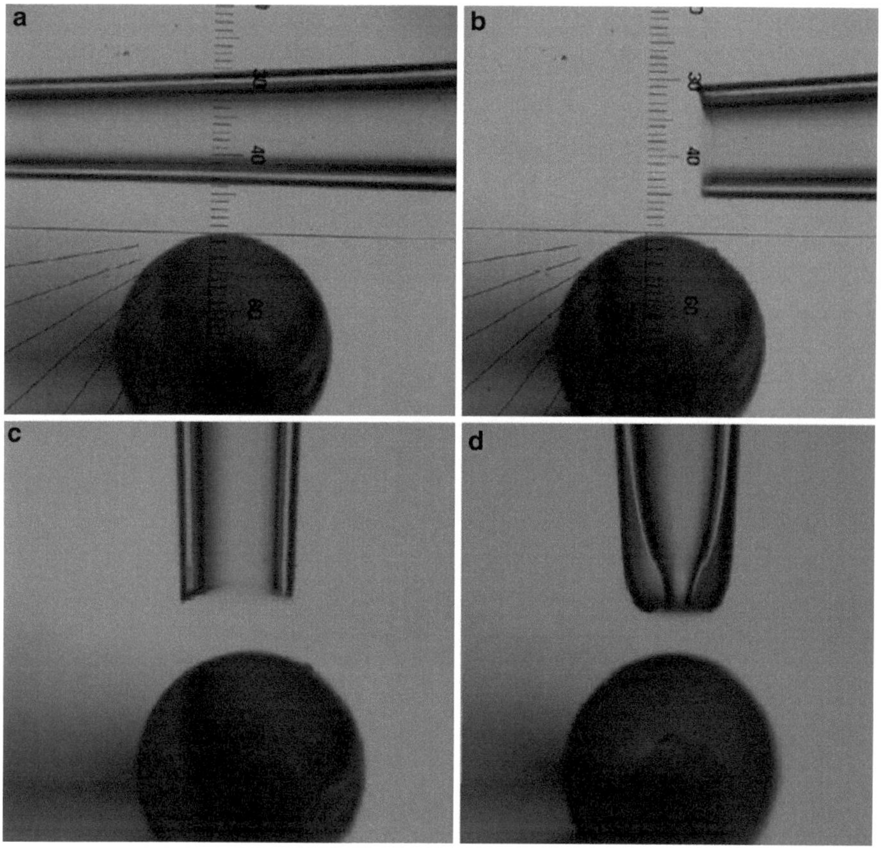

Fig. 3. Holding pipette construction. (**a**) Pulled glass capillary before cutting. (**b**) Glass capillary after cutting. (**c**) Pipette before polishing. (**d**) Pipette after polishing

starts to melt, the pipette will bend downward. Be careful not to expose the pipette tip to the flame, as the small diameter of the glass at that point will result in complete melting. Remove the pipette from the flame when a 150° angle is formed.

3.2.2. Enucleation and Transfer Pipettes

Enucleation and transfer pipettes are forged out of 1 mm OD × 0.75 mm ID glass capillaries (*see* Note 2). These pipettes are beveled and generally have a spike to facilitate penetration of the oocyte's zona pellucida (Fig. 4). The two pipettes only differ in their diameter, with the enucleation pipette being 15–18 μM ID and the transfer pipette being slightly larger than the donor cells that will be transferred (usually between 20 and 35 μM).

1. Pull the glass to generate a gradual and long taper.
2. Set up the pipette in the microforge horizontally.
3. Position the region where you want to cut the pipette on top of the glass ball, but without touching it.

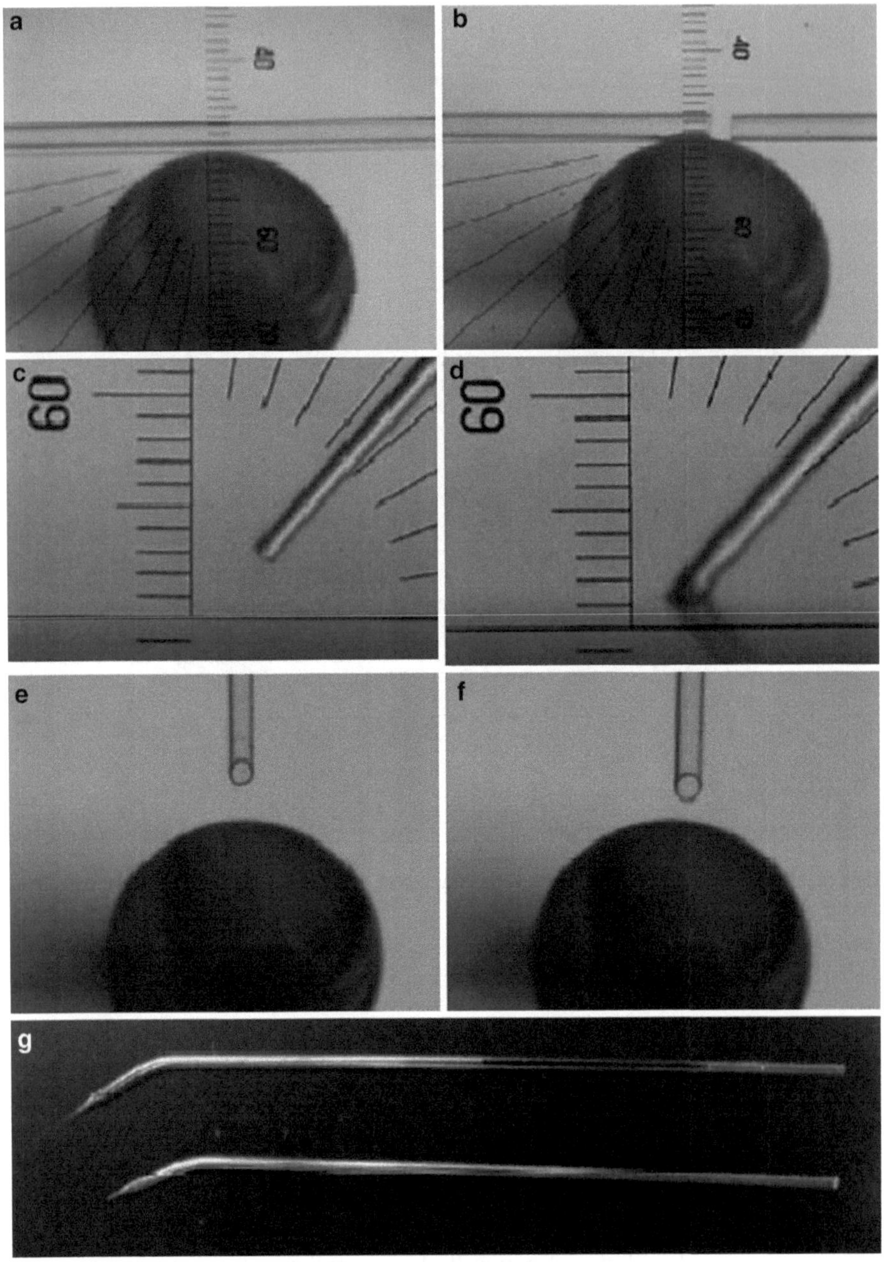

Fig. 4. Enucleation and transfer pipette construction. (**a**) Pulled glass capillary before cutting. (**b**) Glass capillary after cutting. (**c**) Pipette before grinding. (**d**) Pipette being ground. (**e**) Ground pipette before spike construction. (**f**) Pipette with spike. (**g**) Transfer pipette (*top*) and holding pipette (*bottom*) after bending

4. Set the heat level on the microforge to 40–50%.
5. As you heat the filament, the glass ball will move because of filament dilation. As the filament is heated touch the glass bead with the pipette. The pipette will start to melt and at that point remove the heat. The return of the filament to its

original position will break the pipette. If the pipette remains attached to the glass bead, moving the pipette or the filament slightly should break it.

6. To make the bevel place the pipette on the grinder at 45° angle. Use a permanent marker to identify the position of the bevel by marking the top face of the pipette; this will indicate the longer point of the bevel.
7. With the grinding stone wet and rotating at 80–100% its maximum speed, lower the pipette until it makes contact with the grinder and a slight inflection in the pipette is noticed.
8. Grind until the inflection disappears, and the bevel is complete through the pipette tip, approximately 1–2 min.
9. Remove the pipette from the grinder and proceed to wash according to pipette type. Washing at this step ensures that small glass chips that result from grinding will not attach to the pipette during spike generation or pipette bending.
10. Enucleation pipettes can be washed with only 70% ethanol to remove most glass chips that may be present. Connect a 20-mL syringe to the pipette by a silicon tubing of the appropriate size to make an air tight connection. Dip the pipette into a 50-mL tube containing 70% ethanol and first blow some air to remove liquid left from the grinding process. Then aspirate 70% ethanol by holding back the syringe embolus for 5–10 s and applying a pressure of 5–10 mL. Then blow the aspirated liquid until air bubbles come out of the pipette tip.

Transfer pipettes are washed more thoroughly to eliminate residual glass chips. Using the same technique as indicated above, wash the pipette two times in Chromerge, and then twice in each of three tubes containing ddH_2O and one containing 70% ethanol.

1. Then, a spike is added to the pipette to increase its sharpness.
2. Mount the capillary onto the microforge in a vertical position with the bevel mark facing you.
3. Set the heat to 30–40% maximum.
4. While heating, use the micromanipulators in the microforge to barely touch the top of the glass bead.
5. Once the tip of the pipette starts to melt, move the pipette rapidly and straight away from the glass bead, creating the small spike. Practice and trial and error will be needed to find the perfect combination and articulation of movements required to make a good spike without distorting the shape of the pipette opening.

6. Finally, bend the pipette as indicated for holding pipette above. When bending, make sure that the marking that indicates the location of the bevel is facing you directly.

3.3. Nuclear Transfer

3.3.1. Preparation of Culture Dishes for Embryo Manipulation

1. Prepare four 35-mm petri dishes with five 50 μL drops of KSOM medium covered with mineral oil (these will be used to place the stained oocytes, enucleated oocytes, transfer oocytes, and fused oocytes). Place one drop in the center of the dish with the remaining ones surrounding it. The center drop will be use to rinse the oocytes before placing them in the other drops.
2. Add 10 μL of DMAP stock to 1 mL of embryo culture medium. Once DMAP stock is thawed it needs to be heated to become water soluble. This can be done by briefly placing the aliquot in a boiling waterbath for a minute or less. Prepare one or two 35-mm petri dishes with 50-μL drops of DMAP medium covered with mineral oil.
3. Add 5 μL of HOECHST stock to 1 mL of embryo culture medium and prepare one 35-mm petri dish with 5–6 50-μL drops covered with mineral oil.
4. For final embryo culture, prepare dishes with 100-μL drops of KSOM medium. To prepare the drops, place 30 μL of medium in the center of each well of a four-well dish. Then, cover with oil. Finally, add 70 μL of medium to complete the 100 μL and add oil as necessary to completely cover the drop.
5. Allow all plates to equilibrate in the incubator for at least 8 h before use.
6. Prepare oocyte manipulation medium by adding 200 μL of FBS and 1.5 μL of CB stock to 800 μL of HH medium.

3.3.2. Preparation of Recipient Oocytes

Sixteen to seventeen hours after onset of oocyte maturation the cumulus cells are removed from the oocyte by vortex agitation:

1. Warm up HH medium in waterbath or dry heat block.
2. Thaw a vial of hyaluronidase solution.
3. Transfer the oocytes from maturation medium into the tube containing hyaluronidase using a P200 micropipette (up to 300 oocytes can be processed together).
4. Let the cumulus oocyte complex sediment and remove supernatant, leaving behind approximately 150 μL of medium in the tube.
5. Vortex for 5 min at speed 7 (Daigger, Vortex Genie 2 or equivalent).
6. Using a P1000 micropipette rinse the tube two times with 1 mL of HH medium and place the medium in a dish.

7. Slowly and gently rock the plate to concentrate the oocytes in the center of the medium drop.
8. Transfer the oocytes to a 100 μL drop of clean HH medium and gently disperse the oocytes. Then, rock the plate to collect the oocytes in the middle of the drop. Repeat until oocytes are cleaned of granulosa cells (3–5 times).
9. Select MII oocytes by presence of a polar body. Moving the mirror in the light-path of the stereomicroscope will generate high contrast that will aid in identifying the polar bodies. At 17 h postmaturation, at least 50% of oocytes should be at metaphase II stage.
10. Transfer MII and immature oocytes to drops of KSOM medium containing Hoechst 33342. Immature oocytes will continue to mature and can be selected for presence of a polar body after the first group is enucleated. Immature oocytes will be stained at this stage and after selection can be directly processed for enucleation without having to restain.
11. Incubate for 10–15 min.
12. Transfer stained oocytes to KSOM drops and place in incubator until use.

3.3.3. Setup of Manipulation Chamber and Tools

Bovine oocyte micromanipulation is carried out in plastic petri dishes. We typically use the lid of a 100-mm petri dish since the borders are lower, allowing for easier access of micropipettes. Both enucleation and cell transfer drops can be prepared in the same dish. Varied configuration of drop locations can be used. The one described below is only one of the possible variations. Enucleation is done in medium containing Cytochalasin B, to destabilize the cytoskeleton and prevent oocyte lysis, and FBS as a surfactant to minimize cell stickiness. Cell transfer is performed in HH medium without Cytochalasin B.

1. Place 4–5 drops of 50 μL enucleation medium in a row starting at the top of the plate.
2. In an inverted Y shape, place three drops of HH medium at each side. Using an inverted Y shape allows rotating the plate and being able to manipulate every drop without disturbing the other drops (Fig. 1b).
3. Cover the drops with 15 mL of mineral oil.
4. Transfer the plate to the microscope stage.
5. Mount the holding pipette on the left micromanipulator. Adjust the pipette so the tip is parallel to the plate surface. Lower the holding pipette into the oil and then adjust so it is located in the center of the field of view. The holding pipette will not be moved during the manipulation process.

6. Backfill the enucleation pipette with Fluorinert using a syringe and tygon tubing.
7. Connect the enucleation pipette to the hydraulic microinjector and mount it on the right side manipulator. Ensure that no air bubbles are in the line or in the pipette, since this will result in poor flux control. To avoid air bubbles, the oil from the hydraulic microinjector should form a small protruding drop out of the tip before inserting the pipette loaded with Fluorinert.
8. Lower the pipette into the oil. Fluorinert drops will be released from the pipette. Adjust the microinjector pressure to stop the Fluorinert flow but avoid aspirating mineral oil into the enucleation pipette.
9. Move the pipette into an enucleation drop and readjust microinjector pressure.

3.3.4. Oocyte Enucleation

1. Transfer a group of oocytes (30–50) to the enucleation drop.
2. Place the oocytes in the top side of the drop, with the holding and enucleation pipettes at about the middle of the drop. In this way, intact oocytes will be aspirated from the top group, enucleated, and released to the lower side of the drop, with minimal movements of the microscope stage or the holding pipette.
3. Aspirate an oocyte with the holding pipette. Lower or lift the holding pipette to ensure that the oocyte being manipulated is in the same focal plane as the others. For best control, the oocyte should be touching the Petri dish but not compressed against it.
4. Focus the microscope in the middle of the oocyte, where the ZP is in best focus. If the manipulators are leveled with the stage, focus should not need continuous readjustments.
5. Using the enucleation pipette rotate the oocyte until the polar body is visualized and located at the 4 O'clock position (Fig. 5).
6. Use brief exposure to UV light to confirm that the metaphase plate is in proximity of the polar body and make the required adjustments of the oocyte to ensure that both polar body and metaphase plate are in the same focal plane. This is accomplished by rotating the oocyte instead of changing the focus of the microscope. Although initially controlling the oocyte position may be difficult, with practice it should become second nature. Not having to adjust the microscope focus during each enucleation allows you to keep the left hand on the injector and the right hand on the manipulator at all times, therefore achieving greater control and precision.

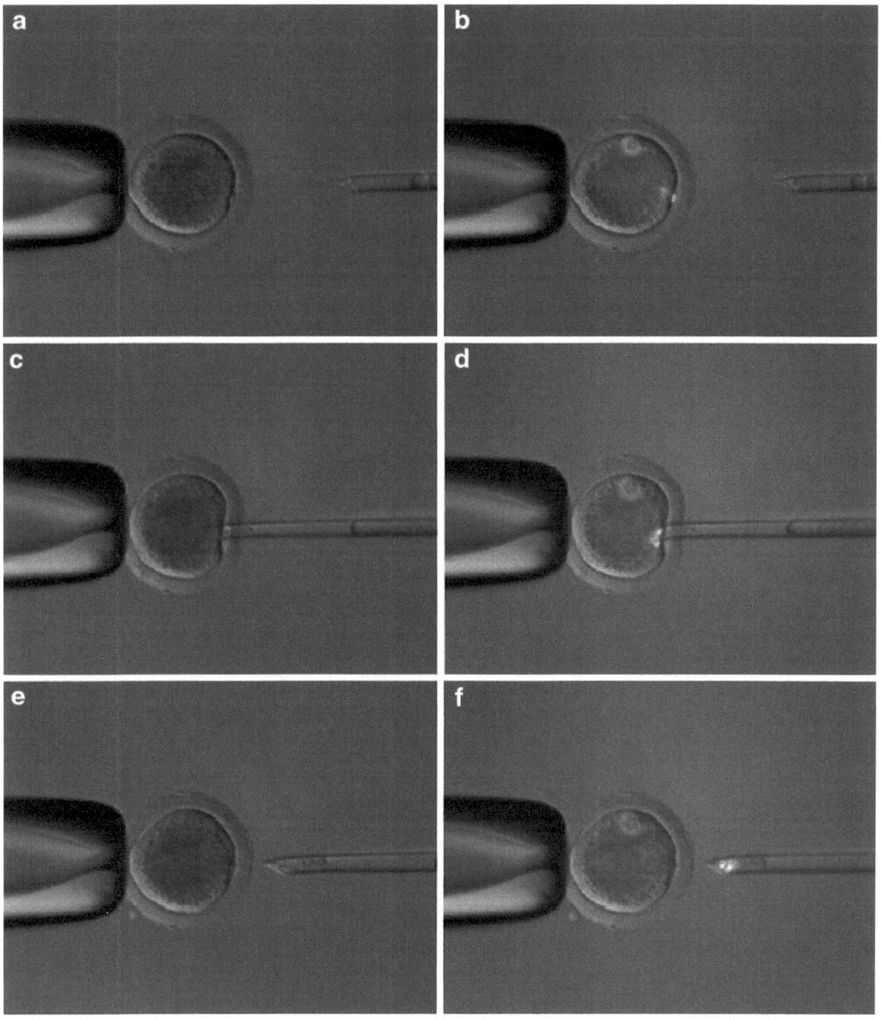

Fig. 5. Oocyte enucleation. (**a**) Situation of the polar body to 4 O'clock position. (**b**) Fluorescent illumination showing the MII plate next to the polar body. (**c**) Penetration of the ZP with the enucleation pipette and aspiration of PB and surrounding cytoplasm. (**d**) Brief illumination with fluorescence to confirm that the MII plate is being aspirated. (**e**) Removal of enucleation pipette. (**f**) Brief illumination with fluorescence to confirm enucleation

7. Penetrate the zona pellucida with the enucleation pipette.
8. Aspirate the polar body and continue with the neighboring oocyte cytoplasm where the metaphase plate with the chromosomes should be located.
9. Briefly activate the UV light to confirm that the chromosomes are being aspirated. Remove the least amount of cytoplasm possible. It is not necessary to discard the aspirated cytoplasm from the enucleation pipette after each enucleation, but it is recommended to do so after each group of oocytes.
10. In some cases, the metaphase plate will be located far from the polar body. In this situation, the polar body should be positioned

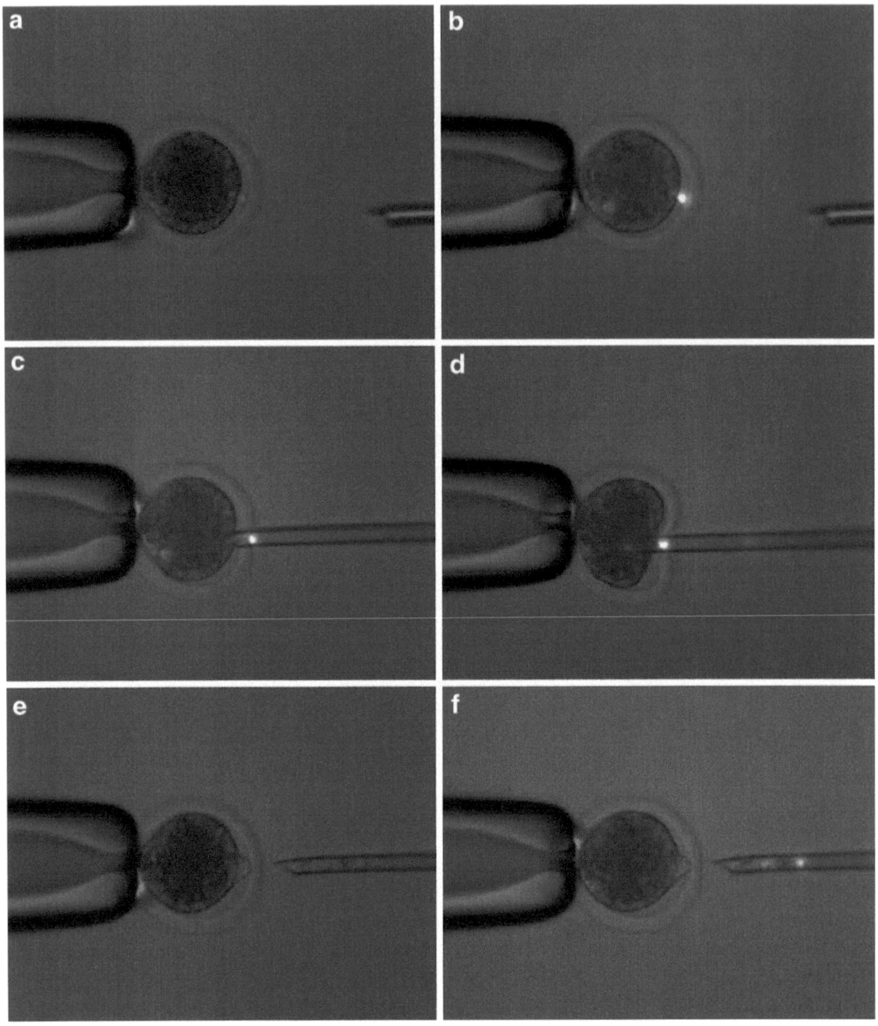

Fig. 6. Oocyte enucleation when MII plate is far from PB. (**a**) Situation of the polar body to 4 O'clock position. (**b**) Fluorescent illumination showing MII plate in the same focal plane of PB. (**c**) Penetration of the ZP with the enucleation pipette and aspiration of PB. (**d**) Advancement of pipette and aspiration of MII plate. (**e**) Removal of enucleation pipette. (**f**) Brief illumination with fluorescence to confirm enucleation

at 4 O'clock with the metaphase plate in focus anywhere from 4 to 9 O'clock positions. Then, aspirate the polar body first and move the enucleation needle toward the metaphase while slowly aspirating to remove the metaphase plate (Fig. 6).

11. Position an intact oocyte on top of the enucleated oocyte and using the enucleation needle push the enucleated oocyte downward. The negative pressure in the holding pipette will suck the intact oocyte which will be then ready for enucleation.
12. Repeat the process until all oocytes in the batch are enucleated.
13. Transfer the enucleated oocytes to KSOM drops, rinsing at least once.

3.3.5. Cell Transfer

Different cell types and at different stages of the cell cycle can be successfully used for bovine cloning (*see* Note 3). We describe the use of fibroblasts cultured using standard cell culture methods and synchronized in G0 stage of the cell cycle by contact inhibition. For this, we plate 100,000 cells in a well of a four-well dish 5–7 days before manipulation. We describe below the preparation of cells for nuclear transfer.

1. Rinse donor cells with PBS.
2. Add 200 µL of Pronase solution and incubate for 5 min.
3. Add 1 mL of HH medium and pipette up and down to release the cells from the plate. Transfer to a 1.5-mL tube and spin down the solution at 250 × *g* for 5 min.
4. Remove medium and add 200 µL of HH, keep the cells at 37°C until used.
5. Rotate the manipulation dish so one set of transfer drops are located on top.
6. Replace the enucleation pipette with a transfer pipette filled with Fluorinert.
7. Add 5 µL of cell suspension to the drop located on top.
8. Transfer a group of 40–50 enucleated oocytes to the drop below the drop containing cells.
9. Raise the holding pipette so it does not contact the drop containing the cell suspension.
10. Load cells into the transfer pipette tip – usually 10–15 cells. Select cells of smaller size with smooth edges. Avoid larger cells that could be at or past S phase of the cell cycle.
11. Move the stage to the drop containing the oocytes and lower the holding pipette.
12. Transfer a single cell into the perivitelline space of the oocyte. Transfer can be made by piercing the zona pellucida (making a new hole) or by using the same hole previously made during enucleation (*see* Note 4). To increase fusion efficiency, the cell should be transferred to the region where the oolema is in closer proximity to the zona pellucida (Fig. 7).
13. Return the oocyte–cell couplets to the incubator in KSOM medium, rinsing at least once.

3.3.6. Oocyte–Cell Fusion

1. Place the fusion chamber in a 100-mm petri dish on the inverted microscope.
2. Connect the chamber to the fusion machine.
3. Prepare a microneedle by pulling a glass capillary, cutting at ~10 µm OD and finally melting the tip to seal the opening. Connect the needle to the micromanipulator.

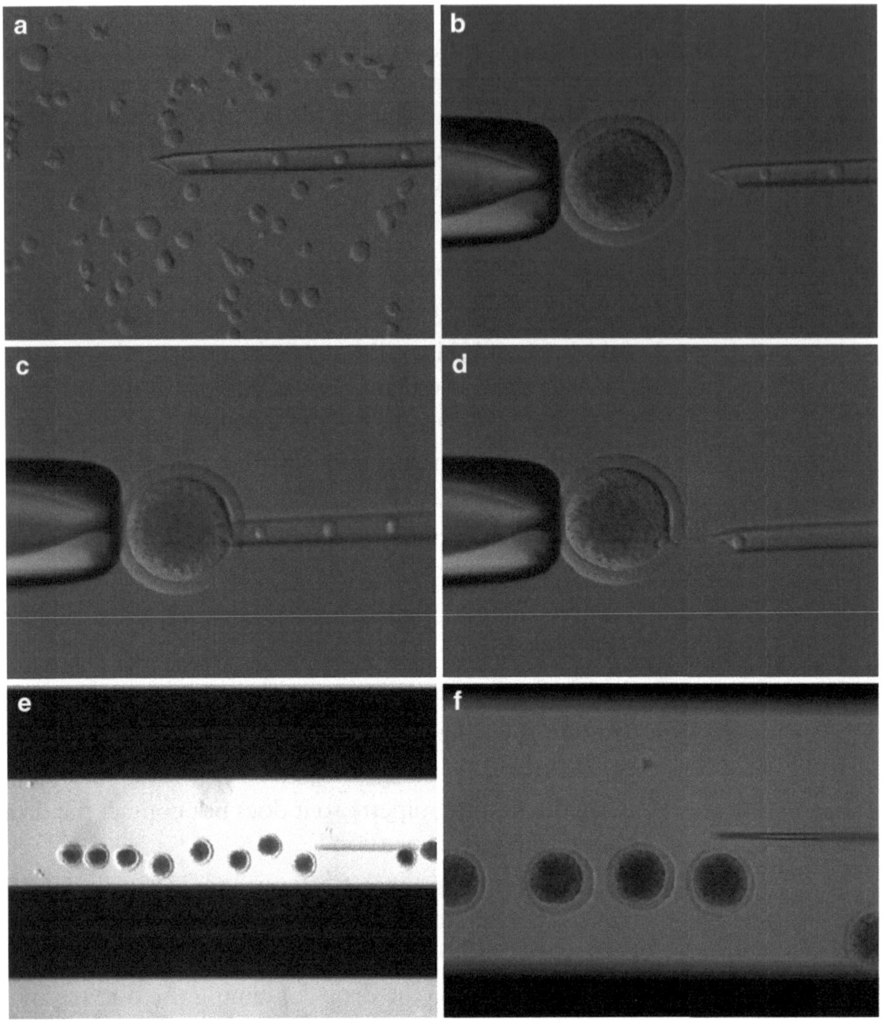

Fig. 7. Cell transfer and electrofusion. (**a**) Collection of cells in the transfer pipette. (**b**) Enucleated oocyte before transfer. (**c**) Penetration of ZP with transfer pipette and deposition of cell between oocyte and ZP. (**d**) Retraction of transfer pipette. (**e**) Oocyte–cell couplets being oriented for electrofusion (low magnification). (**f**) Oocyte–cell couplets being oriented for electrofusion (high magnification), note cell localization toward the bottom electrode

4. Add 20 mL of Sorbitol fusion medium covering chamber and electrodes. If air bubbles are present in between the electrodes, a P1000 pipette can be used to flush medium between the electrodes, thus removing the air bubbles.
5. Prepare a dish for equilibration of the NT couplets, consisting of 6–8 100 μL drops of HH medium and three 100 μL drops of fusion medium.
6. Place 50–60 NT units (oocyte and somatic cell) in one HH drop.

7. Take groups of NT units (10–15) and rinse them through the three drops of fusion medium until they remain at the bottom of the dish.
8. Place the NT units between the electrodes and align them using the microneedle, positioning the somatic cell toward one of the two electrodes (Fig. 7).
9. Once all oocytes are properly aligned, deliver one pulse of 2 kV/cm for 15 μs (*see* Note 5).
10. Immediately after the pulse is delivered remove the NT units from the fusion chamber and wash them five to six times in HH medium. Manipulate the couplets very gently to avoid cell lysis.
11. Return the NT units to the incubator in KSOM medium, rinsing at least once.
12. After 30–60 min transfer the oocytes to an HH drop and confirm fusion by evaluating the presence/absence of the somatic cell under a stereoscope.

3.3.7. Oocyte Activation (see Note 6)

1. Add 1 μL of Ionomycin stock to 1 mL of HH medium and prepare several 100-μL drops on a petri dish.
2. 2–4 h after fusion, and 23–25 h after onset of oocyte maturation, place the NT units in Ionomycin drops for 4 min.
3. Rinse three times in HH medium.
4. Transfer the NT units to equilibrated KSOM medium containing 2 mM of 6-DMAP.
5. Incubate for 4 h.
6. Rinse NT units thoroughly five times in HH medium. Special care should be taken to avoid taking too much medium from one dish to the next, as residues of DMAP in the culture medium will impair proper development.
7. Transfer to embryo culture medium.

3.4. Embryo Culture

1. Culture embryos in groups of 30 in 100-μL drops of KSOM medium under oil at 38.5°C, humidity to saturation, 5% CO_2, 5% O_2, and 90% N_2.
2. 48 h after activation cleavage rate is recorded and noncleaved embryos removed from culture.
3. 72 h after activation, the culture medium is supplemented with serum by adding 5 μL of FBS to each medium drop. Alternatively, the embryos can be transferred into fresh KSOM medium containing 5% FBS.
4. On day 7 of culture, blastocysts can be recovered and transferred to synchronized recipients.

5. For analysis of different stages of embryonic development, PN, 2-, 4-, 8-, 16-cell, morula, and blastocyst stage embryos can be collected at 12, 23, 34, 64, 72, 120, and 168 h post-activation, respectively.

4. Notes

1. Oocytes located at the edge of the plate cannot be well visualized. To minimize the number of oocytes that go to the edges add first the HH medium to the center of the plate and spread it throughout but avoid contact with the walls of the dish. Then, add the oocytes and mix by swirling gently with a pipette. Allow the oocytes to sediment for a few seconds and then, using the manipulation pipette, extend the medium to touch the walls of the dish. The oocytes will remain in their original location as long as this is done gently.
2. Pipettes should be made in advance. It is recommended that several pipettes be available at the time of micromanipulation to have as replacements in case of breakage, blockage, or bad pipettes. Pipettes can be stored in a small box (microscope slide box, pipette tips box, 25-cm petri dish, etc.) where they can be set on a strip of plaster, or foam with slits.
3. Synchronization of the cell cycle of the donor cell and the oocyte is important to maintain the correct ploidy of the embryo. Generally diploid cells are transferred to MII oocytes and the extrusion of a pseudo polar body is prevented to maintain ploidy. Diploid cells can be in G_0 (12) or G_1 (38) stages, and different synchronization schemes can be followed. Alternatively, mitotic cells can be transferred to MII oocytes and polar body extrusion is allowed to take place, generating as a result a diploid zygote (39). Live animals have been obtained from all of these methods of cell synchronization.
4. In the interest of speed, no time should be spent on looking for the enucleation hole; the cell transferred by piercing the ZP again.
5. When using a fusion chamber with 0.5 mm gap between electrodes, set the electrofusion machine to 100 V to deliver a 2-kV/cm pulse. The parameters for electrofusion may require optimization. Pulse voltage can vary from 1 to 2.5 kV/cm and pulse duration from 15 to 50 μs. Also, single or multiple pulses can be delivered. Higher pulse intensity and duration may result in extensive cell lysis, while insufficient pulse intensity will result in reduced fusion efficiency. A scheme that induces about 10% embryo lysis generally results in a good compromise of fusion efficiency and embryo lysis.

6. Given that somatic cells do not carry the sperm's ability to induce oocyte activation, it must be done artificially. There are numerous procedures available to artificially activate mammalian oocytes (40). Sperm-induced oocyte activation is mediated by increases in intracytoplasmic calcium, therefore most artificial activation protocols start by inducing an intracytoplasmic calcium rise. Ionophores, such as Ionomycin, allow the passive diffusion of calcium though the plasma membrane and the ER, thus inducing a single large increase in intracellular calcium (41). Because a single calcium rise is not sufficient to completely downregulate maturation promoting factor (MPF) activity and initiate embryonic development, the calcium stimulus is supplemented with protein kinase or protein synthesis inhibitors (42). Among them, DMAP, a broad ser/thr kinase inhibitor, has been successfully used after ionomycin treatment to induce high rates of bovine oocyte activation (43).

Acknowledgments

We present here the most reliable protocol we know for bovine SCNT. It is the product of cumulative knowledge produced by a vast number of researchers. Some of the techniques here described have likely not been previously published in peer-reviewed journals but were transmitted orally from researcher to researcher. We are indebted to those who came before and helped us design this protocol and we offer an apology to those whose work has not been cited when compiling it. We thank all current and past members of the Cellular Reprogramming Laboratory for contributions toward the establishment of this protocol and to Juan Arechaga, Sebastian Canovas Bernabe and Marcelo Goissis for critical reading of the manuscript. Work in our laboratory was supported by USDA–CSREES competitive grant no 2007-04179 and by the Michigan State University Experiment Station.

References

1. Spemann, H. (1914). Ver Deutsch Zool Ges Leipzig (Freiburg) *24*, 216–221.
2. Briggs, R., and King, T. (1952). Transplantation of living nuclei from blastula cells into enucleated frogs' eggs. Proc Natl Acad Sci U S A *38*, 455–463.
3. Gurdon, J.B. (1962). The developmental capacity of nuclei taken from intestinal epithelium cells of feeding tadpoles. J Embryol Exp Morphol *10*, 622–640.
4. Illmensee, K., and Hoppe, P.C. (1981). Nuclear transplantation in *Mus musculus*: developmental potential of nuclei from preimplantation embryos. Cell *23*, 9–18.
5. McGrath, J., and Solter, D. (1983). Nuclear transplantation in mouse embryos. J Exp Zool *228*, 355–362.
6. Willadsen, S.M. (1986). Nuclear transplantation in sheep embryos. Nature *320*, 63–65.

7. Prather, R.S., Barnes, F.L., Sims, M.M., Robl, J.M., Eyestone, W.H., and First, N.L. (1987). Nuclear transplantation in the bovine embryo: assessment of donor nuclei and recipient oocyte. Biol Reprod *37*, 859–866.
8. Stice, S.L., and Robl, J.M. (1988). Nuclear reprogramming in nuclear transplant rabbit embryos. Biol Reprod *39*, 657–664.
9. Prather, R.S., Sims, M.M., and First, N.L. (1989). Nuclear transplantation in early pig embryos. Biol Reprod *41*, 414–418.
10. Kono, T., Tsunoda, Y., and Nakahara, T. (1991). Production of identical twin and triplet mice by nuclear transplantation. J Exp Zool *257*, 214–219.
11. Meng, L., Ely, J.J., Stouffer, R.L., and Wolf, D.P. (1997). Rhesus monkeys produced by nuclear transfer. Biol Reprod *57*, 454–459.
12. Wilmut, I., Schnieke, A.E., McWhir, J., Kind, A.J., and Campbell, K.H.S. (1997). Viable offspring derived from fetal and adult mammalian cells. Nature *385*, 810–813.
13. Cibelli, J.B., Stice, S.L., Golueke, P.J., Kane, J.J., Jerry, J., Blackwell, C., Ponce de Leon, F.A., and Robl, J.M. (1998). Cloned transgenic calves produced from nonquiescent fetal fibroblasts. Science *280*, 1256–1258.
14. Wakayama, T., Perry, A.C., Zuccotti, M., Johnson, K.R., and Yanagimachi, R. (1998). Full-term development of mice from enucleated oocytes injected with cumulus cell nuclei. Nature *394*, 369–374.
15. Baguisi, A., Behboodi, E., Melican, D.T., Pollock, J.S., Destrempes, M.M., Cammuso, C., Williams, J.L., Nims, S.D., Porter, C.A., Midura, P., Palacios, M.J., Ayres, S.L., Denniston, R.S., Hayes, M.L., Ziomek, C.A., Meade, H.M., Godke, R.A., Gavin, W.G., Overstrom, E.W., and Echelard, Y. (1999). Production of goats by somatic cell nuclear transfer. Nat Biotechnol *17*, 456–461.
16. Onishi, A., Iwamoto, M., Akita, T., Mikawa, S., Takeda, K., Awata, T., Hanada, H., and Perry, A.C. (2000). Pig cloning by microinjection of fetal fibroblast nuclei. Science *289*, 1188–1190.
17. Lanza, R.P., Cibelli, J.B., Diaz, F., Moraes, C.T., Farin, P.W., Farin, C.E., Hammer, C.J., West, M.D., and Damiani, P. (2000). Cloning of an endangered species (*Bos gaurus*) using interspecies nuclear transfer. Cloning *2*, 79–90.
18. Loi, P., Ptak, G., Barboni, B., Fulka, J., Jr., Cappai, P., and Clinton, M. (2001). Genetic rescue of an endangered mammal by cross-species nuclear transfer using post-mortem somatic cells. Nat Biotechnol *19*, 962–964.
19. Chesne, P., Adenot, P.G., Viglietta, C., Baratte, M., Boulanger, L., and Renard, J.P. (2002). Cloned rabbits produced by nuclear transfer from adult somatic cells. Nat Biotechnol *20*, 366–369.
20. Shin, T., Kraemer, D., Pryor, J., Liu, L., Rugila, J., Howe, L., Buck, S., Murphy, K., Lyons, L., and Westhusin, M. (2002). A cat cloned by nuclear transplantation. Nature *415*, 859.
21. Zhou, Q., Renard, J.P., Le Friec, G., Brochard, V., Beaujean, N., Cherifi, Y., Fraichard, A., and Cozzi, J. (2003). Generation of fertile cloned rats by regulating oocyte activation. Science *302*, 1179.
22. Woods, G.L., White, K.L., Vanderwall, D.K., Li, G.P., Aston, K.I., Bunch, T.D., Meerdo, L.N., and Pate, B.J. (2003). A mule cloned from fetal cells by nuclear transfer. Science *301*, 1063.
23. Galli, C., Lagutina, I., Crotti, G., Colleoni, S., Turini, P., Ponderato, N., Duchi, R., and Lazzari, G. (2003). Pregnancy: a cloned horse born to its dam twin. Nature *424*, 635.
24. Gomez, M.C., Pope, C.E., Giraldo, A., Lyons, L.A., Harris, R.F., King, A.L., Cole, A., Godke, R.A., and Dresser, B.L. (2004). Birth of African Wildcat cloned kittens born from domestic cats. Cloning Stem Cells *6*, 247–258.
25. Lee, B.C., Kim, M.K., Jang, G., Oh, H.J., Yuda, F., Kim, H.J., Hossein, M.S., Kim, J.J., Kang, S.K., Schatten, G., and Hwang, W.S. (2005). Dogs cloned from adult somatic cells. Nature *436*, 641.
26. Li, Z., Sun, X., Chen, J., Liu, X., Wisely, S.M., Zhou, Q., Renard, J.P., Leno, G.H., and Engelhardt, J.F. (2006). Cloned ferrets produced by somatic cell nuclear transfer. Dev Biol *293*, 439–448.
27. Kim, M.K., Jang, G., Oh, H.J., Yuda, F., Kim, H.J., Hwang, W.S., Hossein, M.S., Kim, J.J., Shin, N.S., Kang, S.K., and Lee, B.C. (2007). Endangered wolves cloned from adult somatic cells. Cloning Stem Cells *9*, 130–137.
28. Shi, D., Lu, F., Wei, Y., Cui, K., Yang, S., Wei, J., and Liu, Q. (2007). Buffalos (*Bubalus bubalis*) cloned by nuclear transfer of somatic cells. Biol Reprod *77*, 285–291.
29. Berg, D.K., Li, C., Asher, G., Wells, D.N., and Oback, B. (2007). Red deer cloned from antler stem cells and their differentiated progeny. Biol Reprod *77*, 384–394.
30. Polejaeva, I.A., Chen, S.H., Vaught, T.D., Page, R.L., Mullins, J., Ball, S., Dai, Y., Boone, J., Walker, S., Ayares, D.L., Colman, A., and Campbell, K.H. (2000). Cloned pigs

produced by nuclear transfer from adult somatic cells. Nature *407*, 86–90.

31. Wheeler, M.B. (2003). Production of transgenic livestock: promise fulfilled. J Anim Sci *81*, 32–37.
32. Kuroiwa, Y., Kasinathan, P., Matsushita, H., Sathiyaselan, J., Sullivan, E.J., Kakitani, M., Tomizuka, K., Ishida, I., and Robl, J.M. (2004). Sequential targeting of the genes encoding immunoglobulin-(mu) and prion protein in cattle. Nat Genet *36*, 775–780.
33. McCreath, K.J., Howcroft, J., Campbell, K.H., Colman, A., Schnieke, A.E., and Kind, A.J. (2000). Production of gene-targeted sheep by nuclear transfer from cultured somatic cells. Nature *405*, 1066–1069.
34. Lai, L., Kolber-Simonds, D., Park, K.W., Cheong, H.T., Greenstein, J.L., Im, G.S., Samuel, M., Bonk, A., Rieke, A., Day, B.N., Murphy, C.N., Carter, D.B., Hawley, R.J., and Prather, R.S. (2002). Production of alpha-1,3-galactosyltransferase knockout pigs by nuclear transfer cloning. Science *295*, 1089–1092.
35. Wells, D.N., Misica, P.M., Tervit, H.R., and Vivanco, W.H. (1998). Adult somatic cell nuclear transfer is used to preserve the last surviving cow of the Enderby Island cattle breed. Reprod Fertil Dev *10*, 369–378.
36. Lanza, R.P., Dresser, B.L., and Damiani, P. (2000). Cloning Noah's ark. Sci Am *283*, 84–89.
37. Seshagiri, P., and Bavister, B. (1989). Phosphate is required for inhibition by glucose of development of hamster 8-cell embryos in vitro. Biol Reprod *40*, 607–614.
38. Kasinathan, P., Knott, J.G., Wang, Z., Jerry, D.J., and Robl, J.M. (2001). Production of calves from G1 fibroblasts. Nat Biotechnol *19*, 1176–1178.
39. Wakayama, T., Rodriguez, I., Perry, A.C.F., Yanagimachi, R., and Mombaerts, P. (1999). Mice cloned from embryonic stem cells. Proc Natl Acad Sci U S A *96*, 14984–14989.
40. Machaty, Z. (2006). Activation of oocytes after nuclear transfer. Methods Mol Biol *348*, 43–58.
41. Nakada, K., and Mizuno, J. (1998). Intracellular calcium responses in bovine oocytes induced by spermatozoa and by reagents. Theriogenology *50*, 269–282.
42. Alberio, R., Zakhartchenko, V., Motlik, J., and Wolf, E. (2001). Mammalian oocyte activation: lessons from the sperm and implications for nuclear transfer. Int J Dev Biol *45*, 797–809.
43. Susko-Parrish, J.L., Leibfried-Rutledge, M.L., Northey, D.L., Schutzkus, V., and First, N.L. (1994). Inhibition of protein kinases after an induced calcium transient causes transition of bovine oocytes to embryonic cycles without meiotic completion. Dev Biol *166*, 729–739.

Chapter 11

Cell Fusion-Induced Reprogramming

Jeong Tae Do and Hans R. Schöler

Abstract

Genomic reprogramming can be accomplished by five different types of methods: nuclear transfer, cell fusion, in vitro culture, introduction of egg extract, and transduction of transcription factors. We have shown that fusion-induced reprogramming is an efficient method for reprogramming differentiated somatic cells to a pluripotential state (pluripotential reprogramming) – Oct4 gene reactivation occurs within 1–2 days postfusion of somatic cells with pluripotent stem cells. Reactivation of *Oct4* can be monitored by detection of the GFP signal from the *Oct4*-GFP transgene of somatic cells. In the current report, we fused double transgenic (OG2/ROSA26) somatic cells with pluripotent embryonic stem (ES) cells, and demonstrated the presence of the somatic cell genome in all GFP-positive ES-like colony-forming cells, confirming their identity as the cell fusion hybrids.

Key words: Fusion, Hybrids, Reprogramming, Pluripotency, Stem cells

1. Introduction

The function of somatic cells within a given microenvironment is specified by epigenetic information established during embryonic development. The memory of cells gets erased during the process of "reprogramming," and the cellular program reverts to that of an earlier developmental state. In pluripotential reprogramming, the pluripotent state becomes established in somatic cells. In this manner, pluripotent embryonic stem (ES) cells can transfer pluripotency-specific features to somatic cells, leading to erasure of the somatic cell memory (1–3). Pluripotential reprogramming is associated with several cellular characteristics: establishment of pluripotent features in somatic cells: expression of pluripotency-specific markers, inactivation of tissue-specific gene expression, potential to contribute to the development of all three germ layers, and presence of an undifferentiated epigenetic state (4).

S. Ding (ed.), *Cellular Programming and Reprogramming: Methods and Protocols*, Methods in Molecular Biology, vol. 636, DOI 10.1007/978-1-60761-691-7_11, © Springer Science+Business Media, LLC 2010

Fusion-induced reprogramming is considered a useful tool for screening factors involved in cellular reprogramming. Recent studies have shown that overexpression of Nanog (5) and Sall4 (6) could enhance the efficiency of the reprogramming process. In addition, epigenetic modification–related factors, such as G9a (a histone methyltransferase) (7) and Dnmt3a (8), affect the establishment of somatic cell pluripotency through fusion-induced reprogramming.

Cell fusion can be accomplished by two types of methods: electroshock and chemical modifications. Electroshock-mediated fusion has been performed and described by others (9, 10). Here we describe a detailed method for generating cell fusion hybrids by a chemical fusion method using polyethylene glycol (PEG). To monitor the reprogramming of somatic cell genome, $OG2^{+/-}$/ $ROSA26^{+/-}$ double transgenic somatic cells, which carry GFP under the control of the *Oct4* promoter and a *neo/lacZ* transgene that is expressed ubiquitously, were chemically fused with pluripotent embryonic stem (ES) cells. Since Oct4 serves as an essential function in the maintenance of cellular pluripotency, including that of germ cells (11–13), reactivation of *Oct4* expression in somatic cells following cell fusion is indicative of onset of genomic reprogramming (Fig. 1). We therefore determined the gene expression profile and the epigenetic status of isolated cell fusion

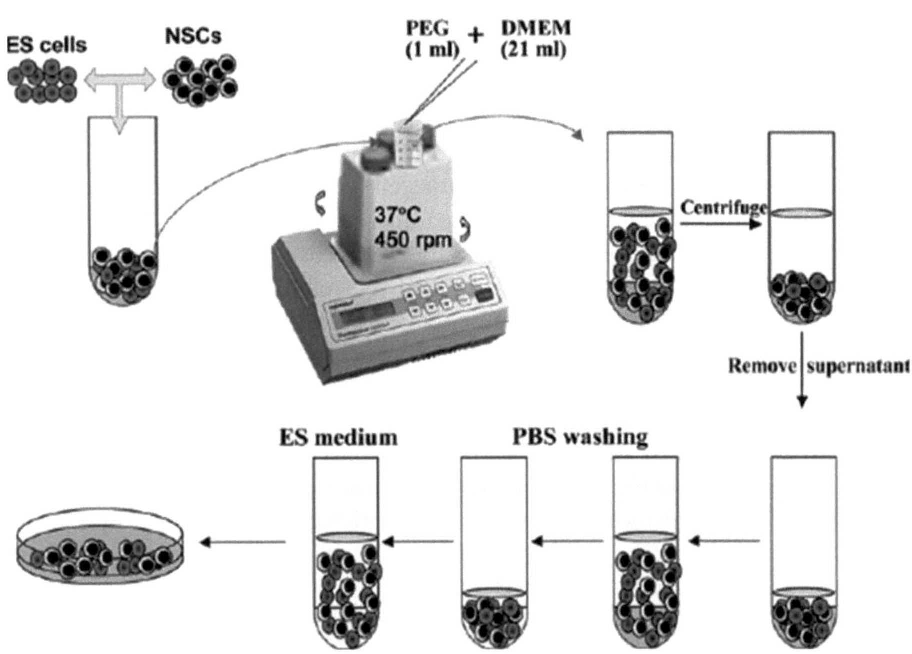

Fig. 1. Schematic illustration of the procedure for cell fusion using polyethylene glycol (PEG). Cell fusion is accomplished by subjecting cells in 50-ml conical tubes to PEG treatment under constant shaking of 450 rpm in an Eppendorf thermomixer equipped with block for 4 × 50 ml conical tubes. Cells are centrifuged, washed with serum-free medium, and subsequently cultured in ES cell culture medium onto a feeder-layered dish

hybrids in an effort to assess whether the somatic cell genome had undergone appropriate reprogramming.

2. Materials

2.1. Cell Culture: ES Cells, Neural Stem Cells, and Mouse Embryonic Fibroblasts

1. Pluripotent fusion partner cell: ES cells (E14 and HM-1 line).
2. Somatic fusion partner cell: neural stem cells (NSCs) or mouse embryonic fibroblasts (MEFs) derived from OG2/ROSA26 double transgenic mice carrying GFP under the control of the *Oct4* promoter and a neo/lacZ transgene that is expressed ubiquitously.
3. MEF culture medium: Dulbecco's Modified Eagle's Medium (DMEM) (Gibco/BRL) supplemented with 15% fetal bovine serum (FBS, Gibco/BRL), and 1× penicillin/streptomycin/glutamine (Gibco/BRL). Store at 4°C.
4. ES cell culture medium: DMEM supplemented with 15% FBS, 1× penicillin/streptomycin/glutamine, 1× nonessential amino acids (Gibco BRL), 0.1 mM β-mercaptoethanol, and 1,000 U/ml leukemia inhibitory factor (LIF; ESGRO®, Chemicon). Store at 4°C.
5. NSC expansion medium: NS-A media (Euroclone) or DMEM/F12 (Invitrogen) supplemented with N2 supplement (Invitrogen), 10 ng/ml epidermal growth factor (EGF, Invitrogen), 10 mg/ml basic fibroblast growth factor (bFGF, Invitrogen), 50 μg/ml BSA (Fraction V; Gibco BRL), 1× penicillin/streptomycin/glutamine, and 1× nonessential amino acids. Store at 4°C.
6. Neurosphere culture medium: DMEM/F12 supplemented with N2 supplement, B27 supplement (Invitrogen), 0.6% Glucose (w/v), 8 mM HEPES (Sigma), 10 ng/ml EGF, 10 mg/ml bFGF, 50 μg/ml BSA, 1× penicillin/streptomycin/glutamine, and 1× nonessential amino acids. Store at 4°C.
7. Phosphate-buffered saline (PBS Ca^{2-} /Mg^{2-}, Gibco/BRL). Store at room temperature (RT).
8. 0.1% Gelatin: Gelatin 2% solution Type B (Sigma) diluted in PBS. Store at RT.
9. 0.25% trypsin and ethylenediamine tetraacetic acid (EDTA) (1 mM) solution from Gibco/BRL. Store at 4°C.
10. EGF and bFGF dissolved in PBS containing 0.1% BSA and immediately frozen in single-use aliquots (10 μl) at –20°C. Store residual solution at 4°C for 2–4 weeks (do not refreeze).

11. Dissociation solution: Hank's balanced salt solution (HBSS Ca^{2-}/Mg^{2-}, Sigma) supplemented with 1.33 mg/ml Trypsin-EDTA, 0.2 mg/ml Kynurenic acid (Sigma), and 2 mM Glucose (Sigma). Store at 4°C.
12. Leukemia inhibitory factor. Store at 4°C.
13. 100× penicillin/streptomycin/glutamine solution. Store at 4°C.
14. 100× nonessential amino acids solution. Store at 4°C.
15. 7.5% BSA solution. Store at 4°C.
16. 1 M HEPES solution (Sigma): to make 8 mM in neurosphere medium, add 320 μl 1 M HEPES solution to 40 ml neurosphere medium. Store at 4°C.
17. 45% Glucose (Sigma) solution: to make 0.6% in neurosphere medium, add 266 μl 45% Glucose to 40 ml neurosphere medium. Store at 4°C.
18. Hank's balanced salt solution (HBSS Ca^{2-}/Mg^{2-} and Ca^{2+}/Mg^{2+}, Sigma). Store at RT.
19. Earle's balanced salt solution (EBSS, Sigma). Store at RT.
20. Kynurenic acid. Store at RT.
21. Sucrose (Sigma). Store at RT.
22. BSA. Store at 4°C.
23. 70 μm cell strainer (Becton Dickinson).
24. 5-cc syringe and 18-gage needle.

2.2. Cell Fusion

1. Polyethylene glycol 1500 (PEG 1500; PEG 50% w/v, Roche). Store in single-use (1 ml) aliquots at 4°C.
2. DMEM without FBS. Store at 4°C.
3. PBS Ca^{2-}/Mg^{2-}. Store at RT.
4. Cell-counting chamber: improved Neubauser Hemocytometer (Assistent, Germany).
5. Fusion equipment: Eppendorf thermomixer equipped with block for 4 × 50 ml conical tubes (Eppendorf).

2.3. Selection of Fusion Hybrid Cells

1. G418 selection medium: ES cell culture medium supplemented with 6 μl/ml Geneticin (50 mg/ml active Geneticin, Gibco/BRL). Store at 4°C.
2. 50× HAT selection medium: dissolve HAT Media Supplement (50×) Hybri-Max™ (Sigma) in MEF culture medium and filter through a 0.22-μm filter. Freeze in single-use aliquots (500 ml) at −20°C.
3. Fluorescence-Activated Cell Sorting: Flow cytometry analyzer: FACSAria cell sorter (Becton Dickinson) installed with FACSDiva software (Becton Dickinson).
4. 70 μm cell strainer (Becton Dickinson).

2.4. Freezing and Thawing of Fusion Hybrid Cells

1. Freezing medium: DMEM medium supplemented with 30% FBS (v/v) and 10% (v/v) Dimethyl sulfoxide (DMSO, Sigma). Store at 4°C for 1 week.
2. Cryopreservation tube (CryoTube vial, Nunc).
3. MEF culture medium: washing medium.
4. Liquid nitrogen tank.

3. Methods

3.1. ES Cell (E14 and HM-1) Culture

1. Prepare feeder-layered dishes for the culture of ES cells and hybrid cells (see Note 1). Thaw MEF cells (passage 4) and culture in 10-cm gelatinized dish in MEF culture medium until they have reached confluence (~2–3 days after thawing). Treat cells with MEF culture medium containing 10 μg/ml mitomycin C (MMC, Sigma) for 2 h at 37°C, 5% CO_2 in air. Remove medium, rinse cells three times with PBS and trypsinize cells. Resuspend cell pellet in MEF culture medium. Split cells in a 60-mm gelatinized dish at a concentration of 1.5×10^5 cells/ml (see Note 2).
2. Thaw one vial of frozen ES cells and plate onto a feeder-layered dish in ES cell culture medium (*see* thawing method described in Subheading 3.6.).
3. Passage ES cells 2 days after thawing. Remove ES cell culture medium, wash cells with PBS, add Trypsin-EDTA, and resuspend cells in MEF culture medium. Centrifuge cells at $200 \times g$ for 5 min, discard supernatant, resuspend cell pellet in ES cell culture medium, and replate cells onto feeder-layered dishes. ES cells are normally split in a ratio of 1:5 to 1:10.

3.2. Preparation of NSCs for Somatic Fusion Partner Cells

1. Recover 12.5- to 16.5-dpc OG2/ROSA26 heterozygous embryos after mating of male $OG2^{+/+}/ROSA26^{+/+}$ homozygous double transgenic mouse with normal CD1 or B6C3F1 mice, or following the mating of OG2 female mice with ROSA26 male mice. Separate male and female embryos according to the morphology of the gonads. Specifically, female cells can be used to investigate the reactivation of the inactive X chromosome during reprogramming.
2. Isolate lateral ventricles of the brain and place them into a 50-ml conical tube.
3. Incubate the isolated tissues in dissociation solution at 37°C for 30 min and homogenize by gentle pipetting.

4. Place a 70-μm cell strainer (Becton Dickinson) into a 50 ml conical tube and filter the homogenized cells through the cell strainer to remove large cell clusters.
5. Centrifuge filtered cells at 200×*g* for 5 min and remove supernatant.
6. Add 0.9 M sucrose (in 0.5× HBSS) to tube, resuspend cells, centrifuge at 750×*g* for 10 min, and remove supernatant.
7. Resuspend cell pellet in 2 ml of neurosphere culture medium, place on top of 10 ml of 4% BSA in EBSS, centrifuge at 200×*g* for 7 min, and remove supernatant.
8. Add neurosphere culture medium to tube, resuspend cells, and replate cells onto 35-mm suspension culture dish (Corning). Brain tissue from a single embryo provides sufficient material for 1 dish.
9. Culture primary neurospheres for 5–8 days before derivation of NSCs.
10. Plate neurospheres onto 60-mm gelatin-coated dishes in NSC expansion medium.
11. Establish NSCs by dissociation and replating onto the gelatin-coated dishes in NSC expansion medium (see Note 3).

3.3. Preparation of MEFs for Somatic Fusion Partner Cells

1. Recover 13.5-dpc ROSA26/OG2 heterozygous embryos. Separate male and female embryos according to the morphology of the gonads. Cut off the embryo head and remove liver, internal organs, and gonads. It is important that the preparation is devoid of any gonadal tissue, as primordial germ cells (PGCs), which reside in the gonads, stain positive for Oct4-GFP and can thus produce a false-positive result for reprogrammed cells.
2. Transfer embryo bodies to a new dish containing fresh PBS.
3. Attach an 18-gage needle to a 5-cc syringe and remove plunger.
4. Place embryo bodies into syringe and add 3 ml of Trypsin-EDTA. Reinsert plunger in syringe and squirt the bodies into a 50-ml conical tube. Incubate at 37°C for 5 min, add 10 ml MEF culture medium, and homogenize the squirted tissues by gentle pipetting.
5. Place 70-μm cell strainer into a clean 50 ml conical tube and filter the homogenized cell through the cell strainer to remove large cell clusters.
6. Centrifuge tube at 200×*g* for 5 min and remove supernatant.
7. Resuspend cells in MEF culture medium and replate onto the gelatin-coated dishes (passage 0).

8. Culture MEF cells at 37°C, 5% CO_2 in air. Change the medium the next day and culture cells for 2 days.
9. After 3 days of primary culture, passage the MEF cells. Remove medium, rinse cells with PBS, trypsinize cells, and resuspend cell pellet in MEF culture medium. Split cells in a 1:3 ratio into a gelatinized dish. Use these MEF cells directly for fusion experiments; freeze the rest for the next experiment (see Note 4).

3.4. Cell Fusion

Schematic illustration of the procedure for cell fusion is shown in Fig. 1.

1. Prewarm 1 ml of PEG and 30 ml of DMEM (in 50 ml conical tube) and place the tubes into Eppendorf thermomixer equipped with block for 4 × 50-ml conical tubes. Set the block at 37°C.
2. Prepare single-cell suspension of ES and somatic cells by trypsinization and washing with PBS.
3. Count cell number of each fusion partner cell using a hemocytometer.
4. Mix ES cells with NSCs or MEFs in a 1:1 or 1:5 ratio in a 50-ml conical tube (see Note 5).
5. Centrifuge cell mixture at 400 × *g* for 5 min and remove supernatant.
6. Centrifuge cell pellet at 400 × *g* for 5 min and completely remove residual medium.
7. Place tube into Eppendorf thermomixer block and vortex the block at 450 rpm (see Note 6).
8. Add 1 ml of a prewarmed 50% PEG 1500 to cell pellet over 1 min.
9. Add 1 ml of prewarmed DMEM over 1 min.
10. Add 2 ml of prewarmed DMEM over 1 min.
11. Add 3 ml of prewarmed DMEM over 1 min.
12. Add 5 ml of prewarmed DMEM over 1 min.
13. Add 10 ml of prewarmed DMEM over 1 min. Total volume of PEG + DMEM is 22 ml.
14. Centrifuge cells at 400 × *g* for 9 min then discard supernatant.
15. Add PBS to tube, gently resuspend cells (so as not to break cell aggregates), centrifuge at 400 × *g* for 9 min, and then discard supernatant.
16. Gently resuspend cells in ES cell culture medium, plate into 6-mm feeder-layered culture dish, and culture in incubator at 37°C, 5% CO_2 in air (see Note 7).

3.5. Selection and Further Culture of Fusion Hybrid Cells

3.5.1. Drug Selection

1. Select fusion hybrid cells in ES cell culture medium containing 300 μg/ml G418 or 1× HAT (the latter when HM-1 ES cells are used) the day following fusion to eliminate unfused ES cells (see Note 8).
2. Examine the *Oct4*-GFP signal under a fluorescent microscope starting on day 1 postfusion.
3. Passage cells every 2 days.
4. After 10–15 days of drug selection, nearly all surviving ES-like colonies are GFP-positive fusion hybrids (Figs. 2 and 3).

3.5.2. FACS Sorting for Oct4-GFP-Positive Cells

1. Oct4-GFP expression is first observed at around 40–45 postfusion (see Note 9). GFP-positive hybrid cells can be sorted at different time point by FACS for specific purposes.
2. Prepare single-cell suspension of cell mixtures from the culture dish by trypsinization and washing with MEF culture medium.
3. Centrifuge cells at 200×*g* for 5 min and remove supernatant.
4. Add 1 ml of ES cell culture medium and resuspend cells.
5. Place a 70-μm cell strainer into a 15-ml conical tube and filter the cells through the strainer to remove large cell clusters.
6. Wash cell strainer with 0.5 ml ES cell culture medium to recover any residual cells.

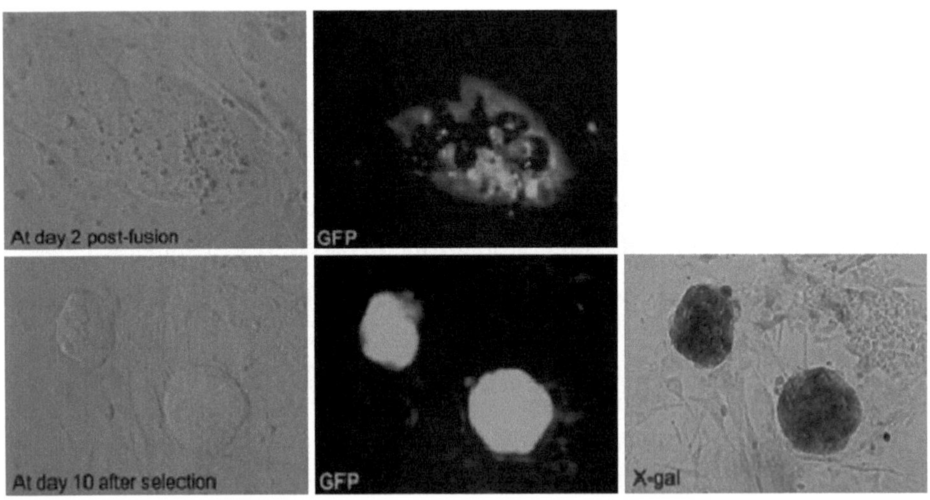

Fig. 2. Reprogramming of NSCs after fusion with ES cells. The reprogramming of the somatic cell genome can be determined by detection of the *Oct4*-GFP signal (i.e., GFP-positive cells). GFP-positive cells appear in a colony at 45 h postfusion (*upper* panels). After G418 selection, fusion hybrid cells can be selected by staining with X-gal for the presence of the NSC-derived lacZ reporter

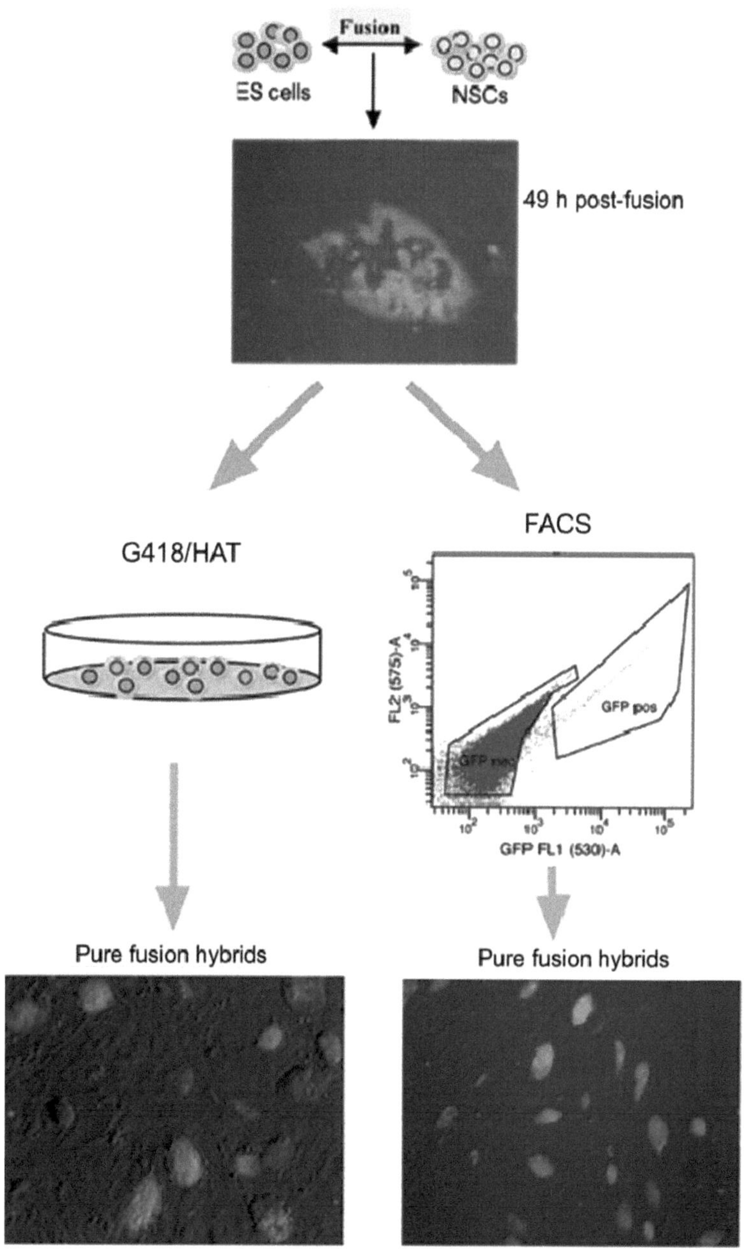

Fig. 3. Two methods are used for selection of fusion hybrid cells. Unfused NSCs and fusion hybrid cells can be selected by treatment with G418. Nearly all ES-like colony-forming cells stain GFP-positive in day-10 selection culture medium. The GFP-positive cells can also be sorted by FACS from day 2 postfusion onward. The proportion of cells that are GFP-positive cells on day 2 postfusion is about 0.01–0.1%

7. Transfer cells to the FACSAria cell sorter.
8. Use fluorescence at 530 and 575 nm to discriminate false-positive auto-fluorescent cells (slightly yellowish-green signal) from the true GFP-positive cells (Fig. 3).

9. Sort GFP-positive cells into a 15-ml conical tube containing 7 ml ES cell culture medium. About 10^6 cells can be sorted into this tube. GFP-positive cells can be sorted directly into lysis buffer for RT-PCR.
10. After FACS sorting, centrifuge cells, resuspend cell pellet in ES cell culture medium, and plate cells onto a feeder-layered dish.

3.6. Freezing and Thawing of Fusion Hybrid Cells

The freezing and thawing procedures apply to ES cells, MEFs, and NSCs.

3.6.1. Freezing

1. Prepare single-cell suspension of cells from the culture dish by trypsinization and washing with MEF culture medium.
2. Centrifuge cells at 200×*g* for 5 min and remove supernatant.
3. Resuspend cell pellet in freezing medium (at room temperature).
4. Transfer 1 ml of cell suspension into each cryotube (2×10^5 cell/ml).
5. Put the cryotube in a styrofoam box and place in a −80°C freezer for 2–3 days.
6. For long-term storage, transfer cryotubes to a liquid nitrogen tank (−196°C).

3.6.2 .Thawing

1. Warm the MEF culture medium to 37°C.
2. Place the cryotube containing frozen cells in a 37°C water bath until the cells have completely thawed. Transfer cell suspension to a 15-ml conical tube containing 9 ml of prewarmed MEF culture medium.
3. Centrifuge cells at 200×*g* for 5 min and remove supernatant.
4. Resuspend cell pellet in ES cell culture medium.
5. Plate cells onto a feeder-layered dish.

4. Notes

1. HM-1 ES cells can be cultured in Glasgow Modified Eagle's Medium supplemented with 15% FBS, 1× penicillin/streptomycin/glutamine, 1× nonessential amino acids, 0.1 mM β-mercaptoethanol, and 1,000 U/ml LIF under feeder-free conditions. However, coculture with feeder cells is better for earlier reprogramming and higher reprogramming efficiency.

2. If the feeder cells are too dense, ES cells do not form nice-looking colonies.
3. During passage of NSCs, expose cells to trypsin for only a very short time (~5 s); NSCs are easily detached by treatment with Trypin-EDTA.
4. During MEF passage, cells become larger. To maintain a homogenous population of small cells, do not split in a ratio higher than 1:4.
5. Although some researchers suggest that using a cell ratio of 1:5 will improve reprogramming efficiency, we have not observed a change in efficiency based on cell ratio used.
6. If an Eppendorf thermomixer equipped with block for 4 × 50-ml conical tubes is not available, use a water bath. Put the tube into the warm water and stir by hand. Keep temperature at 37°C until at least step 13.
7. When embryonic carcinoma (EC) cells are fused with somatic cells, fusion hybrid cells can also be cultured in ES cell culture medium, which will enhance the reprogramming efficiency. Feeder-layered dishes are not necessary.
8. Since unfused ES cells begin to die within 1 day after adding G418 or HAT, the medium should to be changed daily. Washing cells in PBS to remove dead cells is recommended. Unfused ES cells can survive, however, even after more than 15 days of G418 and HAT treatment. Therefore, to obtain a pure population of hybrids, FACS forting or single-cell-cloning is highly recommended. After FACS or single-cell-cloning, addition of selective medium is not necessary. A higher concentration of G418 (up to 500 μg/ml G418) may help to completely remove unfused cells, but it will retard cell proliferation. Although the feeder cells are not neo-resistant, they can be treated with G418 (<300 μg/ml). However, the cells have to be passaged either every day or every other day.
9. In NSCs, derived from GOF18 – which contain all the *Oct4* regulatory elements, such as distal enhancer, proximal enhancer, and proximal promoter – *Oct4*-GFP can be detected within 24 h postfusion. However, the onset of *Oct4*-GFP activation is dependent upon the culture conditions. GFP-positive cells were counted in total fusion mixtures to assess the reprogramming rate. However, this is not an accurate measure of the reprogramming rate, as there were only a few hybrids that had formed from the fusion of ES cells with NSCs; the initial fusion mixtures contain unfused ES cells, unfused NSCs, fused ES–ES cells, fused NSC–NSC cells, and ES–NSC hybrid cells. Therefore, the rate as determined by FACS is more a measure of the fusion rate rather than the actual reprogramming rate. The actual reprogramming rate is

higher than 95%; nearly all NSCs are reprogrammed once correctly fused with ES cells.

10. The reprogramming efficiency varies according to the somatic cell type; i.e. NSCs are more efficiently reprogrammed than MEFs, following fusion with ES cells. Albeit, neurosphere and cumulus cells have a similar reprogramming efficiency (14).

References

1. Do, J.T., Han, D.W., Gentile, L., Sobek-Klocke, I., Stehling, M., Lee, H.T., et al. (2007) Erasure of cellular memory by fusion with pluripotent cells. *Stem Cells.* 25, 1013-1020.
2. Tada, M., Morizane, A., Kimura, H., Kawasaki, H., Ainscough, J.F., Sasai, Y., et al. (2003) Pluripotency of reprogrammed somatic genomes in embryonic stem hybrid cells. *Dev Dyn.* 227, 504-510.
3. Tada, M., Takahama, Y., Abe, K., Nakatsuji, N. and Tada, T. (2001) Nuclear reprogramming of somatic cells by in vitro hybridization with ES cells. *Curr Biol.* 11, 1553-1558.
4. Do, J.T., Han, D.W. and Scholer, H.R. (2006) Reprogramming somatic gene activity by fusion with pluripotent cells. *Stem Cell Rev.* 2, 257-264.
5. Silva, J., Chambers, I., Pollard, S. and Smith, A. (2006) Nanog promotes transfer of pluripotency after cell fusion. *Nature.* 441, 997-1001.
6. Wong, C.C., Gaspar-Maia, A., Ramalho-Santos, M. and Reijo Pera, R.A. (2008) High-efficiency stem cell fusion-mediated assay reveals Sall4 as an enhancer of reprogramming. *PLoS One.* 3, e1955.
7. Ma, D.K., Chiang, C.H., Ponnusamy, K., Ming, G.L. and Song, H. (2008) G9a and Jhdm2a regulate embryonic stem cell fusion-induced reprogramming of adult neural stem cells. *Stem Cells.* 26, 2131-2141.
8. Do, J.T., Han, D.W., Gentile, L., Sobek-Klocke, I., Stehling, M. and Scholer, H.R. (2008) Enhanced Reprogramming of Xist by induced upregulation of Tsix and Dnmt3a. *Stem Cells.* 26, 2821-2831.
9. Pells, S. and McWhir, J. (2004) Studying nuclear reprogramming with cell hybrids. *Methods Mol Biol.* 254, 301-312.
10. Tada, M., Tada, T., Lefebvre, L., Barton, S.C. and Surani, M.A. (1997) Embryonic germ cells induce epigenetic reprogramming of somatic nucleus in hybrid cells. *EMBO J.* 16, 6510-6520.
11. Kehler, J., Tolkunova, E., Koschorz, B., Pesce, M., Gentile, L., Boiani, M., et al. (2004) Oct4 is required for primordial germ cell survival. *EMBO Rep.* 5, 1078-1083.
12. Nichols, J., Zevnik, B., Anastassiadis, K., Niwa, H., Klewe-Nebenius, D., Chambers, I., et al. (1998) Formation of pluripotent stem cells in the mammalian embryo depends on the POU transcription factor Oct4. *Cell.* 95, 379-391.
13. Niwa, H., Miyazaki, J. and Smith, A.G. (2000) Quantitative expression of Oct-3/4 defines differentiation, dedifferentiation or self-renewal of ES cells. *Nat Genet.* 24, 372-376.
14. Do, J.T. and Scholer, H.R. (2005) Comparison of neurosphere cells with cumulus cells after fusion with embryonic stem cells: reprogramming potential. *Reprod Fertil Dev.* 17, 143-149.

Chapter 12

An Improved Method for Generating and Identifying Human Induced Pluripotent Stem Cells

Prashant Mali, Zhaohui Ye, Bin-Kuan Chou, Jonathan Yen, and Linzhao Cheng

Abstract

This chapter describes the methods we use to derive and characterize human induced pluripotent stem (iPS) cells. We describe in order, first our culture techniques for the starting fibroblast populations and methods for retrovirus preparation and concentration. Subsequently, a detailed iPS derivation protocol suitable for human fibroblast populations is discussed using standard retroviral vectors expressing the classic four or three reprogramming genes. Finally, we elaborate a robust technique for monitoring and identification of potential iPS cells through live staining of reprogrammed cells. We also outline steps for characterization of the resulting iPS cell lines.

Key words: Embryonic stem cells, Induced pluripotent stem cells, Retroviruses, Genetic modification, Transformation, Self-renewal, Pluripotency, Embryoid bodies, Teratoma

1. Introduction

Human embryonic stem (hES) cells have the ability to self renew and differentiate into cell types of all germ layers and thus have the potential to serve as an unlimited source for cell-replacement therapy (1, 2). Recent advances in generating induced pluripotent stem (iPS) cells which circumvent the ethical and source issues associated with the derivation of hES cells by directly converting an easily accessible somatic tissue cell to a pluripotent state, have made the dream of making patient-specific pluripotent cell lines and eventually transplantable tissues for therapy a distinct possibility (3–8).

In this chapter, we describe our protocol for the generation of iPS cells from human fibroblasts of adult or fetal origin. A basic proficiency of hES cell culture and genetic modification techniques

S. Ding (ed.), *Cellular Programming and Reprogramming: Methods and Protocols*, Methods in Molecular Biology, vol. 636,
DOI 10.1007/978-1-60761-691-7_12, © Springer Science+Business Media, LLC 2010

is assumed: details of those protocols may be found elsewhere (9, 10). We expect a period of about 6 months to derive and completely characterize a new iPS cell line and this protocol walks the user through the various steps involved therein. We highlight in particular the critical junctures and also areas where user cell specific changes may be desired. While the protocol below for derivation using four transgenes (encoding Oct4, Sox2, Klf4, and Myc) is very robust, derivation using only the first three factors may require further user cell type specific optimization.

2. Materials

2.1. Tissue Culture

1. Dulbecco's Modified Eagles Medium (DMEM), High Glucose.
2. Minimum Essential Medium with Earle's salts (Invitrogen).
3. KNOCKOUT ™ D-MEM: Optimized D-MEM for ES Cells (Invitrogen).
4. Fetal Bovine Serum (Hyclone defined).
5. Knockout Serum Replacement (KSR) (Invitrogen).
6. L-glutamine, (200 mM) (Invitrogen).
7. MEM Nonessential amino acids (NEAA), 10 mM (100×) (Invitrogen).
8. 1× Phosphate-buffered saline (PBS) Ca^{+2} and Mg^{+2} free (Invitrogen).
9. β-mercaptoethanol (EM Science).
10. Penicillin–Streptomycin (100×) (Invitrogen).
11. Antibiotic–Antimycotic solution (100×) (Invitrogen).
12. Basic fibroblast growth factor (bFGF) (Peprotech).
13. 0.05% Trypsin-EDTA (Invitrogen).
14. Collagenase Type IV (Sigma): used at 1 mg/ml in KNOCKOUT ™ D-MEM.
15. Stericup™ (Millipore).
16. Trypan blue 0.4% solution (Invitrogen).
17. Hemocytometer.
18. 6-well, 12-well tissue culture plates (BD).
19. 10-cm tissue culture dish (Corning).
20. Gelatin (Sigma).
21. 25-cm cell scrapers (Sarstedt).
22. Matrigel™ matrix (BD).
23. 5-ml, 15-ml and 50-ml polystyrene tubes (Sarstedt).

24. 10% buffered formalin (Fischer Scientific).
25. Mouse (IgM) anti-human antibody TRA-1-60 (Millipore).
26. Secondary Alexa Fluor 555 anti-mouse IgM (Invitrogen)
27. Polybrene (Sigma).

2.2. Retrovirus Production

1. 293 T cells (ATCC).
2. Opti-MEM I medium (Invitrogen).
3. Lipofectamine 2000 (Invitrogen).
4. Poly-D-Lysine (Sigma).
5. Reprogramming retroviral vectors: pMX-Oct4, pMX-Sox2, pMX-Klf4, and pMX-c-Myc (Addgene).
6. Helper plasmids: one expressing Vesicular stomatitis virus (VSV) G protein such as MD.G, and one expressing MLV (retroviral) gag-pol (Addgene).
7. Centricon (Plus-20, 20 ml from Millipore) with a cutoff 100,000 NMWL.
8. 15-cm tissue culture dish (Corning).
9. Whatman 0.45 micron, cellulose acetate filters (F8677).

3. Methods

3.1. Preparation of Cell Culture Media

3.1.1. Media and Feeder Cells for hES and iPS Cells

Cells are maintained in the standard hES cell culture condition, i.e., KNOCKOUT™ D-MEM, 20% KSR, 1× NEAA, 1× L-Glutamine & 1×-Antibiotic/antimycotics, 0.1 mM β-mercaptoethanol and 10 ng/ml of basic fibroblast growth factor (bFGF, purchased from Peprotech). Mouse embryonic fibroblasts (MEFs) were used as feeder cells or a source of conditioned media as previously described (9, 10).

3.1.2. Media for Human Fibroblasts

We propagate human fibroblasts in DMEM (low glucose) with Earle's Salts, 10% FBS, 1× NEAA, 1× L-Glutamine, and 1×-Antibiotic/antimycotics. Derivation and propagation of human mesenchymal stem cells (MSCs) have been described previously (10). Addition of low levels of bFGF (1 ng/ml) was used for culture of hMSCs. For propagating fibroblasts (IMR90 and hMSCs), the split ratios were never more than 1:4, with cells passaged about every 6 days (see Note 1).

3.2. Retroviral Production and Usage

We have successfully used both lentiviral and retroviral vectors for reprogramming human adult and fetal fibroblasts (6). For retroviral vectors expressing the four standard Yamanaka four factors Oct4 (O), Sox2 (S), Klf4 (K), and c-Myc (M), we used the original pMX-based vectors. These and other retroviral vectors (such as pMIG-based) expressing the four genes are available via Addgene. Transgenes derived from mouse coding sequences are fine for human cell preprogramming, likely due to the high levels of homology between mouse and human genes for the four factors (6). In the following, we describe a detailed and optimized protocol based on the method and pMX-based vectors that Yamanaka and colleagues first described for human cells (3). Making high titer retroviruses is absolutely critical for successful reprogramming of human cells, but cotransduction of ecotropic (mouse) receptor gene is not necessary (3). Others and we instead use VSV.G pseudo-typed retroviral or lentiviral vectors for transducing efficiently both human and mouse cells (4–8). We also detail below a protocol for retroviral production, concentration, storage, and usage at appropriate concentrations. Protocols for the use of lentiviral vectors by a similar method using 293 T cells have been previously published (11).

3.2.1. Production of Recombinant Retroviruses

1. Day 0: Coat 15-cm dishes with 50 μg/ml poly-D-lysine dissolved in PBS (12 ml/dish) for a period of 1 hour. Wash twice with PBS and then dispense 8–10 million 293 T cells in standard DMEM (high glucose) + 10% FBS to a total volume of 20 ml.
2. Day 1: After 24 h the plates should be about 70–80% confluent. Proceed to make the transfection cocktail (one for each viral vector encoding Oct4, Sox2, Klf4, and Myc). Our modified formula (to reduce Lipofectamine and DNA amounts) for cells seeded in a 15-cm plate is as follows: Add 36 μl of Lipofectamine to 1.2 ml of OPTI-MEM-I in a 15-ml polystyrene tube, and incubate for 5 min at room temperature in a 5 ml polystyrene tube. In parallel, mix 24 μg of total DNA (i.e., 3 μg VSV.G, 6 μg of retro-gag/pol, and 15 μg of retroviral vector, see Note 2) into 1.2 ml of OPTI-MEM in another 5 ml polystyrene tube. After 5 min, mix the diluted DNA with diluted Lipofectamine and incubate for 20 min at room temperature. In the mean time, change to fresh media in the 293 T cultures now using DMEM (with 1%FBS) to a total volume of 20 ml. Finally, add the DNA–lipid complexes drop wise onto the 293T cultures. Gently swirl the plates and shake back and forth and sideways to mix uniformly and place dishes in the incubator. Culture overnight.
3. Day 3: After 48 hours, Harvest the supernatant from the plates in 50-ml tubes and store at 4°C. Add fresh DMEM with 1% FBS to a total volume of 20-ml to the plates for another harvest later.
4. Day 4: Harvest the second round of supernatant and collect into the original 50-ml tubes. There should now be a total of about 40 ml supernatant per each vector. Typical titers of

unconcentrated viruses obtained using this procedure are in the range of ~10^5–5×10^5 transducing units/ml. To eliminate cell contamination, filter the supernatant using a 0.45-μm filter unit (low protein binding).

3.2.2. Usage of Recombinant Retroviral Vectors

1. To concentrate viruses by 50–100-folds the Centricon (Plus-20, 20 ml from Millipore) with a cutoff 100,000 NMWL is our method of choice (11). The filter filtration method also helps to reduce the free form of VSV.G proteins that are toxic to target cells. Each device concentrates ~18 ml to ~200 μl each time with a spin for 20–30 min at 3,000 rpm, and repeated loading is fine. Using this procedure starting from a volume of ~40 ml of supernatant we typically concentrate down to about 300–500 μl. Designate this final volume amount as V. A practical guide is provided below to use an appropriate amount of concentrated viruses without the need to measure the viral titers precisely.
2. We recommend the use of freshly made viruses for transduction, but if it is not possible, we make small aliquots and store them at −80°C where it is stable for many years (see Note 3).
3. Using the appropriate amount of virus is critical for successful reprogramming. We find that both too much and too little virus amount adversely affects reprogramming efficiency. Consequently our recommendation for the appropriate dosage based on the above viral production protocol is as follows: If V is the total amount of virus obtained per viron type per 15-cm plate (see step 1), then we recommend as a starting amount V/12 for each virus per 100 Kcells/10 cm^2 all dissolved together in a total volume of 2 ml to be transduced. The ratio of retroviral viruses per cell (per each vector) is typically ~4–5. It gives at least 70–80% transduction efficiency, as measured by a compatible GFP-expressing viral vector in both IMR90 fetal fibroblasts and adult MSCs (the virus amount for the latter may need to be increased to V/8). In case transduction efficiency is too poor we recommend increasing the ratio to V/10, V/8 or V/4 of each virus type, until a positive reprogramming result is obtained. This is one of the reasons why concentrated viruses are used.

3.3. Basic Reprogramming Procedure

The overall reprogramming procedure is highlighted in Fig. 1a. Specific details are as follows.

1. Day-1: In general, fibroblasts are seeded at a density in the range of 50–100 K/10 cm^2. For this protocol we will consider the specific case for 100 K cells per one well of a six-well

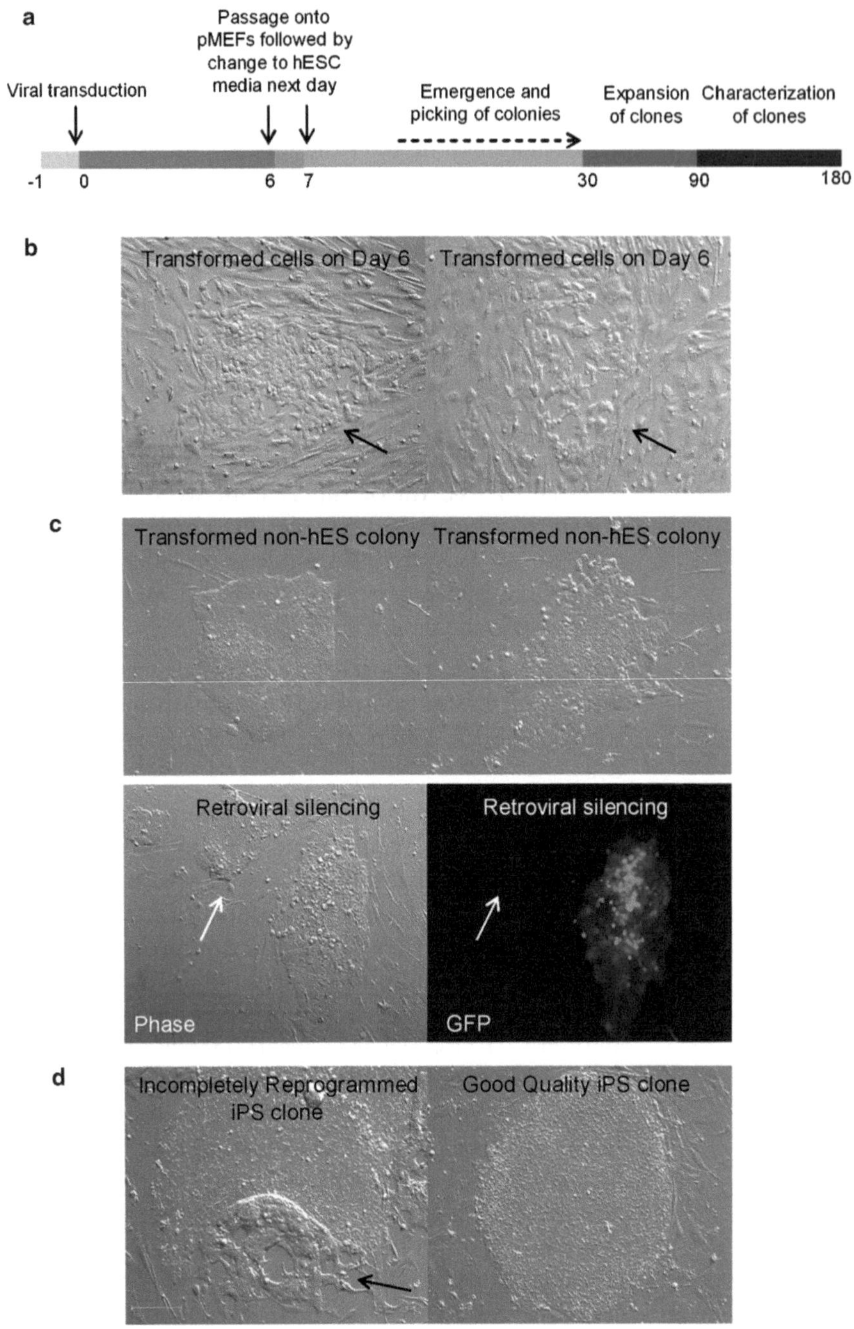

Fig. 1. (**a**) Timeline for the overall iPS cell derivation protocol is outlined. (**b**) By Day 6, "transformed" cell clusters are visible due to retroviral vector-mediated gene expression. (**c**) As transforming/reprogramming proceeds, a myriad of colonies are observed that are visible starting around day 9-12. Two examples of transformed but non-hES-like colony morphologies is highlighted. If the retroviruses also co-express GFP then Silencing of retroviral-mediated transgene expression, assayed by a loss of GFP expression, is a good indication of "correct transformation" or reprogramming. (**d**) Not all hES-like colonies that are picked (based on morphology) grow and expand equally well. Among those that proliferate, one cannot assume that they are clonal either genetically or epigenetically. Some clones self renew normally after picking, resembling hES cell colonies (*right*). However, others may occasionally bud-off transformed cells (*left*), likely due to either incomplete reprogramming or contaminated cells near the hES-like colony originally picked. These latter clones usually do not show good growth and differentiation ability in the longer term.

plate. It is important that cells must be dispensed evenly across the well (see Note 4).

2. Day 0: Cells are transduced using a combination of the retroviruses as per the concentrations recommended in the retroviral usage section. The retroviral amount for the four-factor (OSKM) and three-factor (OSK) conditions is the same. Pre-mix the viruses with the standard fibroblast media and 6 μg/ml polybrene to a total of 2 ml per well before dispensing.
3. Day 1: Supplement the wells with an additional 1 ml of fibroblast media. Note that the concentration of polybrene now is reduced to 4 μg/ml and the total media per well is 3 ml.
4. Day 2: Aspirate the retrovirus containing media and add fresh 3 ml fibroblast media to the well.
5. Day 4: Aspirate the old media off the well and add fresh 3 ml media as specified on day 2. The cells should be proliferating well by now, and will begin to get confluent. We find that increased cell–cell contact and continued proliferation during these initial days is favorable for the reprogramming process.
6. Day 5: Plate irradiated MEF feeder cells into gelatin-coated dishes. For the above specific case, we recommend one six-well plate, and two 10-cm dishes per one well of reprogramming cells. The former plate serves as a monitoring dish useful for day-to-day observations and analysis of efficiency and other assays during the course of reprogramming. The larger 10-cm dishes are a convenient format for the purpose of colony picking.
7. Day 6: By now the reprogramming well will be confluent with cells, and occasional clusters of transformed cells will be visible (Fig. 1b). Several regions of the well will show cells growing rapidly without any contact inhibition (see Note 5). On this day the cells are passaged onto MEFs and the procedure is outlined next.

   First, prepare the plates preseeded with MEF feeder cells as follows. Change the media in the MEFs plates or dishes (after inspection) with fresh fibroblast media at 2 ml per well of six-well plates (or 12 ml per 10-cm dish).

   Next, aspirate the media off the reprogramming well, and wash once with PBS and add 1 ml trypsin to the well. In about 5–10 min, the cells will detach and breakdown into small clumps. Add 2 ml of fresh fibroblast media to the well to neutralize the trypsin. Pipet up and down to mix the cells and breakdown any remaining clumps.

   When using the OSKM condition proceed as follows: dilute the cell suspension (3 ml) by transferring them into a new tube containing an additional 7 ml of media to bring the total to 10 ml. Dispense 3 ml of the total volume of cells into each

10-cm dish, and 0.5 ml/well of a six-well plate with feeder cells (*see below*). Note that the split ratio used here is effectively 1:20 (see Note 6). Typically this corresponds to 10–25 K/10 cm^2 density of cell plating.

When using the OSK condition proceed as follows: the 3 ml of harvested cells will be transferred into a new tube. Single cell suspension will be dispensed 0.5 ml/well of six-well plates with feeder cells. Note that the split ratio used here is effectively 1:6 (see Note 6).

Finally, disperse the cells evenly and place into the incubator for overnight attachment of the cells.

8. Day 7: Aspirate off the fibroblast media and directly change it to hES cell culture media. Use 3 ml/well for one well of a six-well plate and 18 ml for the 10-cm dishes.
9. Day 9: Aspirate off the old media and add fresh hES cell culture media at the amounts mentioned on Day 7. Carefully scan the plates to check for transformation characterized by growth of cell clumps. Although colonies of "transformed" cells may emerge at day 9 and onward (Fig. 1c, see Note 7), however most of these epithelial colonies are typically not correctly reprogrammed cells. The candidate iPS cells emerge a little later and resemble hES cells in both morphology and expression of pluripotency markers such as TRA-1-60. (Fig. 2b).
10. Days 11 and 13: Repeat the feeding and cell-monitoring procedure from day 9.
11. Day 14: Recommended but optional: Collect the reprogramming cells from one well of a six-well plate to do FACS analysis for TRA-1-60 expression after trypsin digestion. We recommend repeating this procedure again on Days 18, 21, 24, and 27. Detection of a positive population at these time points and an increase in their percentages over time reflects that reprogrammed cells have emerged (refer Fig. 2a).
12. Day 15: As the MEF feeder cells are now more than a week old, from this day on MEF-conditioned media is used for feeding the reprogramming plates. The feeding and cell monitoring procedure remains the same as on Day 9. It is normal to observe both an increase in size of some transformed cell clusters as well as a loss of cells by apoptosis in others. A lot of the transformed cells are semiadherent and can get dislodged and reattach in different parts of a reprogramming plate. Consequently, especially when running multiple reprogramming experiments it is highly recommended that aspirating tips and feeding pipettes should be changed between different experimental conditions.
13. Days 17, 19: Repeat the feeding and cell-monitoring procedure from Day 15. Closely monitor, in particular, if the media color changes to even a pale yellow during these days. If so, from then on, it is important to switch to daily media changes

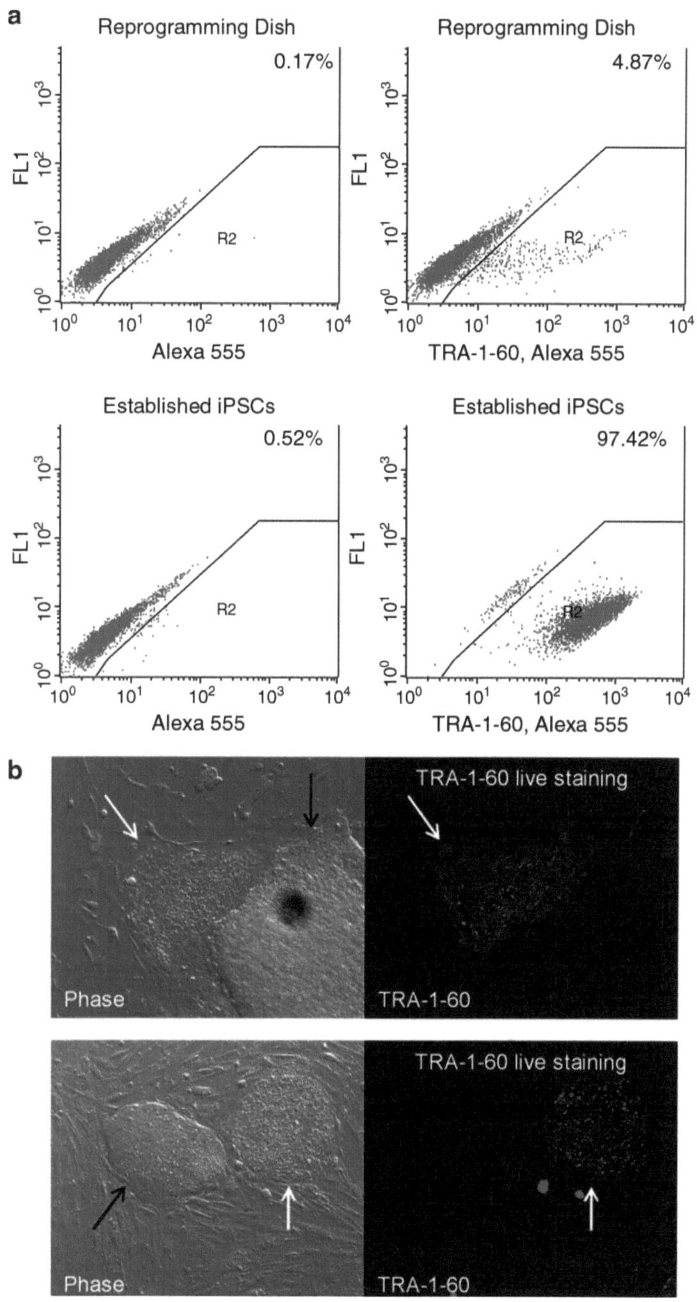

Fig. 2. (**a**) TRA-1-60 expression analysis by FACS is a useful method for measuring the kinetics of reprogramming. The *upper* panel gives the pattern of TRA-1-60 expression in cells from the whole reprogramming dish as compared to a fully reprogrammed iPS clone (*lower panel*). (**b**) Further, TRA-1-60 live staining is a very efficient way for distinguishing hES-like colonies (*white arrow*) from transformed cell clusters (*black arrows*). Two representative examples are provided.

instead of the every alternate day procedure. When changing media daily only 2 ml per well of a six-well plate or 12 ml per 10-cm dish may be used. If, however, one observes that even with this daily feeding the media gets acidic too soon, we recommend increasing the media amounts progressively in steps of 1 ml per well of a six-well plate (i.e., 2 ml to 3 ml to 4 ml and so on), and in steps of 6 ml for the 10-cm dish (i.e., 12 ml to 18 ml to 24 ml and so on). Usually a rapid change in media color is a reflection of high confluence of the reprogramming wells and is not a desired phenomenon (see Note 8).

14. Day 21 onward: Repeat the feeding and cell-monitoring procedure from above.

3.4. Human iPS Colony Identification and Picking

Note that in general by Day 21 for most starting fibroblast populations both a large number and varying morphology of colonies are visible in different regions of the plate. This is a reflection of a remarkable mesenchymal to epithelial transition that occurs during the reprogramming process from fibroblasts. However, not all epithelial-like colonies are hES like – which are characterized by a flatter cobblestone like morphology with individual cells clearly demarcated from each other in the colonies. Moreover, a lot of the non-hES-like colonies form closely resembling but not identical compact clusters of cells and are capable of sustained self-renewal and successive passaging for long periods under hES cell culture conditions. This makes identification of the successfully reprogrammed colonies a very critical step in this process. In our hand the most reliable method for selecting a reprogrammed colony is live staining by the TRA-1-60 antibody that also recognizes undifferentiated hES cells.

1. The primary and secondary antibodies are both used at a 1:200 dilution (see Note 9) and are premixed together into hES cell media. After aspirating the existing media off the reprogramming dish we add this antibodies containing media at 1 ml per one well of a six-well plate or 6 ml per 10-cm dish. This amount is sufficient to safely cover the surface of the dishes without the need for a shaker. The plates are then directly placed into the tissue culture incubator for about 1 h (37°C, 5% CO_2).
2. Following this period the media is aspirated, washed once with PBS and finally fresh hES cell medium is added. The plates are left in the incubator for 15 min and then imaged under a standard fluorescence microscope. Successful antibody staining can very specifically delineate reprogrammed colonies from just plain transformed counterparts (refer Fig. 2b), and can be detected for up to 24–36 h. This aspect is particularly useful since it helps in identification and tracking of the candidate iPS colonies both before picking and also the day after they are picked and transferred into a new well.
3. For the purpose of picking we follow this procedure: First, we fabricate our colony-picking tool, which is made by drawing the thin end of a glass Pasteur pipette into a J-shaped fine curve (about 10–50 microns in thickness). For this the pipette is held

over a Bunsen flame, and when the thin end starts to melt, the glass is slowly pulled apart, and just before it melts off completely, it is in a (critical step) quick motion pulled apart and away from the flame. This usually leaves a very thin curved end on one or both of the two separated parts of the pipettes. Second, the TRA-1-60 positive stained colonies are identified under the fluorescence microscope at a 10× magnification. We prefer to select colonies at least 100–500 microns in diameter. These can usually be discerned by the naked eye when the plate is held up to light. Once a positive colony is identified, it is brought to the center of the viewing field and the magnification is switched down to 4×. Third, the dish cover is removed and the picking tool is immersed in the media. Looking through the microscope at this 4× resolution it is easy to view the approach trajectory of tool as it is brought next to the colony, and then it is used to gently scrape the colony off the surface until it is completely detached and floating in solution. At the same 4× magnification, a 10-μl micropipette tip is brought next to the colony and which is then sucked up in a volume no more than 5 μl to avoid carry over of additional floating/dislodged cells in the dish. Fourth, this drawn volume can be either dispensed directly into a single well of a 96-well MEF feeder plate, or (preferred alternative) it is dispensed into 50 μl of hES media in an eppendorf tube, following which a 200-μl pipette tip is used to break the picked colony into 3–5 smaller clumps by a few gentle pipetting motions and then dispensed in a single well of a 96-well plate.

4. The colonies thus picked are allowed to attach for 48 h before media is changed, and subsequently these are treated like normal hES colonies and passaged, expanded, and maintained using standard culture procedures. We recommend picking at least 10 distinct colonies by the end of each reprogramming experiment.

3.5. Initial iPS Clone Expansion and Characterization

Until proven that they are truly pluripotent, these TRA-1-60+ colonies we picked are referred to as hES-like colonies or potential iPS colonies. These potential iPS cell colonies in their early passages must be constantly monitored. Some of the colonies will be more fragile and prone to rapid apoptosis and/or differentiation, while some others may show more robust growth, and while yet some others may occasionally bud-off transformed non-hES-like cells (refer Fig. 1d). All these phenotypes are normal and we observe them on a regular basis, and may characterize, respectively, transient, stable, or unstable reprogramming states of the individual clones. Usually from starting ten colonies we are able to derive four to five stable hES-like colonies that display normal growth patterns and remain a pure population during subsequent expansion. These are the clones we focus on for subsequent characterization.

Typically expansion of a clone from a single colony stage to a confluent one-well of a six-well plate takes about 5–7 weeks. We strongly recommend only 1:1.5 to 1:2 passaging of clones during this initial critical and slow period of expansion, and we often use the 50:50 ratio of plain hES cell medium and MEF-conditioned medium for feeding the cells. Cell passaging is either by mechanical means or by the use of collagenase (but not trypsin), and especially when working with very small wells and colonies, we strongly recommend the use of 1,000-μl pipette tip (and not smaller) for gentle scraping/breaking of colonies during cell splitting.

To facilitate subsequent rapid characterization of the cells, we typically expand the iPS colonies uninterrupted (with only occasional freezing of cells) while using a faction of cells for the following assays (in the order of preference). We estimate that we need cells from a total of 24 wells (in six-well plates) for the assays or tasks listed below.

1. Pluripotency markers (iPS cells from two wells can be replated into smaller well with MEF feeder cells): We fix cells after iPS cells reach an optimal size. Cells can be stained for surface markers such as TRA-1-60 (and/or TRA-1-81) and SSEA4 (and/or SSEA3) or nuclear antigens such as NANOG (human and endogenous) and OCT4. Alkaline phosphatase staining can be done by standard histochemistry.
2. Karyotype analysis (two wells): cells are harvested after appropriate treatments required for karyotyping analysis.
3. DNA and RNA extraction (two wells): Gene expression analysis and fingerprinting is carried out using RNA or DNA derived in this step.
4. Embryoid body (EB) formation (three wells): Cells are collected by collagenase passaging and resuspended in FBS-containing media (standard procedures) and allowed to form EBs for up to 2 weeks. Formation of cystic structures is typically observed within 6–10 days varying from clone to clone. After 2 weeks, the EBs are broken down into smaller clumps using a 200 μl pipet tip and allowed to attach onto gelatin-coated plates for an additional 2 days followed by fixing and staining for the three embryonic germ layers and trophectoderm.
5. Teratoma formation (12–15 wells): Cells are collected from the plates into a 50-ml tube by directly scraping them in their native media using a cell-scraper. After spinning down, the cell pellet is resuspended on ice (critical step) in 400 μl of a 1:1 mixture of matrigel and knockout DMEM and collected in an eppendorf tube and stored on ice. This volume of suspension is suitable for intramuscular injection into the hind limb of two immunodeficient mice (200 μl each). We

prefer to use immunodeficient mice with further reduced NK activities, such as SCID/Beige or $RAG^{-/-}IL2RG(\gamma_c)^{-/-}$ mice. Palpable tumors can be detected as early as 6 weeks postinjection with the improved method, but up to 4 months is also normal.

Typically, full characterization of an iPS clone takes between 5 and 6 months. There is clone to clone variation in differentiation ability and expansion/survival potential, so identification of a good quality normal karyotype clone is a tedious but essential aspect of reprogramming (see Note 10).

4. Notes

1. Poorly growing primary fibroblast cultures can be improved by addition of bFGF and relatively confluent passaging (1:1.5 to 1:2).
2. The overall protocol for production of retroviral vectors by transfection of 293 T cells is similar to that for lentiviral vectors (11). However, the ratio of the gag-pol plasmid and a transducing vector plasmid is different between the two systems. More retroviral vector plasmid should be used (2–2.5-fold) as compared to the MLV (retro-) gag-pol helper plasmid. The total amount of DNA (three plasmids) is kept the same (24 μg).
3. For a short-term storage (up to a week), it is better to store retroviral virions at 4°C. Note that the virions do not survive well after repeated freez and thaw.
4. To ensure cells are uniformly dispersed, especially for smaller wells (i.e., in 12-well plates or smaller), it is recommended that one pre-mix the cells into the total volume of media (to be used per well) before dispensing.
5. Early signs of transformed growth of cells on day 6 are typically a very good sign that the viral transduction procedure is working well.
6. This split ratio can be further increased to up to 1:40 for cells that grow well in hES cell media – for example, in our experience, hMSCs can be passaged at this increased dilution. Again, it is important to assess the growth rate and viability of a user's cell type before making these adjustments.
7. A large number and type of "transformed" (epithelial-like) colonies are typically visible in a reprogramming plate. In the examples provided here (Fig. 1c), the colonies do not have the characteristic hES cell morphology and serve to help the user identify and familiarize oneself with partially or incorrectly reprogrammed cells.

8. Over confluence of cells during the reprogramming procedure typically inhibits both emergence and expansion of the true iPS colonies, and also makes identification and clean picking of colonies difficult. In addition, the cells risk peeling off from the surface over time.
9. A similar approach has been previously described with TRA-1-81 antibody (7). We exclusively used TRA-1-60 antibody that enables the identification of reprogrammed cells from a variety of human cell types. The TRA-1-60 live staining has little adverse effects on the growth of TRA-1-60+ or negative colonies even after multiple rounds of staining. To conserve antibody usage further (1:300 to 1:400) dilutions of the TRA-1-60 antibody may be used too; however, the stained colonies will appear dimmer but still discernible under the microscope.
10. When troubleshooting for failure to obtain *bona fide* iPS cells using the above procedure we recommend that the user checks the following steps (in order of importance): first, the titer of the unconcentrated retroviral supernatant (at least ~10^5 transducing units); second, transduction efficiency of the target cell type (at least 60–80%); third, poor quality of the pMEFs used during reprogramming (can affect reprogramming efficiency by tenfold or even more); and fourth, refractory or senescent nature or late passage of the target cell type (use earlier passage of the cells).

Acknowledgments

The authors would like to thank members of the Cheng lab for helpful discussions and assistance during various aspects of this study. This work was funded in part by a grant (R01 HL073781) and a supplement from NIH.

References

1. Thomson JA, Itskovitz-Eldor J, Shapiro SS, Waknitz MA, Swiergiel JJ, Marshall VS, Jones JM. (1998) Embryonic stem cell lines derived from human blastocysts. *Science* 282(5391):1145-1147.
2. Reubinoff BE, Pera MF, Fong C, Trounson A, Bongso A. (2000) Embryonic stem lines from human blastocysts: somatic differentiation in vitro. *Nat Biotechnol* 18:399-404.
3. Takahashi K, Tanabe K, Ohnuki M, Narita M, Ichisaka T, Tomoda K, Yamanaka S. (2007) Induction of pluripotent stem cells from adult human fibroblasts by defined factors. *Cell* 131: 861-872.
4. Yu J, Vodyanik MA, Smuga-Otto K, Antosiewicz-Bourget J, Frane JL, Tian S, Nie J, Jonsdottir GA, Ruotti V, Stewart R, Slukvin II, Thomson JA. (2007) Induced pluripotent stem cell lines derived from human somatic cells. *Science* 318:1917-1920.
5. Park IH, Zhao R, West JA, Yabuuchi A, Huo H, Ince TA, Lerou PH, Lensch MW, Daley

GQ. (2008) Reprogramming of human somatic cells to pluripotency with defined factors. *Nature* 451:141-146.

6. Mali P, Ye Z, Hommond HH, Yu X, Lin J, Chen G, Zou J, Cheng L. (2008) Improved efficiency and pace of generating induced pluripotent stem cells from human adult and fetal fibroblasts. *Stem Cells* 26:1998-2005.
7. Lowry WE, Richter L, Yachechko R, Pyle AD, Tchieu J, Sridharan R, Clark AT, Plath K. (2008) Generation of human induced pluripotent stem cells from dermal fibroblasts. *Proc Natl Acad Sci U S A* 105:2883-2888.
8. Park IH, Arora N, Huo H, Maherali N, Ahfeldt T, Shimamura A, Lensch MW, Cowan C, Hochedlinger K, Daley GQ. (2008) Disease-specific induced pluripotent stem cells. *Cell* 134:877-886.
9. Dravid G, Hammond H, Cheng L. (2006) Culture of human embryonic stem cells on human and mouse feeder cells. *Methods Mol Biol (Clifton, N.J.)* 331:91-104.
10. Cheng L, Hammond H, Ye Z, Zhan X, Dravid G. (2003). Human adult marrow cells support prolonged expansion of human embryonic stem cells in culture. *Stem Cells* 21:131-142.
11. Ye Z, Yu X, Cheng L. (2008) Lentiviral gene transduction of mouse and human stem cells. *Methods Mol Biol (Clifton, N.J.)* 430:243-53.

Chapter 13

Using Small Molecules to Improve Generation of Induced Pluripotent Stem Cells from Somatic Cells

Caroline Desponts and Sheng Ding

Abstract

Induction of pluripotent stem cells from somatic cells by defined factors was shown to be possible only recently, but already several laboratories have made tremendous strive toward improving and understanding the process. Originally, Oct4, Sox2, Klf4, and cMyc were identified as being the combination of genes necessary to induce reprogramming. It was later shown that cMyc was dispensable; however, in its absence the process was less efficient and took a considerably longer period of time to occur. Furthermore, others have shown that the combination of Oct4, Sox2, Nanog, and Lin28 could also induce reprogramming. One major caveat associated with these techniques remains the need for overexpression of several genes using viral systems. Until very recently, most studies were done using integrating viruses such as retroviruses and lentiviruses. This method ensured that the protein of interested would be expressed at a high concentration and for an adequate period of time necessary to induce reprogramming. Up to date, others have now been able to use different nonintegrative method such as adenovirus and plasmid transfection to induce reprogramming. Furthermore, piggyBac transposition was successfully used to induce reprogramming of murine cells. Most importantly, it was recently published that reprogramming can be induced in the absence of virus, with proteins and small molecules. All of the later methods are appealing since they do not require the integration of the virus or plasmid to exert its effect. However, one avenue that would be all the more therapeutically appealing would be to induce reprogramming in the absence of gene overexpression systems, using small molecules to modulate signaling pathways in the somatic cells. A few molecules have already been identified with the ability to either improve the process or replace one or two of the genes deemed necessary for reprogramming. We have screened successfully for compounds that can replace some of these factors, and share the methods developed following these screens.

Key words: Reprogramming, Pluripotency, Small molecule, BIX-01294,Bayk8644 retroviruses, iPSC

S. Ding (ed.), *Cellular Programming and Reprogramming: Methods and Protocols*, Methods in Molecular Biology, vol. 636, DOI 10.1007/978-1-60761-691-7_13, © Springer Science+Business Media, LLC 2010

1. Introduction

The development of methods that induce pluripotency in somatic cells offers great promise for the treatment of different diseases. It was previously shown that stable genomic integration and high expression of four factors, Oct4/Sox2/Klf4/c-Myc or Oct4/Sox2/Nanog/LIN28, can reprogram fibroblast cells, pre-B cells, and liver and stomach epithelial cells into induced pluripotent stem cells (iPSCs) (1–4). It is now accepted that iPSCs can be generated by viral integration of Oct4/Sox2/Klf4 without c-Myc (5, 6), with the advantage that these iPSC exhibited less tumorigenicity in chimeras and progeny mice (6). Nonetheless, in the absence of c-Myc overexpression, the reprogramming process is much slower and efficiency substantially reduced. Recently, some groups have been able to induce reprogramming of somatic cells in the absence of viral integration, using either plasmid transfection, adenoviruses, or piggyBac transposon (7–10). More importantly, Zhou et al. have shown that it is possible to induce reprogramming of somatic cells using proteins and small molecule, avoiding the use of controversial viruses (11). In this study, Oct4, Sox2, Klf4, and Sox2 proteins were fused to a polyarginine at the C-terminus to facilitate their entry in the target cells. Treatment of the target cells with these proteins in the presence of valproic acid led to the induction of pluripotency in a few of the treated cells (0.006%). These new methods are certainly more appealing since they do not involve viral integration; however, the reprogramming process remains much less efficient as compared with reprogramming in the presence of integrating lentiviruses or retroviruses.

Since the viral integration and gene over expression of Oct4, Klf4, and Sox2 is not a viable therapeutic option, our laboratory has directed its effort in finding small molecules that could replace the presence of virally transduced TFs to find a chemical cocktail that would allow reprogramming of somatic cells in chemically defined conditions. Using a screening approach, we identified a few small molecules with the ability to replace some TF in the induction of pluripotency in different cellular system. In one study, neural progenitor cells (NPCs), which endogenously express *Sox2* (12), were transduced with Oct4 and Klf4 alone (OK) and were successfully reprogrammed to iPSCs; this process was greatly enhanced in the presence of a G9a histone methyltransferase (G9a HMTase) inhibitor, BIX-01294 (BIX) (13, 14). In addition, we observed that BIX could enable reprogramming of NPCs transduced with c-Myc, Klf4, and Sox2, in the absence of Oct4 ectopic expression. Therefore, in this particular system, BIX seemed to compensate for Oct4 overexpression. We later found that small molecules could assist in the induction of pluripotency from mouse embryonic fibroblast (MEF) in the absence of Sox2 and cMyc, when the cells were exposed to a combination of BIX and Bayk8644 (BayK), an L-channel calcium agonist (15, 16). BayK was of interest because it exerts its effect upstream in cell signaling pathways and does not directly cause epigenetic modifications. It is likely that this type of molecule can be exploited to induce reprogramming in a more specific manner than molecules acting directly at the epigenetic level causing DNA or histone modification. Some of these epigenetic modifiers have already been shown to facilitate the reprogramming process, such as BIX (14, 16), 5′(azacytidine (17), and valproic acid (11, 18, 19). Interestingly, valproic acid was shown to assist in the reprogramming of human fibroblast transduced with only Oct4 and Sox2, in the absence of the two oncogene Klf4 and cMyc (19) and was shown to support reprogramming of somatic murine cells in the absence of viral vector, when cells were directly exposed to proteins (11).

This document describes one of the methods developed following our screens. It describes a method to generate iPS cells from murine MEF cells transduced with Oct4 and Klf4 retroviruses and treatment with different chemical compounds. The reprogramming procedure is done on Matrigel-coated plates instead of using feeder cells, in an effort toward developing a chemically defined reprogramming strategy.

2. Materials

2.1. MEF Derivation

1. 129 S2/SvPasCrlf, ROSA26$^{+/-}$/OG2$^{+/-}$ (obtained from Hans R. Schöler, Max Planck Institute for Molecular Biomedicine, Münster, Germany) or OG2 pregnant mice 13.5 days postcoitus (dpc) (B6;CBA-Tg (Pou5fl-EGFP)2Mnn/J; The Jackson Laboratory).
2. Scissors.
3. Forceps.
4. Gauze.
5. PBS without calcium/magnesium.
6. 70% Alcohol.
7. Animal experiments have to be performed according to the Animal Protection Guidelines of your institution and IACUC approval is necessary for this manipulation.

2.2. Cell Culture

1. 6-Well plates (Corning).
2. 96-Well plates, black with clear bottom (Corning).
3. Growth Factor Reduced Matrigel (BD Biosciences, Bedford, MA).

4. PBS without calcium/magnesium.
5. Trypsin 0.05%/5.3 mM EDTA.

2.2.1. MEF Media

1. 2 mM Glutamax.
2. 1× Antibiotic/antimycotic.
3. 10% ES-FBS.
4. DMEM.

2.2.2. Plat-E Cell Media

1. 10% FBS.
2. 20 mM HEPES.
3. 1× Antibiotic/antimycotic.
4. 0.1 mM Nonessential amino acid (NEAA).
5. DMEM.

2.2.3. mES Cell Media

1. 7.5% ES-FBS.
2. 7.5% Knock Out Serum Replacement (KOSR).
3. 2 mM Glutamax.
4. 1× Nucleosides.
5. 1.1 mM b-Mercaptoethanol.
6. 0.1 mM Nonessential amino acid (NEAA).
7. 10^3 U/ml Leukemia Inhibitory Factor (LIF; ESGRO) (Millipore).
8. Knock Out DMEM.

2.2.4. 2× Freezing Media for iPS Cells

1. 20% DMSO.
2. 30% mES cell media.
3. 50% ES-FBS.

2.3. Retroviruses Production and Transduction

1. Vectors were obtained with Addgene; pMXs-Oct3/4, pMXs-Sox2, pMXs-Klf4, and pMXs-c-Myc.
2. PVDF syringe filters (0.45 μm).
3. Syringe sterile, without needle.
4. Polybrene (10 mg/ml) (Millipore/Chemicon).

2.4. Compounds

1. The synthesis and full characterization of compound BIX-01294 was done as previously described (13). However, this compound is now available through different vendors such as Alexis Biochemical (Plymouth Meeting, PA), Sigma-Aldrich (St-Louis, MO), and Stemgent (San Diego, CA).
2. Bayk8644 was purchased from EMD/Calbiochem Biochemical (San Diego, CA).

2.5. Detection and Identification of iPSCs

2.5.1. Alkaline Phosphatase Detection Kit (Millipore)

1. 10% Formalin solution (app. 4% formaldehyde).
2. 10× Tris Buffered Saline (10× TBS buffer).
3. Tween 20.

2.5.2. Primary Antibodies

1. Mouse anti-Oct3/4 antibody (1:200) (Santa Cruz Biotechnology).
2. Mouse anti-SSEA1 antibody (1:200) (Santa Cruz Biotechnology).
3. Rabbit anti-Sox2 antibody (1:200) (Chemicon).
4. Rabbit anti-Nanog antibody (1:500) (Abcam).

2.5.3. Secondary Antibodies

1. Alexa Fluor 488/555 donkey anti-mouse IgG (1:500) (Invitrogen).
2. Alexa Fluor 488/555 donkey anti-rabbit IgG (1:500) (Invitrogen).

2.5.4. Blocking Buffer Composition

1. PBS without calcium magnesium.
2. Triton X-100.
3. Normal Donkey serum (Jackson Immuno Research).
4. Buffer: PBS containing 0.3% Triton X-100 and 6–10% normal donkey serum.
5. DAPI (4′,6-diamino-2-phenyindole, dilactate), 1,000× Stock solution; 1 mg/ml in water.

3. Methods

3.1. MEF Derivation

129 S2/SvPasCrlf or ROSA26$^{+/-}$/OG2$^{+/-}$ MEFs are derived according to the protocol reported on the WiCell Research Institute website: "Introduction to human embryonic stem cell culture methods" (http://www.wicell.org/index.php?option=com_content&task=category&id=310&Itemid=149&sectionid=16). Animal experiments have to be performed according to the Animal Protection Guidelines of your Institute.

1. Thirteen days d.p.c pregnant females are euthanized.
2. The mice are transferred to a biohazard hood and the remaining manipulation should be performed there to prevent contamination.
3. Mice are then placed on a dissecting board on their back.
4. The fur is cleaned with 70% ethanol, an incision is then performed on the ventral section, being careful to not cut through the peritoneum.

5. The instruments are sterilized again in 70% ethanol.
6. Cut through the peritoneum being careful not to cut through the intestine (which would release contamination).
7. The uterine horns are then exposed, removed, and placed in a dish containing sterile PBS, wash three times with fresh PBS.
8. Each embryo is then released from the uterine horns, separated from their embryonic sac and placenta and transferred to a new dish, containing fresh sterile PBS.
9. Wash three times with PBS.
10. At this point, the head, the liver, and the heart (the red visceral tissue) are removed. This allows eliminating major sources of stem cells.
11. The rest of the embryo is then taken to a new dish containing minimal amount of PBS.
12. Count the number of embryo present.
13. At this point, the embryos are minced using scissors or scalpel to obtain as small pieces as possible.
14. The embryos are then exposed to 2 ml of trypsin and minced for another 5 min.
15. Afterward, 10 ml of trypsin is added, pipet up and down and the embryos are incubated for 20 min in 37°C incubator.
16. Halfway through the incubation, take the embryos out and pipet up and down using a 10-ml steri-pipet to ensure a proper dissociation of the cells.
17. The cells are then placed back at 37°C for the remaining time of the incubation.
18. After 20 min, pipet up and down the cells, expose to MEF media to quench the trypsin and transfer to a conical tube.
19. The cells are centrifuge at $300 \times g$ for 5 min, the supernatant is removed and the cells are resuspended in MEF media and placed in tissue culture dishes for expansion.
20. For this manipulation usually, the equivalent of three embryos is resuspended in enough media to be plated in one 75-cm^2 flask (nine embryos would be plated in three 75-cm^2 flask).
21. The cells are then placed in an incubator at 37°C, 5% CO_2.
22. Once the cells reach confluency, between 1 and 3 days, the cells are then collected and frozen in cyrovials.
23. Usually one flask of cells is frozen into three cryovials.

3.2. Retrovirus Production

Virus production and transduction was performed using a modified protocol published previously (3).

pMX-based retroviral vectors for mouse Oct4, Klf4, c-Myc, and Sox2 were obtained from Addgene.

1. Vectors were amplified using MaxiPrep.
2. The viral production and transduction process was performed as described before (20) with some modifications.

Day 0

3. Plat E cells were plated in four 10-cm tissue culture dishes at a density of 0.1×10^6 cell/cm^2.

Day 1

4. Add 1.5 ml of DMEM to four 5-ml polystyrene tubes.
5. Add 30 μl of Fugene6 to each tube.
6. Mix gently, finger tapping and incubate at room temperature (RT) for 5 min.
7. Add 10 μg of the respective plasmid DNA (Oct4, Klf4, c-Myc, or Sox2) to each tubes.
8. Incubate at RT for 20 min.
9. During that time, remove the media from the PlatE cells and add PlatE cells media that does contain no Antibiotic/Antimycotic (see Note 1).
10. Once the 20-min incubation is done, add the transfection mixture dropwise to the PlatE cells and mix, rocking the dish back and forth, to ensure equal distribution.
11. Incubate overnight at 37°C.

Day 2

12. The day after, remove the media containing the transfection mixture off the cells, add 7 ml of PlatE cell media without antibiotic/antimycotic.
13. Place the cells back at 37°C until they reach confluence, which should be less than 8 h later.
14. Once the cells reach confluence, place them at 32°C for 48 h (see Note 2).

Day 4

15. After 48 h, the supernatant from each plate is collected separately and filtered using a syringe PVDF filter (0.45 μm). 800 μl of mixture is added to each well of a 6-well plate containing the target cells (see Section 3.3).

Preparation of transduction mixture for Oct4/Klf4 transduction (*see* Note 3).

400 μl Oct4 supernatant.

400 µl Klf4 supernatant.
0.4 µl Polybrene.

Negative controls
Control 1
800 µl supernatant from pMXs empty vector transfection.
0.4 µl Polybrene.

Control 2
800 µl Fresh media.
0.4 µl Polybrene.

Positive control
200 µl Oct4 supernatant.
200 µl Klf4 supernatant.
200 µl Sox2 supernatant.
200 µl c-Myc supernatant.
0.4 µl Polybrene.

3.3. Transduction of MEF Cells

Day 3

1. OG2 MEF cells are plated on Matrigel (1:50) coated plates at a density of 3.3×10^5 cells/well on 6-well plate in MEF media.

Day 4

2. MEF media is removed.
3. 800 µl of the appropriate transduction mixture is then added to each well.
4. Incubate at 37°C for 12–14 h (usually overnight).

Day 5

5. Remove the supernatant and discard accordingly (do not forget this is a viral mixture and it should be exposed to bleach for a 20-min prior to being disposed of).
6. Add mES cell media in the presence of small molecules (see Note 4). Different combinations of small molecules have shown great efficiency in promoting iPS cell formation.
   (a) Combination A: 1 µM BIX-01294 (Stemgent) (13) and 2 µM Bayk8644 (Stemgent).
   (b) Combination B: 1 µM BIX-01294 and 40 nM RG108 (Stemgent) (13).
7. Put back at 37°C.
8. Refresh the media with compound every 3 days (see Note 5).

3.4. Observation of iPS Cell Generation

1. Considering that day 5 is the day of MEF cells chemical treatment.

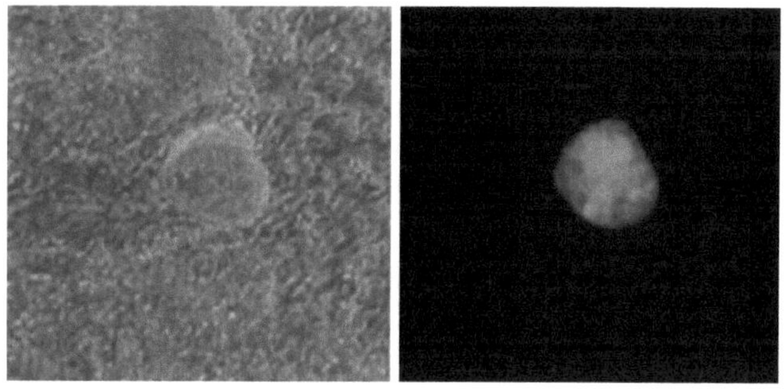

Fig. 1. Example of colonies that develop within 18 days of treatment (**a**) Phase contrast picture of a colony 18 days after MEF cell transduction (×10). (**b**) Picture showing green fluorescence of a colony as an indication of Oct4 promoter activation (×10)

2. In the positive control, colonies will develop within a few days after infection, most of the colonies should be GFP^+ by day 13 (9 days after infection).
3. Oct4/Klf4 with BIX and Bayk (OK2B) treated MEF cells should generate colonies by day 17–18 (12–13 days after treatment with small molecules), GFP^+ colonies should be observed within 14–21 days (day 19–26) (see Note 6).
4. Once the first GFP^+ colonies are observed (Fig. 1), the compound treatment is terminated and cells are then exposed to mES cell media alone or mES cell media supplemented with MEK inhibitor PD0325901 (0.5–1 μM) (16).

3.5. Confirmation of Pluripotency Status of the GFP+ iPS Cell Colonies

Once iPS cell colonies are observed, they are left to expend (for 2–3 days) to a size that would facilitate manipulation.

The day prior to picking colonies, gelatin-coated 96-well plates were inoculated with irradiated MEF cells (CF1) at a density of 25×10^3 cells/well.

Under a biohazard hood, colonies were observed under the microscope and lifted using a sterile glass pipette loop. Each colony was then picked and placed in the well of a 96-well plate containing 20 μl of 0.05% trypsin. The same manipulation was repeated for 4–5 colonies, and then the 96-well plate was placed at 37°C for 3–4 min (see Note 7).

Once the colonies were dissociated, the cells were transferred to the wells of a 96-well plate (coated with gelatin and irradiated MEF cells) containing 200 μl of fresh mES cell media. Usually the dissociated cells of one colony were plated in 5–10 wells of a 96-well plate, depending on the colony size.

3.5.1. For Propagation

A few days later, once they reached a good size (see Fig. 1), GFP^+ colonies were transferred to a 24-well plate (coated with gelatin and irradiated MEF cells) and so on until the cells could be expanded in 6-well plates. At each step, cells at early passage were frozen down.

Cells to be frozen down were resuspended in 500 μl of mES cell media and 500 μl of 2× freezing media was added to the cells, which were then transferred to cryovials. The cryovials were then transferred to a cell-freezing container and kept at −80°C overnight. The day after, the cryovials were transferred to liquid nitrogen tank for long-term storage.

2× Freezing media consist of 50% ES-FBS, 30% mES cell media, and 20% DMSO (for a final concentration of 25% ES-FBS, 65% mES cell media, and 10% DMSO).

3.5.2. iPS Cell Characterization

Several elements have to be assessed to ensure that the iPS cells share the same properties as mES cells.

1. Alkaline phosphatase staining.
2. Immuno-staining with at least these four markers Oct4, Sox2, SSEA-1, and Nanog.
3. mRNA expression of different pluripotency genes (20).
4. Methylation status of promoters for pluripotency genes (16).
5. Micro array analysis in comparison to original MEF cells and mES cell RNA (16).
6. In vitro differentiation toward three germ layers.
7. Most importantly, germline transmission (16).

3.5.3. Alkaline Phosphatase Staining

1. GFP^+ iPS cells were plated on irradiated MEF coated on 96-well back plated with clear bottom.
2. After the colonies reached the desired size, the cells were tested for the presence one of the pluripotency marker, alkaline phosphatase.
3. For alkaline phosphatase detection, staining was performed as suggested by the manufacturer, Millipore.

3.5.4. Immunostaining

1. Media was removed and the cells, were briefly washed once with PBS.
2. Cells were fixed with 4% formalin for 15 min at RT.
3. The formalin was then washed by exposing the cells to PBS three times for 5 min at RT.
4. Cells were then incubated at RT for 1 h in blocking buffer.
5. The blocking buffer was then removed.
6. Primary antibody in blocking buffer (at the concentration mentioned in the Materials see Subheading 2.5) was then added to the cells, incubation was overnight at 4°C.
7. The day after, primary antibody solution was removed and the cells were washed with PBS, three times for 10 min at RT.

8. The secondary antibody solution was then added to the cells, which were incubated at RT for 30–60 min.
9. Afterward, the secondary antibody solution was removed and cells were washed in PBS five times for 5 min at RT.
10. In the third wash, DAPI was added to the PBS (1:1,000) for nuclear staining; this wash was usually performed for 10 min instead of 5 to get a good DAPI staining.
11. Cells were then observed under the fluorescent microscope (Nikon Eclipse TE2000-U/X-cite 120 EXFO microscope equipped with a photometric CoolSnap HQ^2 camera).

4. Notes

1. It is not recommended to wash the cells with PBS to remove the left over media (containing the antibiotic) the PlatE cells tend to detach easily, PBS will cause the cells to detach.
2. After 24 h incubation look at the cells and ensure that the media is not acidic (yellow), if so, add 2–3 ml of media to ensure survival of the cells.
3. The transduction mixture can also be prepared by adding 200 μl of Oct4 supernatant, 200 μl of Klf4 supernatant, 400 μl of media, and 0.4 μl of Polybrene (10 mg/ml), to mimic the viral load of the control treatment.
4. Sometime, better results are achieved by adding MEF media to the cells at this stage and adding the mES cell media with compound on day 6 instead.
5. We sometimes notice that changing the media every 5 days instead of 3 might lead to better results in some cases.
6. It is very unusual to notice colonies that will not turn GFP^+ in these conditions.
7. No more than 4–5 colonies were picked at the same time to ensure that each colony did not remain in trypsin for extended period of time.

Acknowledgments

We would like to thank several people that were instrumental in the development of this method: Yan Shi, Jeong Tae Do, Heung Sik Hahm, and Hans R. Schöler.

References

1. Hanna, J., Markoulaki, S., Schorderet, P., Carey, B. W., Beard, C., Wernig, M., Creyghton, M. P., Steine, E. J., Cassady, J. P., Foreman, R., Lengner, C. J., Dausman, J. A., and Jaenisch, R. (2008) Direct reprogramming of terminally differentiated mature B lymphocytes to pluripotency. Cell 133, 250–264.
2. Takahashi, K., Tanabe, K., Ohnuki, M., Narita, M., Ichisaka, T., Tomoda, K., and Yamanaka, S. (2007) Induction of pluripotent stem cells from adult human fibroblasts by defined factors. Cell 131, 861–872.
3. Takahashi, K., and Yamanaka, S. (2006) Induction of pluripotent stem cells from mouse embryonic and adult fibroblast cultures by defined factors. Cell 126, 663–676.
4. Yu, J., Vodyanik, M. A., Smuga-Otto, K., Antosiewicz-Bourget, J., Frane, J. L., Tian, S., Nie, J., Jonsdottir, G. A., Ruotti, V., Stewart, R., Slukvin, II, and Thomson, J. A. (2007) Induced pluripotent stem cell lines derived from human somatic cells. Science 318, 1917–1920.
5. Aoi, T., Yae, K., Nakagawa, M., Ichisaka, T., Okita, K., Takahashi, K., Chiba, T., and Yamanaka, S. (2008) Generation of pluripotent stem cells from adult mouse liver and stomach cells. Science 321, 699–702.
6. Nakagawa, M., Koyanagi, M., Tanabe, K., Takahashi, K., Ichisaka, T., Aoi, T., Okita, K., Mochiduki, Y., Takizawa, N., and Yamanaka, S. (2008) Generation of induced pluripotent stem cells without Myc from mouse and human fibroblasts. Nat Biotechnol 26, 101–106.
7. Okita, K., Nakagawa, M., Hyenjong, H., Ichisaka, T., and Yamanaka, S. (2008) Generation of mouse induced pluripotent stem cells without viral vectors. Science 322, 949–953.
8. Stadtfeld, M., Nagaya, M., Utikal, J., Weir, G., and Hochedlinger, K. (2008) Induced pluripotent stem cells generated without viral integration. Science 322, 945–949.
9. Kaji, K., Norrby, K., Paca, A., Mileikovsky, M., Mohseni, P., and Woltjen, K. (2009) Virus-free induction of pluripotency and subsequent excision of reprogramming factors. Nature 458, 771–775.
10. Woltjen, K., Michael, I. P., Mohseni, P., Desai, R., Mileikovsky, M., Hamalainen, R., Cowling, R., Wang, W., Liu, P., Gertsenstein, M., Kaji, K., Sung, H. K., and Nagy, A. (2009) piggyBac transposition reprograms fibroblasts to induced pluripotent stem cells. Nature 458, 766–770.
11. Zhou, H., Wu, S., Joo, J. Y., Zhu, S., Han, D. W., Lin, T., Trauger, S., Bien, G., Yao, S., Zhu, Y., Siuzdak, G., Scholer, H. R., Duan, L., and Ding, S. (2009) Generation of induced pluripotent stem cells using recombinant proteins. Cell Stem Cell 4, 381–384.
12. Blelloch, R., Wang, Z., Meissner, A., Pollard, S., Smith, A., and Jaenisch, R. (2006) Reprogramming efficiency following somatic cell nuclear transfer is influenced by the differentiation and methylation state of the donor nucleus. Stem Cells 24, 2007–2013.
13. Kubicek, S., O'Sullivan, R. J., August, E. M., Hickey, E. R., Zhang, Q., Teodoro, M. L., Rea, S., Mechtler, K., Kowalski, J. A., Homon, C. A., Kelly, T. A., and Jenuwein, T. (2007) Reversal of H3K9me2 by a small-molecule inhibitor for the G9a histone methyltransferase. Mol Cell 25, 473–481.
14. Shi, Y., Do, J. T., Desponts, C., Hahm, H. S., Scholer, H. R., and Ding, S. (2008) A combined chemical and genetic approach for the generation of induced pluripotent stem cells. Cell Stem Cell 2, 525–528.
15. Schramm, M., Thomas, G., Towart, R., and Franckowiak, G. (1983) Novel dihydropyridines with positive inotropic action through activation of Ca2+ channels. Nature 303, 535–537.
16. Shi, Y., Desponts, C., Do, J. T., Hahm, H. S., Scholer, H. R., and Ding, S. (2008) Induction of pluripotent stem cells from mouse embryonic fibroblasts by Oct4 and Klf4 with small-molecule compounds. Cell Stem Cell 3, 568–574.
17. Mikkelsen, T. S., Hanna, J., Zhang, X., Ku, M., Wernig, M., Schorderet, P., Bernstein, B. E., Jaenisch, R., Lander, E. S., and Meissner, A. (2008) Dissecting direct reprogramming through integrative genomic analysis. Nature 454, 49–55.
18. Huangfu, D., Maehr, R., Guo, W., Eijkelenboom, A., Snitow, M., Chen, A. E., and Melton, D. A. (2008) Induction of pluripotent stem cells by defined factors is greatly improved by small-molecule compounds. Nat Biotechnol 26, 795–797.
19. Huangfu, D., Osafune, K., Maehr, R., Guo, W., Eijkelenboom, A., Chen, S., Muhlestein, W., and Melton, D. A. (2008) Induction of pluripotent stem cells from primary human fibroblasts with only Oct4 and Sox2. Nat Biotechnol 26, 1269–1275.
20. Takahashi, K., Okita, K., Nakagawa, M., and Yamanaka, S. (2007) Induction of pluripotent stem cells from fibroblast cultures. Nat Protoc 2, 3081–3089.

Chapter 14

Reprogramming of Committed Lymphoid Cells by Enforced Transcription Factor Expression

Huafeng Xie, Catherine V. Laiosa, and Thomas Graf

Abstract

Reprogramming of committed cells from one lineage to another is possible in the hematopoietic system using enforced expression of transcription factors. Here we describe methods to convert committed B and T progenitors into macrophages. In order to obtain a labeled population of starting cells, we employ a lineage ancestry system using a cross between lineage-specific Cre recombinase mice and Rosa26 reporter mice. After infection of these well-defined cell populations with a transcription factor containing retroviral vector that also harbors an infection marker, cells are cultured under conditions permissive for both lymphoid and myeloid development. Multicolor flow cytometry is then used to monitor changes in marker expression on the cell surface reflecting changes in cellular identity. These protocols may be modified to trace cellular reprogramming induced by other transcription factors and in other cellular contexts.

Key words: Reprogramming, Hematopoiesis, Transcription factor, Retrovirus, Lineage commitment

1. Introduction

Hematopoietic stem cells (HSCs) are capable of differentiating into at least ten different types of hematopoietic cells including B and T lymphocytes, erythrocytes, megakaryocytes, granulocytes, and macrophages (1, 2). Multiple developmental pathways have been identified by which HSCs and other multipotent hematopoietic progenitors diversify into distinct lineages (3). Transcription factors have emerged as key determinants in this diversification, altering cell fates by establishing lineage-specific gene expression programs and repressing lineage inappropriate programs (3, 4). As revealed by gain and loss of function approaches, transcription factors are capable of imposing alternative fates on otherwise restricted progenitors and reprogramming lineage committed cells (5–7).

S. Ding (ed.), *Cellular Programming and Reprogramming: Methods and Protocols*, Methods in Molecular Biology, vol. 636,
DOI 10.1007/978-1-60761-691-7_14,

These findings suggest that the balance between these instructive factors in multipotent progenitors drives lineage choice. In addition, enforced expression of transcription factors using multiple traceable viruses enables the elucidation of synergistic and antagonistic transcription factors interactions during cell fate determination. Here we describe protocols used to reprogram committed B and T lymphocytes into myeloid-like cells by enforced expression of myeloid transcription factors C/EBPα and/or PU.1 (8, 9). These protocols can serve as a general template for designing new experiments in which a defined hematopoietic cell population is isolated, exposed to exogenous factors in vitro and then monitored for phenotypic changes using flow cytometry or other techniques.

For both B and T cell experimental systems, the following basic outline was followed: (1) transient transfection of packaging cell lines with retroviral vectors to produce ecotropic retroviral supernatants; (2) isolation of cell population of interest from bone marrow, spleen, or thymus of mice; (3) infection of primary cells of interest with retroviruses; (4) coculture of infected primary cells with stromal cells; and (5) analysis of infected cells using multicolor flow cytometry. For most experiments, we isolated cell populations from "lineage ancestry" mice, allowing the starting cell population to be traced throughout the reprogramming process. For B cell experiments, to generate labeled B lineage cells, we used CD19-Cre Rosa26R-EYFP mice in which CD19 drives the expression of Cre recombinase leading in the bone marrow or spleen to the irreversible activation of EYFP after excision of a stop cassette at the constitutive *ROSA26R* locus (10–12). For T cell experiments, we used Lck-Cre Rosa26R-EYFP mice that label T cell progenitors in the thymus at the pro-T (DN3/DN4) stage of differentiation (10, 12, 13).

2. Materials

2.1. Transient Transfection of Packaging Cell Lines with Retroviral Vectors to Produce Ecotropic Retroviral Supernatants

1. Phoenix ecotropic packaging cells (14).
2. Retroviral vectors expressing gene of interest and co-infection marker (Fig. 1).
3. Cell culture reagents: DMEM (Invitrogen), fetal calf serum (FCS; Hyclone), Penicillin–streptomycin, 1×PBS, Trypsin–EDTA (all for Invitrogen), 10-cm cell culture dishes (BD Falcon).
4. Transfection reagents: Optimem media, Lipofectamine reagent, PLUS reagent (all from Invitrogen).

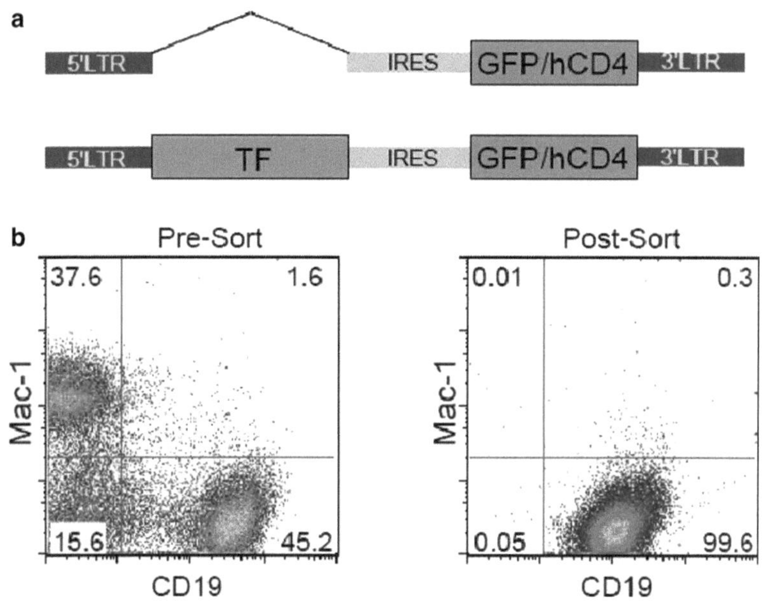

Fig.1. Retrovirus vector constructs and cell purification. (**a**) Constructs of retrovirus vectors used in cell reprogramming assays. 3′ and 5′ LTR; IRES element; virus infection indicator, GFP or truncated human CD4; transcription factor cDNA. (**b**) Purification of CD19+ B lineage cells by MACS. The *left* panel shows an FACS profile of presorted bone marrow sample stained with anti-CD19 (B lineage marker) and Mac-1 (myeloid marker) antibodies. The *right* panel shows an FACS profile after sorting of CD19+ cells. Numbers indicate percentage of cells in the corresponding gate

2.2. Isolation of Cell Population of Interest from Bone Marrow or Thymus of Mice

2.2.1. Lineage Tracer Mice

1. CD19-Cre Rosa26R-EYFP mice.

   $CD19^{Cre/wt}Rosa26R\text{-}EYFP^{fl/wt}$ mice were generated by crossing $CD19^{Cre/Cre}Rosa26R\text{-}EYFP^{fl/fl}$ mice and $Rosa26R\text{-}EYFP^{fl/fl}$ mice (Fig. 2).

2. Lck-Cre Rosa26R-EYFP mice.

   Lck-Cre Rosa26R-EYFP mice were generated by crosses of $Lck\text{-}Cre^{+/+}$ transgenic mice (13) (available at Taconic, also see Note 1 – background differences) and $Rosa26R\text{-}EYFP^{fl/fl}$ knock-in mice to produce $Lck\text{-}Cre^{+/-}$ $Rosa26R\text{-}EYFP^{fl/wt}$ progeny (Fig. 2). Animals were genotyped by PCR as described (10) and in some cases peripheral blood was analyzed by FACS to verify lineage-specific labeling (see Note 2 – FACS phenotyping).

2.2.2. Isolation of B lineage Cells from Bone Marrow

1. AutoMACS (Miltenyi Biotech) or similar instrument.
2. 10-cm culture dish (BD).
3. 15-ml conical tubes (BD).
4. Mortar and Pestle (Sigma).
5. 40 μm cell strainer (BD).
6. PBS (BD/Invitrogen).

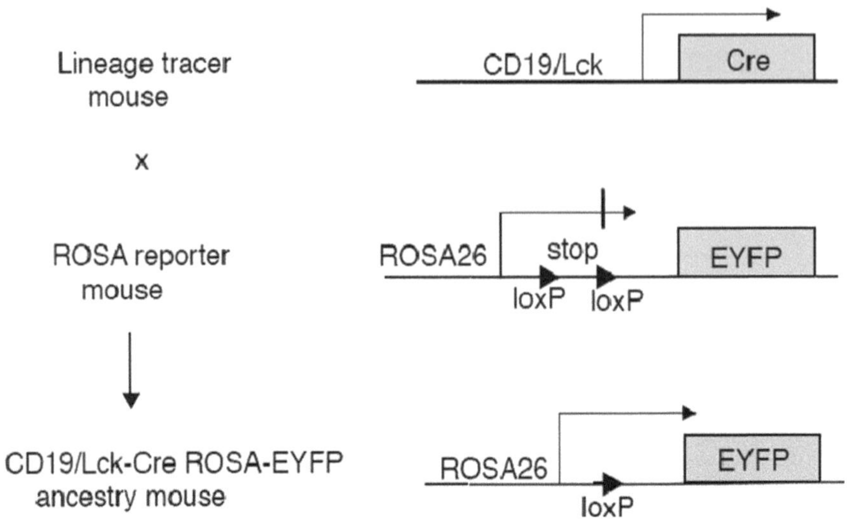

Fig.2. Strategy outlining the generation of lineage ancestry mice. Mice expressing Cre recombinase in either B lineage cells (CD19 Cre, (11)) or T lineage cells (Lck-Cre (13)) were crossed with ROSA26 reporter mice (12). These mice contain a knock-in of EYFP into the ROSA 26 locus whose expression is blocked by a stop cassette flanked by loxP sites. In the resulting lineage tracer mice Cre recombinase activates EYFP in either B or T lineage cells by removing the stop cassette. EYFP, because it is under the control of the ubiquitously expressed ROSA26 promoter, remains expressed even if the cells switch lineages

7. 0.5 M EDTA (BD/Invitrogen).
8. Isolation buffer: 1× PBS, 4% FCS.
9. Fc-Block (BD/eBioscience).
10. Antibodies: Bio-CD19, Streptavidin-PE, PE-CD19, APC-Mac-1 (BD/eBioscience).
11. Streptavidin-microbeads (Miltenyi Biotec).
12. MACS Running buffer.
13. MACS Rinse buffer.
14. 70% Ethanol.

2.2.3. Isolation of Pre-T cells from Thymus

See Subheading 2.2.2.

2.3. Infection of Primary Cells of Interest with Retroviruses

1. 6- or 12-Well tissue culture and non-tissue-culture-treated plates.
2. Retronectin (Takara).
3. Polybrene (Sigma): stock concentration 8 mg/ml (1,000×).
4. Lipofectamine transfection reagent (Invitrogen).
5. PLUS reagent (Invitrogen).

2.4. Coculture of Infected Primary Cells with Stromal Cells

1. B cell cocultures with S17 stromal cells.
2. T cell cocultures with S17 stromal cells or OP9-delta-like 1 stromal cells.
3. Cytokines: SCF, Flt3L, IL-7, M-CSF, IL-3 (all from R&D Systems or Peprotech).

Lyophilized proteins were reconstituted in 1× PBS and 0.1% human BSA and stored at 10 μg/ml at −80°C.

2.5. Analysis of Infected Cells Using Multicolor Flow Cytometry and Fluorescence Microscopy

1. Enzyme-free cell dissociation buffer (Invitrogen).
2. 40 μm cell strainer (BD Falcon).
3. FACS buffer: 1× PBS, 2% FCS.
4. Fluorochrome/biotin-conjugated antibodies (BD Pharmingen).
5. DAPI (Molecular Probes): stock concentration 0.2 mg/ml (1,000×).
6. MoFlo cell sorter (DAKO) or similar instrument.
7. LSRII flow cytometer (Becton Dickinson) or similar instrument.
8. FloJo FACS file analysis software.

3. Methods

3.1. Transient Transfection of Packaging Cell Lines with Retroviral Vectors to Produce Ecotropic Retroviral Supernatants

Retroviral supernatants were prepared by transient transfection of Phoenix ecotropic packaging cells (15). One day in advance, Phoenix ecotropic cells were plated at 5×10^6 cells per 10-cm dish in DMEM media with 10% FCS and without antibiotics. Retroviral vector DNA was prepared using the Wizard PureFection DNA Purification Kit or Endo-free Maxiprep Kit and 10–15 μg of DNA were complexed for transfection with PLUS and Lipofectamine reagents according to the manufacturer's instructions. Complexes were added to the cells in Optimem media or DMEM without serum and incubated for 5 h at 37°C. Media was replaced with DMEM media with FCS and antibiotics and cultured for 48 h. Retroviral supernatants were harvested, filtered through a 45-micron syringe filter, and frozen rapidly on dry ice. Frozen supernatants were stored at −80°C for up to 24 months.)

1. Day 1: Plate Phoenix ecotropic cells at 5×10^6 cells per 10 cm dish in DMEM media with 10% FCS and without antibiotics.
2. Day 0: Transfect plated Phoenix ecotropic cells with retroviral vectors using Lipofectamine reagents and PLUS reagents according to the manufacturer's instructions and incubate for 5 h at 37°C with 5% CO_2.
3. 5 h after transfection replace medium with DMEM media with FCS and antibiotics and culture for 24 h at 37°C with 5% CO_2.

4. Day 1: Transfer the cells to a 32°C incubator with 5% CO_2.
5. Day 2: Harvest retroviral supernatants, filter through a 45-µm syringe filter, and freeze rapidly on dry ice. Store frozen supernatants at –80°C for up to 24 months.
6. Add fresh medium to the cells and incubate for 24 h at 37°C with 5% CO_2 followed by 24 h at 32°C with 5% CO_2.
7. Day 4: Harvest the second batch of virus supernatants, filter, and store at –80°C.

3.2. Isolation of Cell Populations of Interest from Bone Marrow, Spleen or Thymus of Mice

3.2.1. Lineage Tracer Mice

Use of labeled primary cells derived from lineage tracer mice largely eliminated the concern that the resulting reprogrammed cells were derived from a small population of contaminating myeloid progenitors. In addition, it allowed enriched (rather than sorted) populations of cells to be used for experiments since only labeled cells were used for analysis after gating on $EYFP^+$ cells by flow cytometry (described below). If specific populations were desired (such as $CD44^-CD25^+$ DN3 cells), $EYFP^+$ cells were sorted using a Moflo cell sorter.

1. CD19-Cre Rosa26R-EYFP mice.

   CD19-$Cre^{Cre/wt}$ Rosa26R-$EYFP^{fl/wt}$ mice were derived from crosses as described in Subheading 2.
2. Lck-Cre Rosa26R-EYFP mice.

   Lck-$Cre^{+/-}$ Rosa26R-$EYFP^{fl/wt}$ mice were derived from crosses as described in Subheading 2.

3.2.2. Isolation of B Cells from Bone Marrow

$CD19^+$ cells were isolated with autoMACS after labeling them with magnetic microbeads. This method enables enrichment of a large amount of $CD19^+$ cells in a relative short time with a purity close to pure. Figure 1b shows FACS plots of bone marrow cells before and after enrichment for $CD19^+$ cells in a typical experiment.

1. Euthanize with CO_2 5-week-old $CD19^{Cre/wt}$Rosa26R-$EYFP^{fl/wt}$ mice.
2. Cut open the skin from the belly, tear away the muscle from the hind legs and collect the bones in PBS on ice.
3. Pool the hind leg bones and transfer to a sterile mortar. Add 1 ml of isolation buffer and grind the bone with prechilled mortar and pestle.
4. Add 10 ml of isolation buffer to the mortar and pipette for 10–15 times to make a single cell suspension.
5. Transfer the cells to 40 µm filters sitting on a 50-ml conical tube. Wash the mortar once with 10 ml isolation buffer and transfer to the filter.
6. Spin the cells down at 4°C at 300 × g for 7 min.

7. Aspirate the supernatant and wash the cell pellet once with 10 ml of isolation buffer. Pass the cells through a 40-μm filter again and collect the cells in 15-ml conical tubes.
8. Count the cells using hemocytometer.
9. Spin down the cells and resuspend the pellet in 3×10^7 cells/ml.
10. Save some total BM cells as controls for FACS analysis of sorted cells and for a control culture of total bone marrow cells. To the rest of the cells add 0.1 μg Fc-block (anti-CD16/CD32)/million cells and incubate for 10 min on ice.
11. Directly add 0.1 μg Biotin-CD19 antibody/million cells to the cell suspension and mix well. Incubate on ice for 20 min with occasional mixing.
12. Wash the cells twice with PBS plus 2 mM of EDTA. Filter the cells through 40 μm cell strainer at the second wash.
13. Spin down the cells at 4°C at $300 \times g$ for 7 min and resuspend the pellet in 10^7 cells/90 μl of PBS with 2 mM EDTA.
14. Add 10 μl of Streptavidin-microbeads per ten million cells to the cell suspensions. Mix well and incubate in a 4°C refrigerator for 20 min. Mix once during the incubation.
15. Wash the cells twice with isolation buffer. Filter the cells through 40 μm cell strainer at the second wash.
16. Pellet the cells at 4°C at $300 \times g$ for 7 min and resuspend them at 2×10^8 cells/ml in isolation buffer.
17. Run the cells in an autoMACS using the program POSSELD, discarding the negative fraction from outlet port neg 1 and collect cells from outlet port pos 2, which are CD19 positive cells.

3.2.3. Isolation of Pre-T Cells from Thymus

Thymi were dissected from 4- to 5-week-old mice and immediately rinsed in isolation buffer to remove any RBCs. Thymocyte suspensions were prepared by pressing the organ through a 70-micron strainer premoistened with 10 ml of isolation buffer, using a syringe plunger. Single cell suspensions were obtained by repeated pipetting of cells followed by passage through a 40-micron strainer. To avoid cell clumping during centrifugation, an equal volume of culture media containing FCS was added and the centrifugation was performed at 4°C. Red blood cell lysis was not routinely performed on thymocyte preparations since most RBCs are eliminated by rinse step. Cells were resuspended in FACS buffer for staining. Double negative (DN) T cells were enriched from thymocyte suspensions from Lck ancestry mice by lineage depletion by magnetic-activated cell sorting (MACS) using biotin-conjugated antibodies (Abs) for lineage markers (see Note 4), Streptavidin-microbeads, and an autoMACS cell separator using the DEPLETES program. If autoMACS is not available, lineage positive cells may also be depleted using biotin-conjugated Abs and SpinSep dense particles according to the manufacturer's instructions. Either technique produces a DN-enriched sample which can be used for further experiments or subsequent rounds or purification. If pure populations are desired (for example, for analysis of individual DN subsets or when using thymocytes derived from nonancestry mice), then further cell sorting of enriched samples can be carried out by labeling the cells with fluorochrome-conjugated antibodies for Streptavidin (for remaining lineage$^+$ cells), c-kit, CD44 and CD25 and sorting each subset individually.

3.3. Infection of Primary Cells of Interest with Retroviruses

All retroviral infections were performed by combining primary cells with ecotropic retroviral supernatants on plates precoated with Retronectin at 25 μg/ml. Five hours prior to infection, Retronectin-coated plates were preloaded with 2 ml of retroviral supernatant and incubated at 32°C. For infection, cells were resuspended at $5–20 \times 10^5$ cells/ml in complete RPMI medium containing 10% FCS, 2 mM glutamine, 20 mM HEPES, 1× penicillin/streptomycin, and 5.5×10^{-5} mM b-mercaptoethanol and mixed at a 1:1 ratio with retroviral supernatants. Polybrene was added at a final concentration of 8 μg/ml. After aspiration of retroviral supernatants from the Retronectin-coated plates the cell–virus mixture was deposited onto preloaded plates. The plates were centrifuged at 700 × *g* for 60 min and then incubated at 32°C for 3 h. After infection, cells were removed by gentle pipetting, washed with medium, and resuspended for culture at $1–5 \times 10^5$ cells/ml in RPMI containing SCF, Flt3L, IL-7, IL-3, and M-CSF (5–10 ng/ml). Cells were deposited onto stromal cell layers (S17, OP9, or OP9DL1, see below) and incubated at 37°C.

1. Coat non-tissue-culture-treated six-well plates with Retronectin at 25 μg/ml according to the manufacturer's instructions.
2. Five hours prior to infection, preload Retronectin-coated plates with 2 ml of retroviral supernatant and incubate at 32°C.
3. Resuspend the purified cells in up to 5×10^5 cells/ml in RPMI containing 10% FCS, 2 mM glutamine, 20 mM HEPES, 1× penicillin/streptomycin, 5.5×10^{-5} mM – mercaptoethanol and cytokines.
4. Aspirate the virus supernatant from the Retronectin-treated plates and add 2 ml of cell suspension to each well of the virus preloaded plate.
5. To each well add 2 ml of virus supernatant.
6. Add polybrene to a final concentration of 8 μg/ml to the wells.
7. Mix well by gently pipetting. Seal the plate with tape and centrifuge at 700 × *g* for 60 min.

8. Remove the tape and transfer the plate to a 32°C incubator and incubate for 3–6 h.
9. Remove the cells by gentle pipetting and transfer to a 15-ml conical tube.
10. Pellet the cells at 4°C at 300 × *g* for 7 min.
11. Wash once with medium, resuspend at $1–5 \times 10^5$ cells/ml in RPMI containing SCF, Flt3L, IL-7, IL-3, and M-CSF (5–10 ng/ml) and seed in culture.
12. Deposit cells onto stromal cell layers and incubate at 37°C until they are harvested for analysis by gentle pipetting.

3.4. Coculture of Infected Primary Cells with Stromal Cells

3.4.1. B Cell Cocultures with S17 Stromal Cells (16)

1. S17 stromal cells were cultured in MEM medium containing 5% FCS, 2 mM glutamine, and 1× penicillin/streptomycin.
2. Irradiate (with 3,000 rad) S17 cells and plate at 2×10^5 cells/well (12-well plate) or 5×10^5 cells/well (6-well plate). Aliquots of irradiated cells can be frozen in liquid nitrogen for future use and thawed 1 day prior to setting up reprogramming experiment.
3. Add up to 5×10^5 virus-infected cells to a pair of duplicate wells in a 12-well plate. Add 2.5×, 1.25×, 0.625×, 0.625×10^5 cells to duplicate well pairs 2–5. Wells with the most cells are analyzed 48 h after plating, the rest at 24 h intervals.
4. Aspirate 0.8 ml of media from cultures at day 3 and replenish with 1 ml complete growth medium.

3.4.2. T Cell Cocultures with S17 Stromal Cells or OP9-Delta-Like 1 Stromal Cells (17)

S17 stromal cells were prepared exactly as described for B cells. For OP9-delta-like 1 stromal cells, cells were cultured in αMEM medium containing 20% FCS, 2 mM glutamine, and 1× penicillin/streptomycin. For cocultures, OP9-DL1 cells were plated at 1×10^5 cells/well (12-well plate).

3.5. Analysis of Infected Cells Using Multicolor Flow Cytometry and Fluorescence Microscopy

To harvest cells from cocultures, remove nonadherent cells by gently repeated pipetting of media. Add 1 ml of enzyme-free cell dissociation buffer to plate containing adherent cells and incubate at 37°C for 5–7 min. Add 2 ml of culture media and collect cells with repeated pipetting and passing them through a 40-micron cell strainer prior to centrifugation at 300 × *g* for 7 min.

3.5.1. Cell Harvesting from Cocultures

1. Remove nonadherent cells by gently repeated pipetting of culture media and pass through a 40-μm filter.
2. Add 1 ml of enzyme-free cell dissociation buffer to wells containing adherent cells and incubate at 37°C for 5–10 min until the stromal cells dislodge from the well.
3. Break up cell clumps by repeated pipetting. Collect the cells and pass them through a 40-μm filter.

4. Wash the well once with staining buffer and pool the cells.
5. Spin down the cells at 300 × *g* for 7 min.
6. Resuspend the pellet in 100 μl of staining buffer with 0.1 μg Fc-block.

3.5.2. Antibody Labeling

Antibody labeling was performed in 100 μl of FACS buffer for samples containing less than 2×10^6 cells. For larger cell numbers, cells were resuspended at 10^7 cells per 500 ml. Prior to antibody staining, cells were treated with Fc-block and incubated on ice for 10 min. Fluorescence-conjugated antibodies were added at a predetermined concentration and incubated on ice (protected from light) for 20–30 min. Staining was followed by a wash in a large volume (at least fivefold) of FACS buffer and centrifugation at 300 × *g* for 7 min. Prior to FACS analyzes, dead cells were excluded by DAPI staining at a final concentration of 0.2 μg/ml.

1. Resuspend the pellet in 100 μl of staining buffer with 0.1 μg Fc-block and incubate 10 min on ice.
2. Add desired antibodies at predetermined concentrations to the cell suspensions. Mix well and incubate on ice for 20 min with occasional vortex.
3. Wash the cells with FACS buffer with at least ten times the volume of cells suspension.
4. Pellet the cells by centrifugation at 300 × *g* for 7 min. Resuspend the pellet with 200–300 μl of FACS buffer. Leave the cells on ice until analysis.
5. Prior to FACS analyzes, exclude dead cells by DAPI staining, using a final concentration of 0.2 μg/ml.

3.5.3. Flow Cytometry

Analytical flow cytometry was performed using an LSRII cytometer (Becton Dickinson) configured with 488 nm, 633 nm, and UV lasers. This system allowed for detection of multiple fluorochromes from the 488-nm laser (EGFP or FITC, PE, PerCP, PECy7) and 633-nm laser (APC and APC-Cy7) as well as DAPI staining from the UV laser. For simultaneous detection of EGFP and EYFP, custom filter/dichroic mirror sets (Chroma) were fitted for the 488 nm laser. EYFP was detected using 530LP dichroic mirror with 550/30 filter. EGFP was detected using 505LP dichroic mirror with 510/20 filter. Spectral compensation and data acquisition was carried out using BD Diva software.

3.5.4. Data Analysis

For analysis of reprogramming assay, flow cytometry was used to detect expression of the lineage tracer (EYFP), co-infection markers (EGFP and/or hCD4), and cell surface markers of interest. Figure 3 shows examples of B and T lymphocytes reprogrammed by C/EBPα. Expression of lymphoid and myeloid markers were

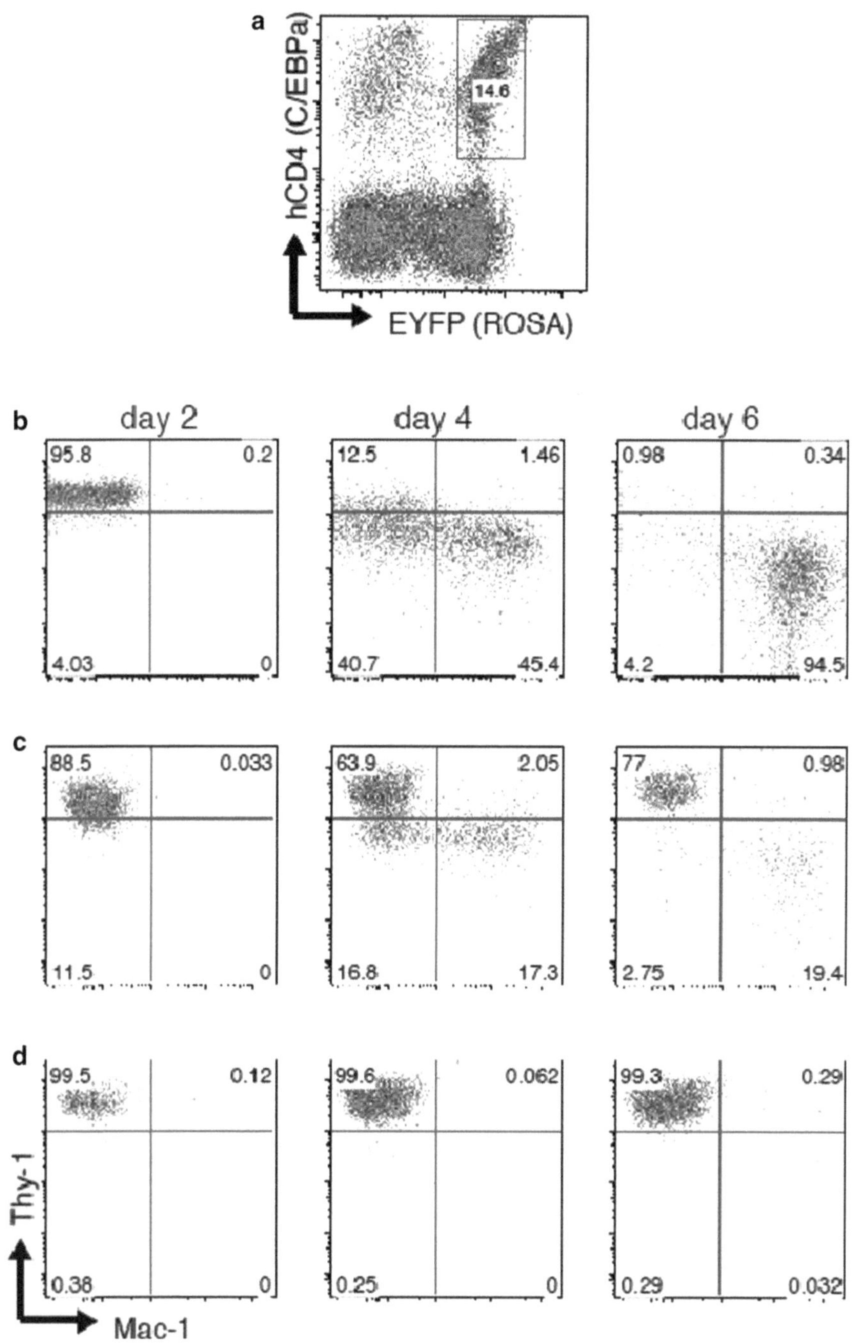

Fig.3. Reprogramming of T cell precursors by C/EBPα and influence of microenvironment. (**a**) FACS profile of C/EBPα virus-infected T cell precursors from T cell lineage tracer mice, gated on the hCD4/EYFP double positive population. This population was further analyzed in (**b**) and (**c**) for the expression of Thy-1 (T cell marker) and Mac-1 (myeloid marker) expression at the indicated days after infection. (**b**) C/EBPα infected T cell precursors cultured on S17 (**b**) and (**c**), on OP9-delta stroma. (**d**) Control virus-infected cells sorted from another infection cultured on OP9 delta stroma. The OP9 delta stroma cells produce a ligand that activates Notch in T cell precursors and partially prevents the induced reprogramming (Data from Laiosa et al. (8))

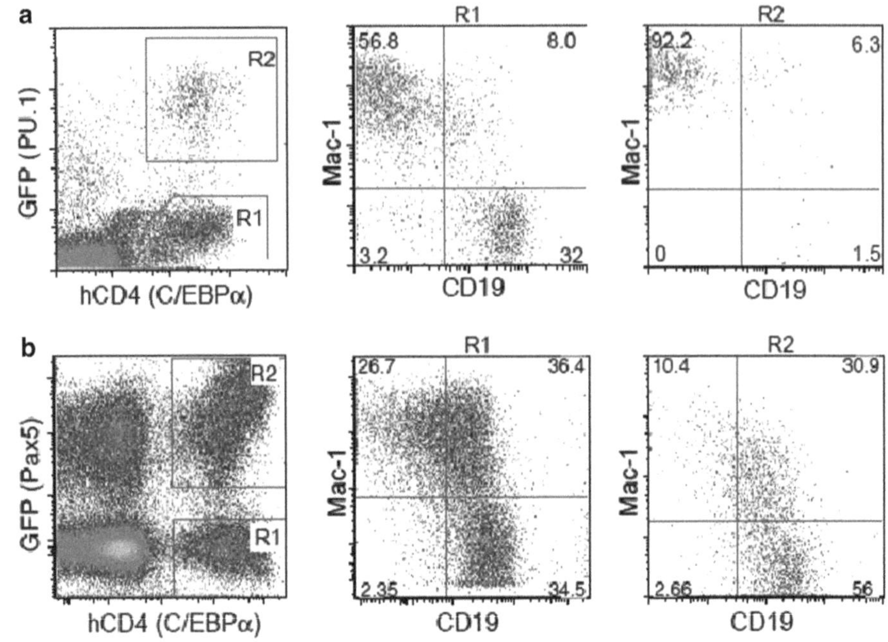

Fig. 4. Synergistic and antagonistic transcription factor interactions during cell reprogramming. (**a**) CD19+ B lineage cells were co-infected with PU.1 and C/EBPα viruses and populations gated that either represent cells infected with only C/EBPα viruses (R1) or both C/EBPα and PU.1 viruses (R2) 4 days after infection. The profiles under R1 and R2 show the expression of CD19 and Mac-1. (**b**) Similar as in (**a**) but comparing cells infected with C/EBPα virus only and cells infected with both C/EBPα and Pax5 viruses 3 days after infection (Data from Xie et al. (9) and unpublished)

analyzed in C/EBPα virus-infected EYFP+ (lineage ancestry marker) population (R1). Gates were set based on antigen expression of control virus-infected cells or by using fluorescence minus one (FMO) controls stained with isotype control antibodies. All data analyzes were performed using FlowJo software (Tree Star).

3.6. An Assay for the Detection of Synergistic and Antagonistic Transcription Factor Interactions

Using the protocols described above, the combined expression of retroviruses carrying different transcription factors that are traceable by coexpressed infection markers, permits to detect both synergistic and antagonistic interactions between two factors. An example of such an experiment is shown in Fig. 4, in which B cells from the bone marrow were co-infected with either C/EBPα-hCD4 virus plus PU.1-GFP virus, or with C/EBPα-hCD4 virus plus Pax5-GFP virus. The FACS analysis demonstrates that the effect of C/EBPα in reprogramming B cells is enhanced by co-infection with PU.1 (day 4 data was shown in the figure) (9) and inhibited by co-infection with the B cell master regulator Pax5 (day 3 data was shown in the figure).

4. Notes

1. For Lck ancestry mice, differences in timing of EYFP labeling of DN subsets was seen when using Lck-Cre mice derived from mixed or C57BL/6 backcrossed animals (available from Taconic). In mixed backgrounds, up to 70% of DN3 cells are labeled with EYFP while that number decreases to 10–20% in C57BL/6 backcrossed animals. In both backgrounds, more than 80% of DN4 cells are labeled with EYFP.
2. For unknown reasons, the Lck-Cre transgene is activated in the germline or early blastocyst of some animals, leading to the EYFP labeling in multiple tissues when animals were crossed with Rosa26R-EYFP mice. In crosses between Lck-$Cre^{+/+}$ transgenic mice and Rosa26R-$EYFP^{ki/ki}$ knock-in mice this was seen in less than 10% of animals screened. To eliminate these animals, peripheral blood samples were screened by FACS to ensure labeling of only $CD4^{+}$ and $CD8^{+}$ cells in the periphery.
3. Lineage antibodies used to deplete thymocyte suspensions of mature lineage cells were as follows: CD3 (clone 145-2C11, T cells), CD4 (clone GK1.5, T cells), CD8α (clone 53-6.7, T cells), B220 (clone RA3-6B2, B cells), CD19 (clone 1D3, B cells), Mac-1/CD11b (clone M1/70, macrophages), Gr-1 (clone RB6-8C5, granulocytes), Ter119 (TER-119, erythroid cells), I-A/I-E (clone 2G9, dendritic cells).
4. When higher infection efficiency is required, cells can be infected with the retrovirus supernatant for two rounds. Carefully aspirate the virus/media mixture without removing the cells and add 2 ml of fresh medium with cytokine. Incubate the cells in 37 for 3–6 h before adding 2 ml of retrovirus supernatant for a second round of infection.
5. For infection of cells with more than one virus, preload Retronectin-coated plate with the same volume of each virus and incubate for 3–5 h at 32°C before removing the supernatant. The total volume of virus suspension should not exceed the volume of the cell suspension.

References

1. Morrison, S.J., N. Uchida, and I.L. Weissman, (1995) *The biology of hematopoietic stem cells.* Annu Rev Cell Dev Biol, **11**: p. 35-71.
2. Orkin, S.H. and L.I. Zon, (2008) *Hematopoiesis: an evolving paradigm for stem cell biology.* Cell, **132**(4): p. 631–44.
3. Orkin, S.H., (2000) *Diversification of haematopoietic stem cells to specific lineages.* Nat Rev Genet, **1**(1): p. 57–64.
4. Sieweke, M.H. and T. Graf, (1998) *A transcription factor party during blood cell differentiation.* Curr Opin Genet Dev, **8**(5): p. 545–51.
5. Cobaleda, C. and M. Busslinger, (2008) *Developmental plasticity of lymphocytes.* Curr Opin Immunol, **20**(2): p. 139–48.
6. Graf, T., (2002) *Differentiation plasticity of hematopoietic cells.* Blood, **99**(9): p. 3089–101.

7. Akashi, K., (2005) *Lineage promiscuity and plasticity in hematopoietic development.* Ann N Y Acad Sci, **1044**: p. 125–31.
8. Laiosa, C.V., et al., (2006) *Reprogramming of committed T cell progenitors to macrophages and dendritic cells by C/EBP alpha and PU.1 transcription factors.* Immunity, **25**(5): p. 731–44.
9. Xie, H., et al., (2004) *Stepwise reprogramming of B cells into macrophages.* Cell, **117**(5): p. 663–76.
10. Ye, M., et al., (2003) *Hematopoietic stem cells expressing the myeloid lysozyme gene retain long-term, multilineage repopulation potential.* Immunity, **19**(5): p. 689–99.
11. Rickert, R.C., J. Roes, and K. Rajewsky, (1997) *B lymphocyte-specific, Cre-mediated mutagenesis in mice.* Nucleic Acids Res, **25**(6): p. 1317–8.
12. Srinivas, S., et al., (2001) *Cre reporter strains produced by targeted insertion of EYFP and ECFP into the ROSA26 locus.* BMC Dev Biol, **1**: p. 4.
13. Lee, P.P., et al., (2001) *A critical role for Dnmt1 and DNA methylation in T cell development, function, and survival.* Immunity, **15**(5): p. 763–74.
14. Kinsella, T.M. and G.P. Nolan, (1996) *Episomal vectors rapidly and stably produce high-titer recombinant retrovirus.* Hum Gene Ther, 7(12): p. 1405–13.
15. Swift, S., et al., (2001) *Rapid production of retroviruses for efficient gene delivery to mammalian cells using 293T cell-based systems.* Curr Protoc Immunol, **Chapter 10**: p. Unit 10 17C.
16. Collins, L.S. and K. Dorshkind, (1987) *A stromal cell line from myeloid long-term bone marrow cultures can support myelopoiesis and B lymphopoiesis.* J Immunol, **138**(4): p. 1082–7.
17. Schmitt, T.M., et al., (2004) *Induction of T cell development and establishment of T cell competence from embryonic stem cells differentiated in vitro.* Nat Immunol, **5**(4): p. 410–7.

Chapter 15

Reprogramming of B Cells

César Cobaleda

Abstract

Cellular reprogramming is an interplay between the original starting cell's plasticity and the (epi)genetic mechanisms used to drive this cell towards a new fate. Our capacity to reprogram mature cells into progenitors thus greatly depends on the inherent physiological plasticity of the initial cell. B lymphocytes possess a high degree of plasticity revealed both during their normal development and under experimental conditions in the laboratory. In this chapter, we discuss the biology of B cell plasticity in the context of physiology and pathology and we provide a specific practical example of this plasticity in a protocol describing the dedifferentiation of mature B cells into multipotential progenitors that can afterwards be reprogrammed into alternative lineages like T cells or macrophages.

Key words: B cells, Reprogramming, Plasticity, Dedifferentiation, Pax5, Stem cells

1. Introduction

1.1. The Many Ways to Reprogramming

Plasticity is a pre-requisite for reprogramming. If cell fate was fixed once it has been established, it would then be impossible to alter developmental programs, either in a "desirable" way (experimental reprogramming for regenerative medicine) or in an unwanted one (cancer is, after all, a deviation from the normal physiological developmental path). The recent advances in reprogramming technology (1–3) are forcing a re-definition of the concept of plasticity itself; if any cell can be reprogrammed under certain conditions, then any cell could be potentially considered as "plastic". In population genetics, phenotypic plasticity is the capacity of an organism with a given genotype to change its phenotype in response to changes in the environment (4). One of the key points in the definition of physiological (or cell-intrinsic) plasticity relies precisely in the capacity of changing cell fate without altering the genotype. In this context, the importance of

S. Ding (ed.), *Cellular Programming and Reprogramming: Methods and Protocols*, Methods in Molecular Biology, vol. 636, DOI 10.1007/978-1-60761-691-7_15, © Springer Science+Business Media, LLC 2010

epigenetic modifications in maintaining cell identity adds a new layer of complexity to the definition of plasticity, and it becomes difficult to separate epigenetic identity from the cellular one, since both are aspects of the same reality, i.e. the phenotype. In this way, any reprogramming will involve epigenetic reprogramming, since it implies the change towards a new cellular identity. Broadly speaking, there are four different approaches to achieve cellular reprogramming:

1. Reprogramming by nuclear transfer (5–7) does not involve changes at the genetic level but, on the other side, the whole nucleus is taken away from its normal environment, in such a way that the new cellular phenotype is "imposed" on the nucleus, thus forcing it to change accordingly its patterns of gene expression to again "correspond" to its new milieu. This method does not in principle require knowledge of the molecular mechanisms that are responsible for the reprogramming process.
2. Reprogramming differentiated cells to induced-pluripotency stem cells (iPS cells) has been achieved by genetically altering the target cells by introduction of stem cell-specific genes that take the control and can transdifferentiate a differentiated cell into a pluripotent one (1, 3, 8, 9). More recently, systems have been developed that either avoid stable genetic changes in target cells (transient vectors) or directly make use of drugs that can help to force reprogramming, reducing the need of genetic transfer (3, 10–13). This approach relies on the knowledge acquired about the factors that are essential for the establishing and maintenance of the stem cell characteristics.
3. Lineage reprogramming by ectopic expression of transcription factors, without the need of reverting cells back to a pluripotent stage (14). This has been achieved in several systems, like the transdifferentiation of committed B lymphocytes to macrophages mediated by C/EBPα (15) (see below) or the conversion of adult pancreatic exocrine cells to insulin-secreting β cells after in vivo adenoviral delivery of *Ngn3*, *Pdx1*, and *Mafa* transcription factors (14, 16). This method requires an accurate understanding of the transcriptional mechanisms in charge of maintaining the identity and function of both the starting cells and the cells that one intends to obtain at the end of the reprogramming process.
4. Lineage reprogramming by loss of transcription factors. All the above-mentioned reprogramming approaches are based on the ectopic expression of reprogramming factors. These factors must establish the new epigenetic program and remove at the same time the original one. This implies that the deletion

of factors that maintain the original epigenetic state of the starting cells must facilitate their reprogramming into new cell types. This is the case of the protocol provided in this chapter, where the deletion of a single factor responsible for maintaining the identity and function of a given cell type (the transcription factor *Pax5* in B cells) allows the cells to revert to the status of an earlier multipotential progenitor (17), as is discussed in detail below.

1.2. B Cells as the Starting Material: Plasticity During Normal B Cell Development

Whatever the way it can be achieved, it is clear that the medical potential of cellular reprogramming for regenerative medicine is enormous, and that the ideal starting cells must be as plastic as possible (i.e. more easy to convert into other cells types with a minimal amount of manipulation). From this perspective, a concept that can be more objectively evaluated is that of physiological plasticity, i.e. the capacity of a cell type to give rise to other specialized cells different from itself, under normal physiological conditions, as a part of their normal developmental programs in the organism. From this point of view, stem cells (each one with its different specific potential, from multipotent to unipotent) are plastic, while terminally differentiated cells are not. But between these two extremes there are several intermediates whose intrinsic biological properties can make them more or less prone to change their fate.

B cells are a perfect example of different degrees of physiological plasticity during normal development (besides the plasticity that can be artificially "forced" in the laboratory) (18). Haematopoietic stem cells (HSCs) in the bone marrow (BM) have the capacity to generate all blood cell types. They initially give rise to multipotent progenitors that partially differentiate into common myeloid progenitors (CMPs) or lymphoid-primer multipotent progenitors (LMPPs). Erythrocytes and megakaryocytes arise mainly from CMPs, while myeloid development proceeds either form CMPs or LMPPs (19). The early lymphocyte progenitors (ELPs) arise from LMPPs upon the expression of the recombination activating genes *Rag1* and *Rag2*. From ELPs, common lymphoid progenitors will be generated that can give rise to T, B, or NK cells (20, 21). The plasticity of these uncommitted progenitors is dependent on their capacity to express basal levels of lineage-specific genes in a process known as multilineage priming (22, 23). This expression allows the progenitors to respond to inductive cues that would take them further into specific lineages. In the case of B cells, signalling is provided by IL7, in combination with the three transcription factors E2A, EBF1, and Pax5 (24–26). E2A and EBF1 activate the expression of B lymphoid genes at the onset of B cell development but the commitment to the lineage is controlled by Pax5, which possesses the dual capacity of repressing the transcription of B-lineage-inappropriate genes and activating the expression of B-cell-specific genes (27, 28). This reprogramming of gene expression restricts the broad developmental capacity of uncommitted progenitors to the B cell pathway. Following Pax5 expression, pro-B cells are only able to differentiate along their unidirectional path to mature B cells. This dual action of Pax5 was demonstrated in seminal experiments with pro-B cells derived from *Pax5* knockout mice (29). In these mice, B cell development is blocked at the pro-B cell stage, and these *Pax5*$^{-/-}$ pro-B cells can be grown in culture in the presence of IL7 and stromal feeder cells. It was shown that *Pax5*$^{-/-}$ pro-B cells behave as multipotent progenitors because they express multilineage genes that allow them to be programmed into most of the haematopoietic lineages under the appropriate conditions. All these developmental options are shut down by reintroduction of Pax5 that actively represses non-B cell genes (25, 29).

Developmental plasticity is thus present at the earliest stages of B cell differentiation, as a consequence of the necessary requirements for lineage commitment. The plasticity of CLPs can also have additional physiological roles. For example, intraperitoneal injection of LPS into mice leads to the depletion of B cells in BM with corresponding increases in macrophage and DC numbers (30). CLPs express high levels of the Toll-like receptor 2 (TLR2) and lower levels of TLR4. In this way, the TLR2 ligand Pam3CSK4 or TLR4 ligand LPS can efficiently redirect the developmental fate of CLPs towards DC differentiation, even under lymphoid culture conditions (SCF, Flt3L, and IL-7), which normally promote B cell development. CLPs can therefore directly sense microbial products by TLR signalling, altering their normal differentiation potential to participate in the rapid replenishment of innate immune cells during infection (30).

At the other end of B cell differentiation, the *raison d'etre* of mature IgM^{+}IgD^{+} B cells is to become plasma cells upon encounter with the corresponding specific antigen. In order for this terminal differentiation to happen, *Pax5* and its transcription program must be down-regulated (27, 28, 31). This process is initiated by the engagement of membrane BCR with its specific antigen, triggering a signalling cascade that finally leads to the up-regulation of the master transcriptional regulator of the plasma cell identity, Blimp1 (32, 33). Mature B cells and plasma cells have very different transcriptional programs, which are controlled in a mutually exclusive manner by Pax5 and Blimp1, respectively (27, 28, 31, 34). Many of the genes that are expressed in plasma cells are also transcribed in uncommitted lymphoid progenitors, and are not compatible with B cell development or function. As these genes are required again for terminal differentiation into plasma cells, they cannot be irreversibly repressed by stable epigenetic modifications. This is the function of Pax5: On one side, Pax5 maintains B cell transcriptional program and B cell identity and, on the other side, it allows for a simple mechanism (*Pax5* repression) of eliminating this identity when full transcriptional reprogramming is necessary to generate a plasma cell. This molecular explanation is likely the reason why mature B cells retain a high degree of plasticity, and why this is essentially dependent on a single gene, *Pax5*, whose expression is linked to B cell identity. This high degree of plasticity, together with the facts that B cells are a long-lived population in the blood system and they experience a profound reprogramming in order to terminally differentiate into a very different cell type, had lead some to consider B cells (and T cells too) as a special kind of single-lineage (unipotent) stem cells (35).

The plasticity inherent to normal B cell development has been pushed in the laboratory beyond its physiological limits (18) and has allowed to skew the developmental potential of all different stages of differentiation. For example, the latent myeloid differentiation potential of CLPs has allowed re-directing them towards the granulocyte–monocyte or megakaryocyte–erythroid lineages by ectopic cytokine signalling or ectopic transcription factor expression, in examples of lineage diversion (18). At later stages, committed $CD19^+$ B cells can be transdifferentiated into macrophages by retroviral expression of the myeloid transcription factor C/EBPα (15) in the presence of myeloid cytokines.

1.3. The Dark Side of Plasticity: Tumoral Reprogramming

Plasticity, as many biological properties, also has its dark side, which reveals itself in the form of cancerous differentiation. Cancer is a deviation of the normal processes of differentiation in which a new lineage is created, with new properties and characteristics, but to a great degree parallel to normal lineages. According to this view, and as we mentioned at the beginning, if cell fate could not be changed then cancer would not happen. Over the last few years, results have accumulated providing support for the cancer stem cell (CSC) theory that proposes that tumours are stem cell-based tissues. According to this theory, tumours consist of different types of cells. On one side, tumours contain the more differentiated cells that are unable to propagate the cancer. On the other side, there is a variable percentage of cancer-maintaining cells with stem cell properties (CSCs) that are the ultimate responsible for generating the tumour mass and that are the only ones that can transplant the cancer (35–37). The precise percentage of these cells within the tumour will most probably vary considerably among different tumours, from very few cells to many or perhaps even most of the cancer cells. Whatever the percentage, this view of cancer as a stem cell-driven tissue poses the problem of identifying the cancer cell-of-origin, the first one suffering the oncogenic alteration(s) and the one who is going to generate the cancer-maintaining cell. It is a well-known fact that many different biological functions have to be altered (38) in order for the tumours to arise (proliferation potential, apoptosis evasion, etc.). From this point of view, the plasticity of the target cell and its responsiveness to the oncogene reprogramming effects is also a decisive factor in tumour generation. A given differentiated cell can only give raise to tumour cells if the specific cancer-inducing alteration is capable of conferring stem cell capabilities to this precise cell type. This has been shown in the mouse for several human translocations, suggesting that, in human cancers, stemness can be a new attribute that arises because of the oncogene activity in a certain type of mature cell: some oncogenes, like MOZ-TIF2 (39) MLL-AF9 (40, 41), MLL-ENL (42), or MLL-GAS7 (43) are able of conferring stem-like properties to committed target cells, and they can generate CSCs when transfected into committed haematopoietic progenitors. However, other oncogenes, like BCR-ABLp190, are not able to confer these stem properties (39). So, clearly, different routes must exist to allow differentiated cells to revert to a progenitor-like condition under pathological circumstances. One example of this possibility is the β-catenin pathway alteration in granulocyte–macrophage progenitors in chronic myelogenous leukaemia (44). In this context, the oncogenic defect does not have to immediately generate a CSC, but might give rise to a precancerous cell that, depending on the conditions, can rest and be hidden, can differentiate, either normally or aberrantly (like the chronic phase of CML) or can end up generating the CSCs (45). In AML1–ETO leukaemia, the translocation can be detected in normal stem cells in long-term remission, suggesting that they are not leukaemic per se, but some of their cellular descendants do become tumorigenic with time (46). These pre-leukaemic cells have also been found in TEL–AML1-associated childhood pre-B acute lymphoblastic leukaemias (47). The reversion from differentiated cells to a progenitor-like condition under pathological circumstances can be experimentally shown in *Cd19-cre* $Pax5^{F/-}$ mice in which *Pax5*-deleted mature B cells develop into aggressive progenitor cell lymphomas, which are indistinguishable in their gene expression pattern from uncommitted $Pax5^{-/-}$ pro-B cells but carry rearrangements at the immunoglobulin heavy and light chain loci indicating that they must originate by dedifferentiation from late B cell developmental stages (17). This dedifferentiation model also correlates with human cancers, since cases of transdifferentiation of follicular B cell lymphoma (FL) into myeloid histiocytic/dendritic cell (H/DC) sarcoma have been described where the initial and final tumours are clonally related and contain the same immunoglobulin rearrangements and identical *IgH–BCL2* translocation breakpoints (48).

1.4. Taking Advantage of Pax5-Dependent B Cell Plasticity: Reprogramming B Cells into Early Progenitors

We have discussed the different physiological, pathological, and experimental manifestations of B cell plasticity. The protocol that we are presenting in this chapter describes a different demonstration of mature B cell plasticity with broad theoretical and practical implications: the in vivo dedifferentiation of mature B cells form peripheral lymphoid organs to *Pax5*$^{\Delta/\Delta}$ pro-B cells that are multipotential haematopoietic progenitors homed in the BM (17). This reprogramming process is induced solely by the deletion of *Pax5* in mature B cells. The dedifferentiated *Pax5*$^{\Delta/\Delta}$ pro-B cells can then seed the thymus of recipient *RAG2*$^{-/-}$ mice and generate T cells, and can also be redirected towards other lineages like macrophages in the presence of the appropriate stimulus. All these newly re-differentiated lineages carry the immunoglobulin heavy and light gene rearrangements as a hallmark indication of their past mature B cell nature. There are two important differences in this experimental approach with respect to the others mentioned above. The first one is that this method is what we could call a "passive" reprogramming (versus "active" ones). Indeed, in this method the only genetic alteration is the loss of a gene (*Pax5*). In other methods, ectopic induction of signalling cascades or transcription factor expression imposes an active signal that pushes the cell to differentiate towards a defined fate. In the method described here, the elimination of *Pax5* removes the B cell identity but it does not force the cell towards any desired end product. Also, the in vivo environment of a *RAG2*$^{-/-}$ host with empty lymphoid niches in the peripheral haematopoietic compartments allows the cells to "choose" the fate without imposing any. This is why dedifferentiation can occur along the same lineage towards an early multipotent *Pax5*$^{\Delta/\Delta}$ progenitor. In order for this process to occur, survival of *Pax5*-deleted B cells must be ensured by the ectopic expression of a *Bcl2* transgene.

The other main difference of this experimental procedure with respect to other reprogramming approaches relies on the distinction between transdifferentiation versus dedifferentiation. The C/EBPα-induced conversion of committed B cells into macrophages happens via non-physiological cellular intermediates expressing markers belonging to both B cells and macrophages in a classical transdifferentiation manner (15). Also the recently described conversion of mature B cells into induced pluripotent stem cells (iPS) by forced expression of transcription factors Oct3/4, Sox2, Klf4, and c-Myc in combination with Pax5 downregulation can be considered as a transdifferentiation process, even if the net result is the generation of early progenitors from mature cells (49). Indeed, this conversion also involves the passage through intermediate cell states that present a re-activation of genes related to stem cell renewal and maintenance, but not yet pluripotency, an incomplete repression of lineage-specific transcription factors and incomplete epigenetic remodelling, including persistent DNA hypermethylation (11). Independently of this fact, these results show that mature B cells can only be reprogrammed provided that the Pax5-dependent transcriptional program controlling B cell identity is disrupted by knocking down Pax5 expression, confirming the central role of Pax5 in maintaining the identity of mature B cells. The protocol presented here shows the capacity of mature B cells to spontaneously revert to a physiologically well-defined differentiation state (pre-pro-B cells) in the absence of any imposed signal, in a process triggered by the loss of a single transcription factor (17).

In summary, our capacity to reprogram mature cells into progenitors greatly depends on the inherent physiological plasticity of the starting cell and the molecular mechanisms that we can use to drive this cell towards a new fate. B lymphocytes possess a high degree of plasticity revealed both during their normal development and under experimental conditions in the laboratory. This plasticity is dependent on a single gene, *Pax5*, whose selective removal allows mature B cells to dedifferentiate, serving as a source of multipotential plastic progenitors. This finding has important implications for our understanding of the molecular mechanisms of maintenance of the identity of differentiated cells and how these mechanisms can be used for regenerative medicine.

2. Materials

1. The following genetically modified mouse alleles are necessary in a C57Bl6/J Ly5.2^{+} background.
   a. Pax5 conditional knockout allele (*Pax5-floxed*) (50).
   b. *CreED-30* transgenic mice constitutively expressing a 4-Hidroxy-tamoxifen (OHT)-inducible form of the Cre recombinase (51).
   c. *Eμ-bcl2-36* transgenic mice line (52).
2. *RAG2*$^{-/-}$ *Ly5.1*$^{+}$ recipient mice for transplantations (TACONIC, model 000461-M: B6.SJL(129S6)-*Ptprc*^{a}/BoCrTac-*Rag2*^{tm1Fwa} N10).
3. Dissection material (Fine Science Tools Inc.).
4. FACS and MACS buffer. FACS: PBS pH 7.5 + 1% heat-inactivated foetal calf serum. MACS (Milteny Biotech): PBS pH 7.2, 0.5% BSA, 2 mM EDTA.
5. CD16/CD32 Fc-block solution (PharMingen).
6. MACS anti-PE microbeads (Milteny Biotech) (see Note 1).
7. Anti-IL7Rα monoclonal antibody (see Note 1).

8. Monoclonal antibodies: anti-B220 (RA3-6B2), CD3ε (145-2C11), CD4 (H129.19), CD8a (53-6.7), CD11b/Mac1 (M1/70), CD11c (HL3), CD19 (1D3), CD21 (7G6), CD22 (Cy34.1), CD23 (B3B4), CD25 (PC61), CD40 (FGK45.5 or 3/23), CD49b (DX5), CD72 (K10.6), CD90.2/Thy1.2 (53-2.1), CD93/AA4.1 (PB.493), CD95/Fas (Jo2), CD117/c-Kit (2B8), CD127/IL7Rα (A7R34), CD138 (281-2), Flk-2/Flt3 (A2F10.1), F4/80 (CI:A3-1), Gr1 (RB6-8C5), IgD (1.19), Igκ (187.1), Igλ (R26-46), IgM (M41-42), Ly5.1 (A20), Ly5.2 (104), M-CSFR (AFS98), MHCII (M5-114), pre-BCR (SL156), TCRβ (H57-597), and Ter119 (TER-119) (see Note 1).
9. Nordion Gammacell 1000 Research Irradiator.
10. Casy Cell Counter (Schärfe System GmbH).
11. FACS sorter.

3. Methods

1. Breed the mice to generate *Pax5*$^{F/F}$*CreED-30 Eμ-bcl2* compound mice. Given the fact that *Eμ-bcl2* mice have a shorter life span due to their tendency to develop B-cell lymphomas, a convenient breeding strategy is to cross *Pax5*$^{F/F}$ *CreED-30*×*Pax5*$^{F/F}$ *Eμ-bcl2* (50–52).
2. To avoid the presence of early stages of B cell development in the final mature B cell populations, donor mice must be pretreated with anti-IL7Rα antibody (Fig. 1). This treatment blocks B cell development leading to the loss of pro-B, pre-B, and immature B cells in the BM and spleen (53). The purified anti-IL-7Rα antibody A7R34 (1 mg in 150 μl of PBS) has to be injected in the tail vein of *CreED-30 Pax5*$^{F/F}$ *Eμ-bcl2 Ly5.2*$^{+}$ mice, starting 8–14 days before sacrificing the mice. Immobilize the mice for tail vein injection and massage the tail to facilitate blood flow. Inject 150 μl of PBS (1 mg Ab) per mouse in the caudal vein using an insulin syringe with a 27-G needle. Repeat this injection every second day until the sacrifice (5–7 injections/mouse) (see Note 2).
3. Once mice are 6–8 weeks old and have been pre-treated with anti-IL7Rα Ab as described above, sacrifice them according to regulatory standards and dissect spleen and lymph nodes. At least four mice should be pooled in order to end up with enough pure mature B cells at the end of the procedure (see Note 3). Keep the cell suspensions in ice or at 4°C all throughout the separation procedure. Single cell suspensions of the spleens are prepared by passing them through Falcon Cell

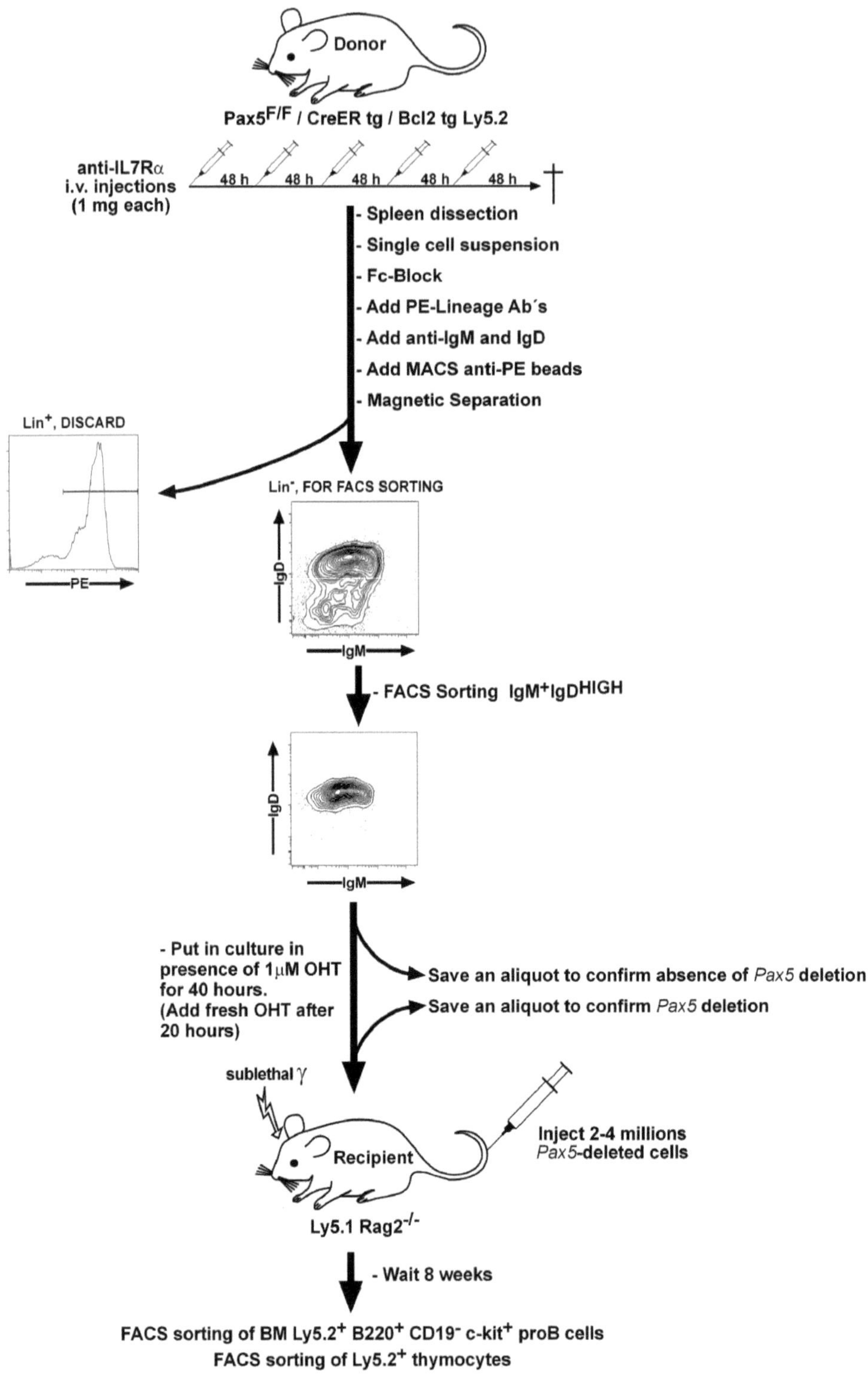

Fig. 1. Flowchart of the mature B cell dedifferentiation experiment. See text for a detailed description

Strainers, in MACS buffer. Resuspend every spleen in a total of 50 μl in a Falcon tube (to avoid clumping) and count using Casy Counter TTC. Discard erythrocyte numbers for calculations (<5 microns size cells). Pellet cells (1,500 rpm, 5 min, 4°C) and resuspend them at an approximate concentration of 40×10^6 cells/ml in MACS buffer. Pool all the spleen suspensions (see Notes 4 and 5).

4. Add Fc-Block at a 1/100 dilution to prevent unspecific binding to Fc receptors. Incubate 10 min at 4°C. Add the antibodies for the Lineage cocktail (CD3ε, CD4, CD8α, CD11c, CD49b, CD93, Gr1, c-Kit, Mac1, TCRβ, Ter119, and Thy1.2) (anti-T, anti-NK, anti-early B, anti-myelo-erythroid), all of them PE-labelled (see Note 6). Add FITC-anti-IgM and APC-anti-IgD antibodies. Incubate 30 min at 4°C. Add MACS anti-PE beads and eliminate all PE-labelled cells using Milteny LS columns or AutoMacs ("depletes" program) according to the manufacturer's instructions (see Note 7). Keep an aliquot for cytometric analysis of lineage depletion efficacy (see Note 8).
5. Pellet the cells, resuspend them in FACS buffer, and filter them through the strain-caps of BD FACS tubes (to avoid clogging the cytometer). Proceed to sort the cells as IgM^+ IgD^{HIGH} (Fig. 1). Sort the cells directly into heat-inactivated Foetal Calf Serum. Once sorted, re-analyze them to confirm purity. Ninety-nine percent or more purity is required for excluding any possible contamination (see Note 9). Pellet and resuspend purified cells in culture medium (see below) at a density of approximately 5×10^5 per ml. Check cells under the microscope. They should be round, uniform, and shiny.
6. Culture the cells for 40 h in Iscove's modified Dulbecco's medium (IMDM) supplemented with 10% FCS, 50 μM β-mercaptoethanol, 1 mM glutamine, PenStrep, and 1 μ 4-hydroxytamoxiphen (OHT, Sigma). After the first 20 h in culture, add a new fresh amount of OHT (1/1,000) from the stock (see Note 10).
7. After 40 h in culture, recollect the cells, count then in Casy Counter and resuspend them in PBS for injection. Resuspend the cells at $2\text{–}4 \times 10^6$ in 150 μl (see Note 11).
8. Injection of *Pax5*-deleted mature B cells into 8–12-week-old *Rag2*$^{-/-}$ *Ly5.1*$^+$ recipient mice. Twenty-four hours prior to the injection, recipient mice have to be γ-irradiated with a sublethal dose (4 Gy) in a Nordion Gammacell 1000 Research Irradiator. Immobilize the mice for tail vein injection and massage tail to facilitate blood flow. Inject 150 μl of PBS containing $2\text{–}4 \times 10^6$ cells per mouse in the caudal vein using an

insulin syringe with a 27-G needle, carefully and slowly aspirating and injecting the cell suspension (see Note 12). After injection, maintain mice with 1.14 gr/l neomycin in the drinking water.

9. Monitor the mice every second day after injection. Sacrifice recipient mice 8 weeks or more after injection, together with healthy controls, both of Ly5.1$^+$ and Ly5.2$^+$ genotypes, and also a *Rag2*$^{-/-}$ mouse control (see Note 13). Collect all main haematopoietic organs and prepare single cell suspensions for flow cytometry (see Note 14). Stain controls with the same staining mixes (see below) as experimental mice in order to clearly discriminate donor Ly5.2$^+$ versus recipient (Ly5.2$^-$, Ly5.1$^+$) cells. Dedifferentiated pro-B cells of donor origin can be FACS-sorted from the BM of transplanted *Rag2*$^{-/-}$ *Ly5.1*$^+$ mice as Ly5.2$^+$CD19$^-$B220$^+$c-Kit$^+$ cells after staining with FITC-anti-Ly5.2, PE-anti-CD19, PE-Cy5-anti-B220, and APC-anti-c-Kit antibodies. Donor-derived DP thymocytes can be sorted from the thymus of transplanted *Rag2*$^{-/-}$ *Ly5.1+* mice as Ly5.2$^+$CD4$^+$CD8a$^+$ cells after staining with FITC-anti-Ly5.2, PE-anti-CD4, and APC-anti-CD8a antibodies.

10. Sorted Pax5$^{\Delta/\Delta}$ Ly5.2$^+$c-Kit$^+$B220$^+$ pro-B cells can be cultured on γ-irradiated ST2 feeder cells in IMDM medium containing 2% heat-inactivated FCS, 0.03% (w/vol) primatone RL, 1 mM glutamine, PenStrep, 50 mM β-mercaptoethanol, 1% supernatant of rIL-7-secreting J558L cells, 2.5% supernatant of rFlt3L-producing SP2.0 cells and 2% supernatant of SCF-secreting CHO cells. These cells can afterwards be induced to differentiate into different haematopoietic lineages (29, 54).

11. For example, to induce macrophage differentiation, maintain pro-B cells on the M-CSF-producing ST2 cells in the absence of IL-7, SCF, and Flt3L for 10 days, and then induce terminal macrophage differentiation by adding extra recombinant mouse M-CSF (25 ng/ml) for 1 week (see Note 15).

12. Study of immunoglobulin rearrangements to confirm the mature B cell origin of dedifferentiated pro-B cells or their re-dedifferentiated descendants. The detailed description of the analysis of *IgH* V(D)J or *Igκ* and *Igλ* genes VJ rearrangements falls out of the scope of this chapter. For a description and lists of specific primers and protocols, *see* refs. 17, 55–57 and Note 16.

4. Notes

1. Large amounts of several of these reagents are required when several mice's spleens are pooled. Be sure that you have enough of them beforehand.
2. The efficiency of the treatment will be checked by analyzing the absence of pro-B cells in the spleen and BM of the mice after sacrifice.
3. The same protocol is followed from both spleen- or lymph node-derived samples.
4. Lymph node suspensions tend to re-aggregate. If this happens, pass them through the strainer only before applying to the sorting machine.
5. BM from the mice is also collected to check for the efficiency of the anti-IL7Rα treatment in eliminating early B cell stages, by staining an aliquot with PE-Cy5-anti-B220 and APC-anti-c-Kit antibodies (pro-B cells, B200$^+$, c-Kit$^+$), or PE-Cy5-anti-B220, PE-anti-CD25, and FITC-anti-IgM (pre-B cells, B220$^+$, CD25$^+$, IgM$^-$).
6. All antibody dilutions have to be previously titrated using control samples to determine their optimal dilution to provide the best signal-to-noise ratio.
7. This step typically removes more than 90% of undesired cells from the suspension (i.e. most non-B cells and all non-mature B cells that had not been eliminated with the anti-Il7Rα treatment) (Fig. 1).
8. A variation of the protocol to obtain *Pax5*-deleted mature B cells involves the use of a different Cre deleter line, *Cd19-Cre* (58) in mice carrying a *floxed Pax5* allele and a constitutive *Pax5* knockout allele (59) in the presence of an *Eμ-bcl2* transgene (52). Cre-mediated elimination of *Pax5* in the lymph nodes of *Cd19-cre* $Pax5^{fl/-}$ *Eμ-bcl2 Ly5.2+* mice induces a change in the surface markers of mature B cells that allows the sorting of *Pax5-deleted* $Lin^-CD25^+IgM^+IgD^-$ cells (17, 28, 50) that can afterwards be injected into the $Rag2^{-/-}Ly5.1^+$ recipients for the dedifferentiation experiments.
9. Any T cell contamination will afterwards result in homeostatic expansion of T cells in the peripheral organs of the recipient mice, thus precluding the evaluation of T cell reconstitution outside of the thymus.
10. Prepare the OHT stock 1,000× (1 mM) in ethanol, and store at −20°C. Add to medium immediately prior to mixing with the cells. To avoid killing some cells with the ethanol, mix

previously the required amount of stock solution in an Eppendorf tube with 500 µl of medium, and then add to the plate/s. Before plating cells, save a small aliquot for genotyping, to confirm absence of deletion of the *floxed Pax5* allele in the absence of active *Cre* recombinase (see Note 11).

11. Once in PBS (i.e. without serum) cells are sticky and can be easily lost by adhesion to the tube's walls. Avoid further centrifugations and keep the cells on ice at all times. Keep a small aliquot for PCR testing of the deletion of the *floxed Pax5* allele by the OHT-induced activity of the CreER™ fusion protein encoded by *CreED-30* transgene. Always verify deletion by PCR genotyping (50) to confirm the validity of the experiment.
12. The "dead volume" at the tip of the syringe and base of the needle has to be taken into account when calculating the number of cells and number of mice that can be injected. To avoid too much loss of sample, the same syringe can be used for different mice, changing the needle. Cells tend to sediment in the tube in ice, if kept for too long time, so gently tap the tube to resuspend them, and try to work quickly.
13. After 3–4 weeks dedifferentiation can be observed in some cases, but cellularity is very low and special methods are required to detect the engrafted cells (for example, CFSE-labelling prior to injection of cells, *see* ref. 17).
14. Presence of thymus indicates an efficient dedifferentiation, since in uninjected $Rag2^{-/-}$ mice the thymus is barely visible, if at all.
15. To induce T cell differentiation, cultured $Pax5^{\Delta}/^{\Delta}$ $Ly5.2^{+}c$-$Kit^{+}B220^{+}CD19$– pro-B cells can be re-injected into sublethally irradiated $Rag2^{-/-}Ly5.1^{+}$ recipients as described above. In this case, since dedifferentiation is not a rate-limiting factor anymore, T cell development can already be analyzed 3 weeks after injection.
16. For V(D)J recombination analysis, sorted populations are required to serve as positive and/or negative controls for rearrangements. These populations should be of the utmost purity to guarantee the absence of background and the accuracy of the obtained results. A common problem while analyzing V(D)J rearrangements by PCR is to detect the appearance of many background bands. It might take quite some time to optimize conditions. Well-tested enzymes for the PCR are *Taqs* from the houses Takara or Eppendorf. The enzyme itself might not be so important, but very good reactions can be obtained by using the buffers and dNPTs supplied with the Takara PCR kit (final Mg^{2+} concentration 1.5 mM). Primers are listed in refs. 17, 55. Most of these

primers work correctly at annealing temperatures of 65–62°C, for 1 min and elongations times at 72°C of 1 min and 45 s (so 105 s of total elongation time per cycle). Denaturing time is 1 min at 95°C. Initiate PCR with a hot start for 4 min at 95°C; final elongation 5 min at 72°C. The use of a JH4 reverse primer as reported in ref. 17 tends to create more background noise bands and a less intense band for VDJ1 rearrangements. An alternative is to use a reverse JH3 primer that gives less noise, as described in ref. 55 but then the gel has to be blotted and hybridized with a specific probe afterwards. One important issue is the amount of template DNA. If too much DNA is used, background bands will be amplified. It is a good idea to normalize the DNA amounts before by using the primers for the constant region Cmu5′and Cmu3′ (55). For the V(D)J detection PCR, run 30–35 cycles in order to be able to see the bands in an agarose gel without having to hybridize. It is better to have low amounts of DNA in this range of visibility than too much DNA that produces a lot of background. It is also an essential need to have very pure material as positive and negative controls. Double-sorted (or MACS-sorted and then FACS-sorted) IgM$^+$ cells from spleen should be the positive control. The negative control could be any appropriate control cells pure from any source of B cell contamination. It is essential to have good negative and positive controls that can be trusted in order to interpret the PCR results.

Acknowledgements

I thank Dr. Meinrad Busslinger for critically reading the manuscript and for his helpful comments. Research in my lab is partially supported by FEDER, Fondo de Investigaciones Sanitarias (PI080164), Junta de Castilla y León (SA060A09 and Proyecto de Biomedicina 2009-10), Proyectos Intramurales Especiales (CSIC), MEC OncoBIO Consolider- Ingenio 2010 (Ref. CSD2007-0017) and a Spanish "Ramón y Cajal" investigator grant from the Spanish Ministerio de Ciencia e Innovación.

References

1. Takahashi K., Yamanaka S. (2006) Induction of pluripotent stem cells from mouse embryonic and adult fibroblast cultures by defined factors. *Cell* **126**:663–76.
2. Lowry W. E., Plath K. (2008) The many ways to make an iPS cell. *Nat Biotechnol* **26**:1246–8.
3. Maherali N., Hochedlinger K. (2008) Guidelines and techniques for the generation of induced pluripotent stem cells. *Cell Stem Cell* **3**:595–605.
4. Price T. D., Qvarnstrom A., Irwin D. E. (2003) The role of phenotypic plasticity in driving genetic evolution. *Proc Biol Sci* **270**:1433–40.

5. Byrne J. A., Pedersen D. A., Clepper L. L., Nelson M., Sanger W. G., Gokhale S., Wolf D. P., Mitalipov S. M. (2007) Producing primate embryonic stem cells by somatic cell nuclear transfer. *Nature* **450**:497–502.
6. Hochedlinger K., Jaenisch R. (2006) Nuclear reprogramming and pluripotency. *Nature* **441**:1061–7.
7. Gurdon J. B., Melton D. A. (2008) Nuclear reprogramming in cells. *Science* **322**:1811–5.
8. Wernig M., Meissner A., Foreman R., Brambrink T., Ku M., Hochedlinger K., Bernstein B. E., Jaenisch R. (2007) In vitro reprogramming of fibroblasts into a pluripotent ES-cell-like state. *Nature* **448**:318–24.
9. Okita K., Ichisaka T., Yamanaka S. (2007) Generation of germline-competent induced pluripotent stem cells. *Nature* **448**:313–7.
10. Stadtfeld M., Nagaya M., Utikal J., Weir G., Hochedlinger K. (2008) Induced pluripotent stem cells generated without viral integration. *Science* **322**:945–9.
11. Mikkelsen T. S., Hanna J., Zhang X., Ku M., Wernig M., Schorderet P., Bernstein B. E., Jaenisch R., Lander E. S., Meissner A. (2008) Dissecting direct reprogramming through integrative genomic analysis. *Nature* **454**:49–55.
12. Huangfu D., Osafune K., Maehr R., Guo W., Eijkelenboom A., Chen S., Muhlestein W., Melton D. A. (2008) Induction of pluripotent stem cells from primary human fibroblasts with only Oct4 and Sox2. *Nat Biotechnol* **26**:1269–75.
13. Huangfu D., Maehr R., Guo W., Eijkelenboom A., Snitow M., Chen A. E., Melton D. A. (2008) Induction of pluripotent stem cells by defined factors is greatly improved by small-molecule compounds. *Nat Biotechnol* **26**:795–7.
14. Zhou Q., Melton D. A. (2008) Extreme makeover: converting one cell into another. *Cell Stem Cell* **3**:382–8.
15. Xie H., Ye M., Feng R., Graf T. (2004) Stepwise reprogramming of B cells into macrophages. *Cell* **117**:663–76.
16. Zhou Q., Brown J., Kanarek A., Rajagopal J., Melton D. A. (2008) In vivo reprogramming of adult pancreatic exocrine cells to beta-cells. *Nature* **455**:627–32.
17. Cobaleda C., Jochum W., Busslinger M. (2007) Conversion of mature B cells into T cells by dedifferentiation to uncommitted progenitors. *Nature* **449**:473–7.
18. Cobaleda C., Busslinger M. (2008) Developmental plasticity of lymphocytes. *Curr Opin Immunol* **20**:139–48.
19. Adolfsson J., Mansson R., Buza-Vidas N., Hultquist A., Liuba K., Jensen C. T., Bryder D., Yang L., Borge O. J., Thoren L. A., Anderson K., Sitnicka E., Sasaki Y., Sigvardsson M., Jacobsen S. E. (2005) Identification of Flt3+ lympho-myeloid stem cells lacking erythro-megakaryocytic potential a revised road map for adult blood lineage commitment. *Cell* **121**:295–306.
20. Kondo M., Weissman I. L., Akashi K. (1997) Identification of clonogenic common lymphoid progenitors in mouse bone marrow. *Cell* **91**:661–72.
21. Hardy R. R., Kincade P. W., Dorshkind K. (2007) The protean nature of cells in the B lymphocyte lineage. *Immunity* **26**:703–14.
22. Hu M., Krause D., Greaves M., Sharkis S., Dexter M., Heyworth C., Enver T. (1997) Multilineage gene expression precedes commitment in the hemopoietic system. *Genes Dev* **11**:774–85.
23. Akashi K., He X., Chen J., Iwasaki H., Niu C., Steenhard B., Zhang J., Haug J., Li L. (2003) Transcriptional accessibility for genes of multiple tissues and hematopoietic lineages is hierarchically controlled during early hematopoiesis. *Blood* **101**:383–9.
24. Miller J. P., Izon D., DeMuth W., Gerstein R., Bhandoola A., Allman D. (2002) The earliest step in B lineage differentiation from common lymphoid progenitors is critically dependent upon interleukin 7. *J Exp Med* **196**:705–11.
25. Cobaleda C., Schebesta A., Delogu A., Busslinger M. (2007) Pax5: the guardian of B cell identity and function. *Nat Immunol* **8**:463–70.
26. Nutt S. L., Kee B. L. (2007) The transcriptional regulation of B cell lineage commitment. *Immunity* **26**:715–25.
27. Delogu A., Schebesta A., Sun Q., Aschenbrenner K., Perlot T., Busslinger M. (2006) Gene repression by Pax5 in B cells is essential for blood cell homeostasis and is reversed in plasma cells. *Immunity* **24**:269–81.
28. Schebesta A., McManus S., Salvagiotto G., Delogu A., Busslinger G. A., Busslinger M. (2007) Transcription factor Pax5 activates the chromatin of key genes involved in B cell signaling, adhesion, migration, and immune function. *Immunity* **27**:49–63.
29. Nutt S. L., Heavey B., Rolink A. G., Busslinger M. (1999) Commitment to the B-lymphoid lineage depends on the transcription factor Pax5. *Nature* **401**:556–62.
30. Nagai Y., Garrett K. P., Ohta S., Bahrun U., Kouro T., Akira S., Takatsu K., Kincade P. W.

(2006) Toll-like receptors on hematopoietic progenitor cells stimulate innate immune system replenishment. *Immunity* **24**:801–12.

31. Shapiro-Shelef M., Calame K. (2005) Regulation of plasma-cell development. *Nat Rev Immunol* **5**:230–42.
32. Kallies A., Nutt S. L. (2007) Terminal differentiation of lymphocytes depends on Blimp-1. *Curr Opin Immunol* **19**:156–62.
33. Martins G., Calame K. (2008) Regulation and functions of Blimp-1 in T and B lymphocytes. *Annu Rev Immunol* **26**:133–69.
34. Shaffer A. L., Lin K. I., Kuo T. C., Yu X., Hurt E. M., Rosenwald A., Giltnane J. M., Yang L., Zhao H., Calame K., Staudt L. M. (2002) Blimp-1 orchestrates plasma cell differentiation by extinguishing the mature B cell gene expression program. *Immunity* **17**:51–62.
35. Lobo N. A., Shimono Y., Qian D., Clarke M. F. (2007) The biology of cancer stem cells. *Annu Rev Cell Dev Biol* **23**:675–99.
36. Dalerba P., Cho R. W., Clarke M. F. (2007) Cancer stem cells: models and concepts. *Annu Rev Med* **58**:267–84.
37. Sanchez-Garcia I., Vicente-Duenas C., Cobaleda C. (2007) The theoretical basis of cancer-stem-cell-based therapeutics of cancer: can it be put into practice? *Bioessays* **29**:1269–80.
38. Hanahan D., Weinberg R. A. (2000) The hallmarks of cancer. *Cell* **100**:57–70.
39. Huntly B. J., Shigematsu H., Deguchi K., Lee B. H., Mizuno S., Duclos N., Rowan R., Amaral S., Curley D., Williams I. R., Akashi K., Gilliland D. G. (2004) MOZ-TIF2, but not BCR-ABL, confers properties of leukemic stem cells to committed murine hematopoietic progenitors. *Cancer Cell* **6**:587–96.
40. Somervaille T. C., Cleary M. L. (2006) Identification and characterization of leukemia stem cells in murine MLL-AF9 acute myeloid leukemia. *Cancer Cell* **10**:257–68.
41. Krivtsov A. V., Twomey D., Feng Z., Stubbs M. C., Wang Y., Faber J., Levine J. E., Wang J., Hahn W. C., Gilliland D. G., Golub T. R., Armstrong S. A. (2006) Transformation from committed progenitor to leukaemia stem cell initiated by MLL-AF9. *Nature* **442**:818–22.
42. Cozzio A., Passegue E., Ayton P. M., Karsunky H., Cleary M. L., Weissman I. L. (2003) Similar MLL-associated leukemias arising from self-renewing stem cells and short-lived myeloid progenitors. *Genes Dev* **17**:3029–35.
43. So C. W., Karsunky H., Passegue E., Cozzio A., Weissman I. L., Cleary M. L. (2003) MLL-GAS7 transforms multipotent hematopoietic progenitors and induces mixed lineage leukemias in mice. *Cancer Cell* **3**:161–71.
44. Jamieson C. H., Ailles L. E., Dylla S. J., Muijtjens M., Jones C., Zehnder J. L., Gotlib J., Li K., Manz M. G., Keating A., Sawyers C. L., Weissman I. L. (2004) Granulocyte–macrophage progenitors as candidate leukemic stem cells in blast-crisis CML. *N Engl J Med* **351**:657–67.
45. Chen L., Shen R., Ye Y., Pu X. A., Liu X., Duan W., Wen J., Zimmerer J., Wang Y., Liu Y., Lasky L. C., Heerema N. A., Perrotti D., Ozato K., Kuramochi-Miyagawa S., Nakano T., Yates A. J., Carson W. E., III, Lin H., Barsky S. H., Gao J. X. (2007) Precancerous stem cells have the potential for both benign and malignant differentiation. *PLoS ONE* **2**:e293.
46. Miyamoto T., Weissman I. L., Akashi K. (2000) AML1/ETO-expressing nonleukemic stem cells in acute myelogenous leukemia with 8;21 chromosomal translocation. *Proc Natl Acad Sci U S A* **97**:7521–6.
47. Hong D., Gupta R., Ancliff P., Atzberger A., Brown J., Soneji S., Green J., Colman S., Piacibello W., Buckle V., Tsuzuki S., Greaves M., Enver T. (2008) Initiating and cancer-propagating cells in TEL-AML1-associated childhood leukemia. *Science* **319**:336–9.
48. Feldman A. L., Arber D. A., Pittaluga S., Martinez A., Burke J. S., Raffeld M., Camos M., Warnke R., Jaffe E. S. (2008) Clonally related follicular lymphomas and histiocytic/dendritic cell sarcomas: evidence for transdifferentiation of the follicular lymphoma clone. *Blood* **111**:5433–9.
49. Hanna J., Markoulaki S., Schorderet P., Carey B. W., Beard C., Wernig M., Creyghton M. P., Steine E. J., Cassady J. P., Foreman R., Lengner C. J., Dausman J. A., Jaenisch R. (2008) Direct reprogramming of terminally differentiated mature B lymphocytes to pluripotency. *Cell* **133**:250–64.
50. Horcher M., Souabni A., Busslinger M. (2001) Pax5/BSAP maintains the identity of B cells in late B lymphopoiesis. *Immunity* **14**:779–90.
51. Schwenk F., Kuhn R., Angrand P. O., Rajewsky K., Stewart A. F. (1998) Temporally and spatially regulated somatic mutagenesis in mice. *Nucleic Acids Res* **26**:1427–32.
52. Strasser A., Harris A. W., Cory S. (1991) bcl-2 transgene inhibits T cell death and perturbs thymic self-censorship. *Cell* **67**:889–99.
53. Sudo T., Nishikawa S., Ohno N., Akiyama N., Tamakoshi M., Yoshida H. (1993) Expression

and function of the interleukin 7 receptor in murine lymphocytes. *Proc Natl Acad Sci U S A* **90**:9125–9.
54. Rolink A. G., Nutt S. L., Melchers F., Busslinger M. (1999) Long-term in vivo reconstitution of T-cell development by Pax5-deficient B-cell progenitors. *Nature* **401**:603–6.
55. Fuxa M., Skok J., Souabni A., Salvagiotto G., Roldan E., Busslinger M. (2004) Pax5 induces V-to-DJ rearrangements and locus contraction of the immunoglobulin heavy-chain gene. *Genes Dev* **18**:411–22.
56. Angelin-Duclos C., Calame K. (1998) Evidence that immunoglobulin VH-DJ recombination does not require germ line transcription of the recombining variable gene segment. *Mol Cell Biol* **18**:6253–64.
57. Schlissel M. S., Corcoran L. M., Baltimore D. (1991) Virus-transformed pre-B cells show ordered activation but not inactivation of immunoglobulin gene rearrangement and transcription. *J Exp Med* **173**:711–20.
58. Rickert R. C., Roes J., Rajewsky K. (1997) B lymphocyte-specific, Cre-mediated mutagenesis in mice. *Nucleic Acids Res* **25**:1317–8.
59. Urbanek P., Wang Z. Q., Fetka I., Wagner E. F., Busslinger M. (1994) Complete block of early B cell differentiation and altered patterning of the posterior midbrain in mice lacking Pax5/BSAP. *Cell* **79**:901–12.

Chapter 16

Adult Cell Fate Reprogramming: Converting Liver to Pancreas

Irit Meivar-Levy and Sarah Ferber

Abstract

Regenerative medicine aims at producing new cells for repair or replacement of diseased and damaged tissues. Embryonic and adult stem cells have been suggested as attractive sources of cells for generating the new cells needed. The leading dogma was that adult cells in mammals, once committed to a specific lineage, become "terminally differentiated" and can no longer change their fate. However, in recent years increasing evidence has accumulated demonstrating the remarkable ability of some differentiated cells to be converted into a different cell type via a process termed developmental redirection or adult cells reprogramming. For example, abundant human cell types, such as dermal fibroblasts and adipocytes, could potentially be harvested and converted into other, medically important cell types, such as neurons, cardiomyocytes, or pancreatic β cells. In this chapter, we describe a method of activating the pancreatic lineage and β-cells function in adult human liver cells by ectopic expression of pancreatic transcription factors. This approach aims to generate custom-made autologous surrogate β cells for treatment of diabetes, and possibly bypass both the shortage of cadaveric human donor tissues and the need for life-long immune-suppression.

Key words: Liver, Pancreas, Beta-cells, Transcription factors, Adult cells reprogramming, Insulin production and secretion, Transdifferentiation

1. Introduction

The instructive role of the pancreatic differentiation factor, PDX-1, in activating the pancreatic lineage has been demonstrated in adult liver cells in mice in vivo and human liver cells in vitro using recombinant adenovirus gene delivery (1–7). Insulin producing cells derived from adult liver, produced and processed the hormone, secreted it in a glucose regulated manner and ameliorated hyperglycemia in vivo (1–7). During the last decade, the potential of converting liver into pancreas has been demonstrated by many

S. Ding (ed.), *Cellular Programming and Reprogramming: Methods and Protocols*, Methods in Molecular Biology, vol. 636, DOI 10.1007/978-1-60761-691-7_16,

groups (8–26) in *Xenopus* (9), rodent (1, 2, 11, 12, 25, 27, 28), and human (3, 16, 26). A critical role for PDX-1 in this process was suggested (2, 14, 16, 26, 29), and the therapeutic outcome was significantly improved by coexpression of several factors including MAFA, NGN3, and/or NeuroD with PDX-1 (12, 19, 30). We have learned that the ectopic gene is only a transient trigger for an irreversible process of developmental redirection and demonstrated an obligatory but insufficient role for adult cells dedifferentiation (5).

Based on the ability of pancreatic transcription factors to alter coordinately a comprehensive profile of gene expression in adult liver cells, this approach has also been termed "reprogramming" or transdifferentiation.

In this chapter, we will focus on the activation of pancreatic lineage in adult human liver cells in vitro. As we previously reported, freshly isolated adult human liver cells can be cultured and propagated in vitro and induced to transdifferentiate along an endocrine pancreatic lineage by ectopic expression of pancreatic specific transcription factors (3). Insulin induction by ectopic PDX-1 (with or without additional pancreatic transcription and growth factors) in adult human liver cells is determined at several levels: the activation of an ectopic insulin promoter (Fig. 1.), the induction of endogenous pancreatic specific transcripts, hormones production (Figs. 2. and 4), processing, and secretion (Fig. 4). The activation of the pancreatic phenotype and function was associated with the lost of the host, hepatic phenotype (Fig. 3). The proper function of the newly generated insulin producing cells is analyzed by their capacity to secrete the processed hormone in a glucose regulated manner; coupling between the induced insulin storage compartment and the glucose sensing apparatus in transdifferentiated cells has been demonstrated (Fig. 4). The most stringent level of β-cell-like function is these cells' capacity to ameliorate hyperglycemia for long periods when implanted in diabetic immune-deficient mice (Fig. 5).

2. Materials

2.1. Liver Cells Isolation

1. Hanks Balanced Salt Solution without Calcium or Magnesium (HBSS) supplemented with 20 mM HEPES pH 7.3 and Gentamicin Sulfate 0.5 mg/mL. Adjust pH to 7.4 and keep cold (4°C).
2. Collagenase type I (Worthington Biochemical Corp., NJ), freshly prepared as 0.03% in HBSS supplemented with 1 mM $CaCl_2$.
3. 5 mM Ethylene Glycol Tetraacetic Acid (EGTA) in HBSS. Adjust pH to 7.2 and store in single use aliquots at –20°C.

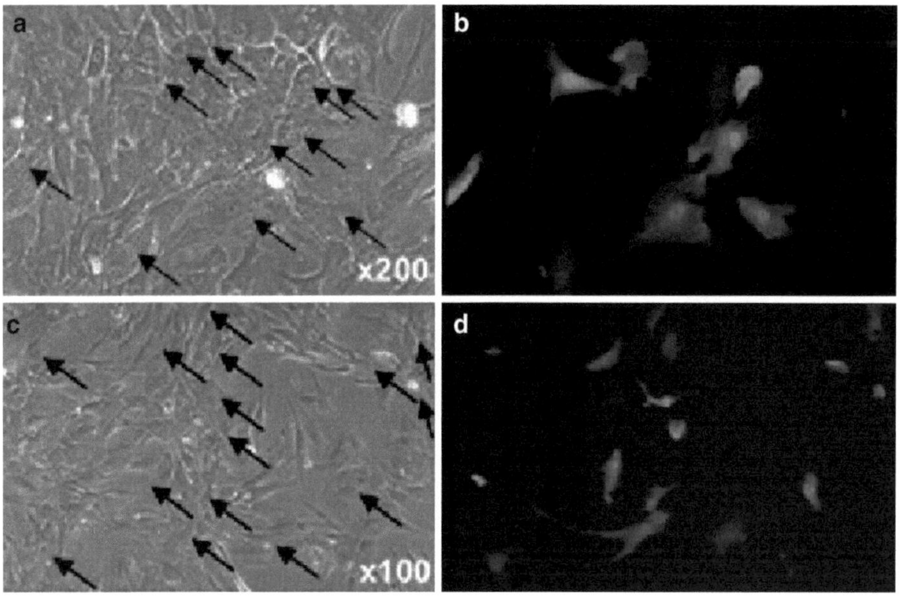

Fig. 1. PDX-1 activates the insulin promoter in human liver cells in vitro. Representative phase contrast morphology (**a, c**), and green fluorescence (**b, d**) of the same field of adult (**a, b**) and fetal (**c, d**) human liver cells, infected by 500 moi *Ad-RIP-GFP* and *Ad-CMV-PDX-1*. *Arrows* indicate the fluorescing cells. Original magnifications, ×200 (**a, b**) and ×100 (**c, d**). Reproduced with permission from Sapir et al. (3)

4. Isolation Stop Solution prepared as HBSS supplemented with 0.2 M EGTA and 10% Fetal Bovine Serum (FBS).
5. Fibronectin-coated plates (3 mg/cm^2) freshly prepared by incubating plates for 15 min at 37°C with 100 μg/mL fibronectin (human) in Dulbecco's Phosphate Buffered Saline (PBS, without Calcium or Magnesium). After additional wash with PBS the fibronectin-coated plates are stored at 4°C until use.

2.2. Primary Cultures of Liver Cells: Maintenance and Treatment

1. Dulbecco's minimal essential medium (DMEM, 1 g/L glucose) supplemented with 10% FBS, 100 U/mL penicillin, 100 μg/mL streptomycin, and 250 ng/mL amphotericin B.
2. Solution of trypsin (0.25%) and ethylenediamine tetraacetic acid (EDTA, 1 mM).
3. Freezing solution: DMEM supplemented with 20% FBS, 10% dimethyl sulfoxide (DMSO) in cryo-containers.
4. Epidermal Growth Factor (EGF; Cytolab Ltd, Israel) is dissolved at 0.5 mg/mL in Dulbecco's Phosphate Buffered Saline (without Calcium or Magnesium) and stored in single use aliquots at –20°C.
5. Nicotinamide (Sigma) is dissolved at 200 mM in culture medium (DMEM) and stored at 4°C.

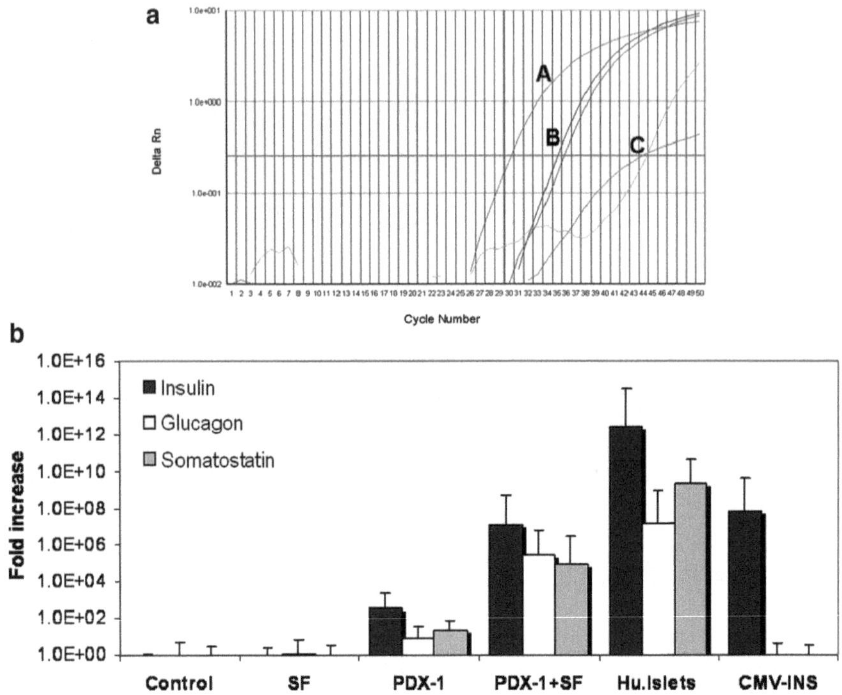

Fig. 2. The promoting effect of Soluble factors on pancreatic hormones gene expression induced by ectopic PDX-1 expression in adult human liver cells. Quantitative real-time RT-PCR analyses of Insulin, Glucagon, and Somatostatin gene expression levels in adult human liver cells treated by *Ad-CMV-PDX-1* with or without soluble factors (Nicotinamide and EGF, are termed as SF) supplementation. (**a**) Comparative expression of the Insulin gene in adult liver cells is presented by a quantitative RT-PCR amplification curve in: human islets (cDNA diluted 1:100, A), *Ad-CMV-PDX-1* and GF treated adult human liver cells (B), or untreated adult liver cells (**C**) (all with similar C_T values for β-actin gene expression). The curves are presented as the Delta Rn (normalized fluorescence units) vs. the cycle numbers of the amplification reaction. (**b**) C_T (threshold cycle) values are normalized to β-actin gene expression within the same cDNA sample, and results are presented as fold of increase (FOI) of the mean ± SD compared with control untreated liver cells (***n*** ≥ 30 for each treatment in five different experiments). Reproduced with permission from Sapir et al. (3)

6. Exendin-4 (sigma) is dissolved at 5 μM in Dulbecco's Phosphate Buffered Saline (without Calcium or Magnesium) and stored in single use aliquots at –20°C.
7. Betacellulin (Recombinant Human, Btc, PeproTech) is dissolved at 4 μM in Dulbecco's Phosphate Buffered Saline (without Calcium or Magnesium) and stored in single use aliquots at –20°C.
8. Activin-A (Recombinant Human, PeproTech) is dissolved at 4 μM in Dulbecco's Phosphate Buffered Saline (without Calcium or Magnesium) and stored in single use aliquots at –20°C. Activin-A is also available as recombinant adenovirus (Ad-CMV-Activin-A).
9. Hepatic Growth Factor (HGF; Recombinant Human, PeproTech) is dissolved at 4 μM in Dulbecco's Phosphate Buffered Saline (without Calcium or Magnesium) and

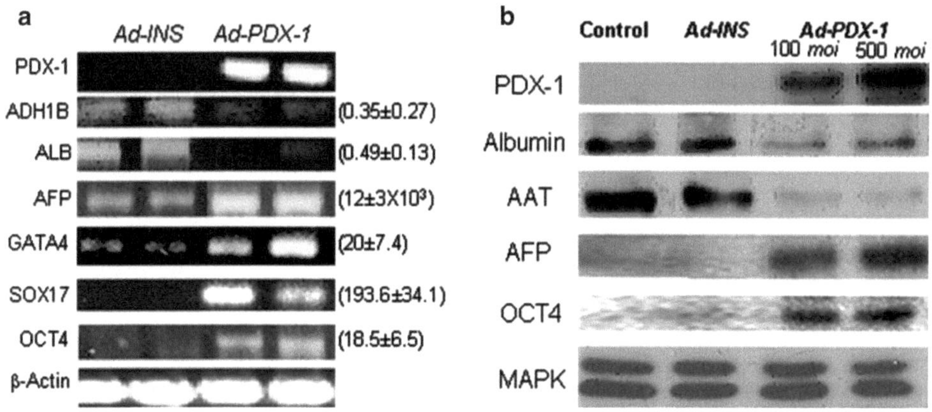

Fig. 3. *PDX-1* suppresses adult hepatic markers gene expression in human liver cells, in vitro. (**a**) Quantitative real-time RT-PCR analyses of human liver cells treated with *Ad-PDX-1*, *Ad-NEUROD1*, *Ad-NKX6.1*, or *Ad-NGN3* for *ALB*, *ADH1B*, *G6PC*, *GLUL*, and *AFP* gene expression. Data presented as relative levels of the mean ± SD compared with *Ad-INS* treated liver cells; $n \geq 8$ in four different experiments; $^{*}p < 0.005$, $^{**}p < 0.01$. *Ad-INS* infected cells serve as both viral infection and the produced pro-insulin control. (**b**) Western blot analyses of albumin, AAT, AFP, and PDX-1 proteins in control untreated (lane 1), *Ad-INS* (lane 2) and *Ad-PDX-1* (100 and 500 moi, lanes 3–4 respectively) treated cells. MAPK serves as protein load control. Representative results, $n = 4$. Reproduced with permission from Meivar-Levy et al. (5)

stored in single use aliquots at –20°C. HGF is also available as recombinant adenovirus (Ad-CMV-HGF).

10. Adenoviruses used: Ad-CMV-GFP, Ad-CMV-Insulin, Ad-CMV-PDX-1, Ad-RIP-GFP, Ad-CMV-NKX6.1, Ad-CMV-PDX-1-VP16, Ad-CMV-NeuroD1, and Ad-CMV-Pax4.
11. Serum-free medium contains DMEM (1 g/L glucose) supplemented with 100 U/mL penicillin, 100 µg/mL streptomycin, and 250 ng/mL amphotericin B with the addition of 10 mg/µL Insulin, 5.5 µg/mL Transferrin, 5 ng/mL, and Selenium (can be purchased as mix ITS (Sigma)).
12. Use BD Falcon™ Cell Culture Dishes or flasks for culturing liver cells.

2.3. Adenovirus Propagation

1. 293 cell culture medium: DMEM (4.5 g/L glucose) supplemented with 10% FBS, 100 U/mL penicillin, 100 µg/mL streptomycin, and 250 ng/mL amphotericin B. For TCID 50, the same medium is used but with only 2% FBS.
2. Solution of trypsin (0.25%) and EDTA (1 mM).
3. Polyethylene Glycol 8000 (PEG, Sigma), autoclave 80 g PEG in 400 mL NaCl (2.5 N) and store at RT. The PEG will be dissolved only by autoclaving the solution. Dissolve the PEG–NaCl solution on heating plate before use.
4. Adenoviruses Saline: 137 mM NaCl, 5 mM KCl, 10 mM Tris–HCl (pH 7.4), and 1 mM $MgCl_2$. Sterilize by autoclaving and store in single use aliquots at 4°C.

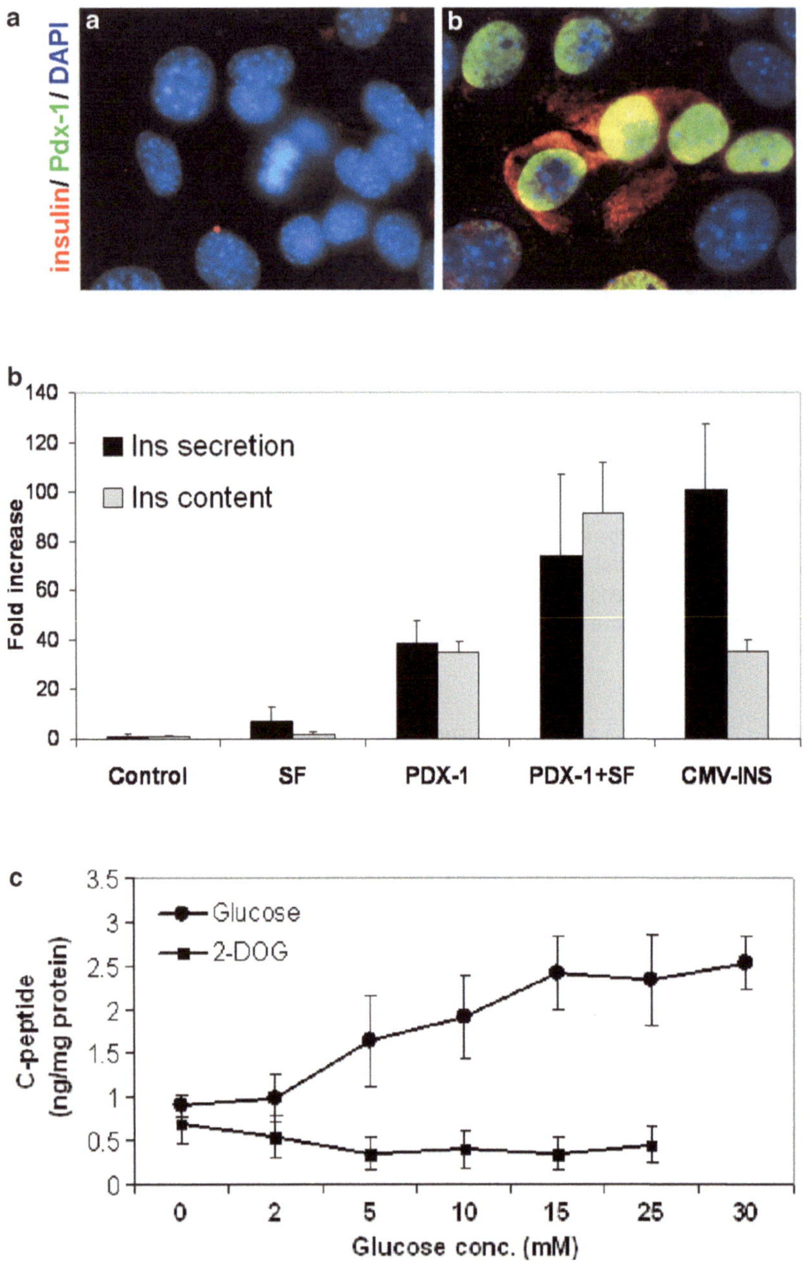

Fig. 4. PDX-1-treated adult human liver cells produce, store, and secrete insulin and C-peptide in a glucose regulated manner. (**a**) Immunofluorescent staining for insulin (cytoplasmatic, *red*) and Pdx-1 (nuclear, *green*) in *Ad-CMV-PDX-1* and GF treated (B) and untreated (A) liver cells in culture. Nuclei are stained in blue (DAPI). Original magnifications, ×600 (A) and ×1,000 (B). (**b**) Adult primary liver cells were treated by *Ad-CMV-PDX-1*, supplemented by GF, and analyzed for insulin content ($n \geq 10$) and insulin secretion ($n \geq 25$) by static incubations for 48 h. *Ad-CMV-hIns* infected cells serve as constitutive human IRI production and secretion control. Results are presented as fold of increase (FOI) of the mean ± SD compared with untreated control liver cells. (**c**) Static incubation of glucose or 2-DOG dose response (0–30 mM) of C-peptide secretion. Results are presented as the mean ± SD; $n = 30$ in four different experiments. Reproduced with permission from Sapir et al. (3)

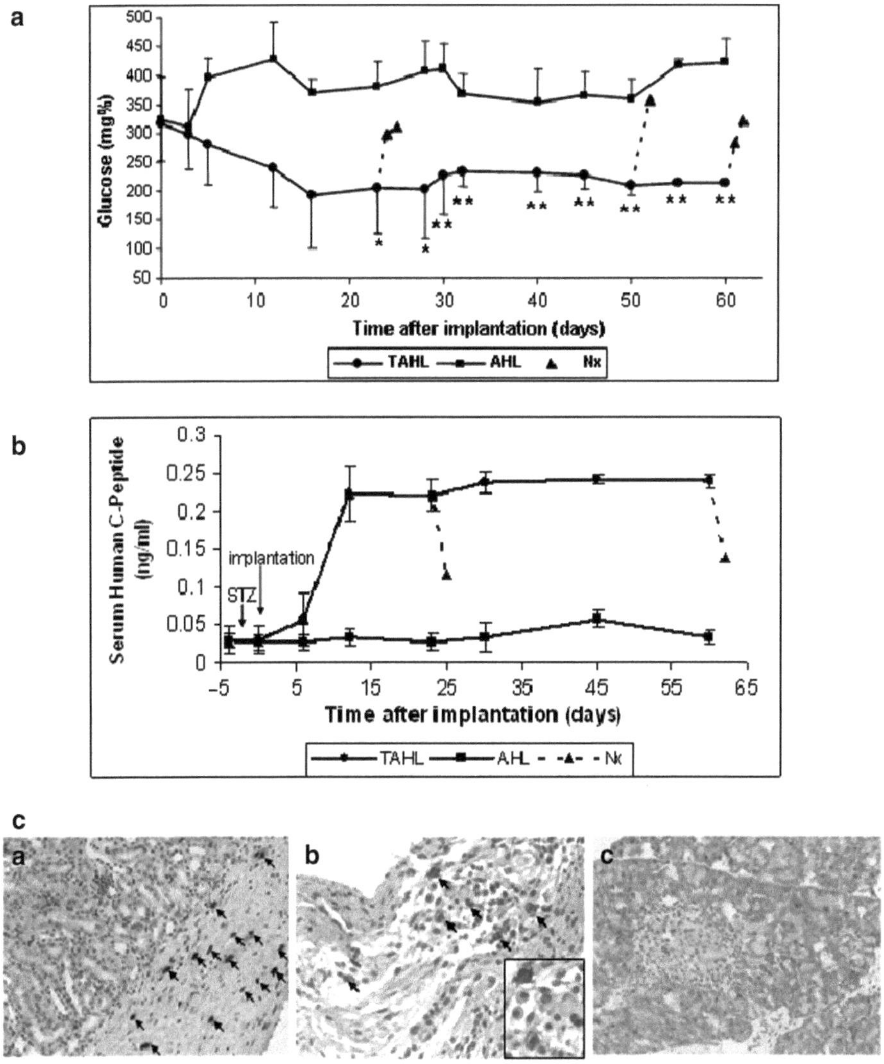

Fig. 5. Transdifferentiated liver cells ameliorate hyperglycemia in NOD–SCID mice. (**a**) Diabetic NOD–SCID mice were implanted under the kidney capsule with 7×10^6 PDX-1 treated adult human liver cells ($n = 15$) or with untreated liver cells ($\boldsymbol{n} = 9$). Glucose levels at the indicated time points after implantation are presented as a mean ± SE in mg%. *Asterisks* denote a significant difference ($^*p < 0.05$, $^{**}p < 0.01$) between the glucose levels of mice implanted by PDX-1-treated liver cells and these implanted by untreated cells. *Dotted lines* denote glucose levels measured after nephrectomy (Nx) at the indicated time points. (**b**) Serum human C-peptide levels in mice implanted by PDX-1 treated liver cells ($n = 10$) or with untreated liver cells ($n = 7$). (**c**) Immunohistochemical analysis of PDX-1 (A) and insulin (B) in the kidney capsule sections, 10 days after transplantation of PDX-1 treated liver cells. Inner panel (B) demonstrates an enlarged magnification of insulin positive liver cells. (**c**) Insulin staining of the same NOD–SCID mouse pancreas. *Arrows* indicate positive cell staining. Original magnifications, ×400 (A, B) and ×200 (C). Reproduced with permission from Sapir et al. (3)

5. CsCl gradient: Light CsCl (0.45 g/mL in 5 mM HEPES pH 7.8) and Heavy CsCl (0.613 g/mL in 5 mM HEPES pH7.8) are stored in 4°C until use. Add 4 mL of light CsCl to Beckman centrifuge tube and then add 4 mL of heavy CsCl very slowly under the light CsCl.

6. Glucose is dissolved at 1 M in KRB solution and stored in aliquots at −20°C.

2.4. Luciferase Assay

There are numerous commercial kits available for Luciferase Assay. We currently use the Luciferase Assay System (Promega, E1500), and elaborate on that protocol. However, any Luciferase Assay protocol can be used to quantify the luciferase activity.

1. Luciferase Cell Culture Lysis Reagent (LCCL), add 4 volumes of water to 1 volume of 5× lysis buffer. Equilibrate 1× lysis buffer to room temperature before use.
2. Luciferase Assay Reagent is prepared by reconstituting Luciferase Assay Substrate with Luciferase Assay Buffer. Reconstituted Luciferase Assay Reagent should be stored in aliquots at −20°C for up to 1 month or at −70°C for up to 1 year. Thaw Luciferase Assay Reagent on ice and mix well before use.

2.5. Isolation of RNA from Human Liver Cells Using TRI Reagent

1. TRI reagent: TRI reagent is a complete and ready-to-use reagent, which is available from different manufactures. Store the TRI Reagent solution at 4°C. The TRI Reagent solution contains a poison (phenol). Use gloves and other personal protection when working with TRI Reagent solution.
2. 1-Bromo-3-Chloropropane (BCP) is a complete and ready-to-use reagent, which is available from different manufactures. Store the BCP solution at 4°C.
3. Prepare 75% ethanol by mixing 25 mL of nuclease-free water with 75 mL 100% ethanol.

2.6. DNase Treatment (DNA-free™, Ambion)

There are numerous commercial kits available for DNase treatment. We currently use the DNase Treatment (DNA-free™, Ambion), and elaborate on that protocol. However, any DNase can be used to degrade the genomic DNA from the RNA sample.

2.7. Reverse Transcriptase Reaction

There are numerous commercial kits available for reverse transcriptase reaction. We currently use the high capacity cDNA reverse transcription kit (Applied Biosynthesis), and elaborate on that protocol. However, any reverse transcriptase enzyme or kit can be use to transcriptase cDNA on the basis of the mRNA.

2.8. Quantitative RT-PCR Analysis Using SYBR® Green

There are numerous commercial SYBR® Green mixes available for Quantitative RT-PCR analysis. We currently use the SYBR® Green mixes from Applied Biosynthesis, and elaborate on that protocol. However, any SYBR® Green mixes can be use for the real-time PCR reactions.

1. SYBR® Green mix is a complete and ready-to-use reagent; store the mix in working aliquots at 4°C.
2. Primers; upon receiving a new primer, reconstitute it in nuclease-free water (100 µM) and store at −20°C. Prepare a primers' working solution of 10 µM in nuclease-free water for the reactions. Primers used: Insulin (NM_000207) F: GCAGCCTTTG-TGAACCAACA, R: CGGGTCTTGGG-TGTGTAGAAGAAG; Glucagon (NM_002054.2) F: CCA-AGATTTTGTGCAGTGGT, R: GGTAAAGGTCCCTTC AGCAT; Somatostatin (NM_001048.3) F: ATGATGCCCT GGAACCTGAAG, R: GCCGGGTTTGAGTTAGCAGAT. Endogenous Control, β-actin (NM_001101) F: TTGCCG_ ACAG- GATGCAGAA, R: GCTCAGGAGGAGCAATG-ATCTT.

2.9. Quantitative RT-PCR Analysis Using TaqMan® Probes

There are numerous commercial TaqMan reaction mixes and TaqMan® probes available for Quantitative RT-PCR analysis. We currently use the TaqMan reaction mixes and TaqMan® probes from Applied Biosynthesis, and elaborate on that protocol. However, any TaqMan reaction mixes and TaqMan® probes can be used for the real-time PCR.

1. TaqMan reaction mix is a complete and ready-to-use reagent; store the mix in working aliquots at 4°C.
2. TaqMan® probes are complete and ready-to-use reagents. Aliquot for a single use, and store at −20°C. The probes used: human β-actin Hs99999903_m1, human Insulin Hs00355773_ m1, human Glucagon Hs00174967_m1, and human Somatostatin Hs00356144_m1.

2.10. Western Blot Analysis

1. Lysis buffer: 0.15 M NaCl, 5 mM EDTA (pH 8), Triton X-100 1%, 10 mM Tris–Cl pH 7.4. The buffer can be aliquoted and stored at 4°C for 2 weeks, or at −20°C for longer periods of time. Before use, add dithiothreitol 5 mM (freshly prepared), 100 mM phenylmethanesulfonyl fluoride (PMSF, from stock solution of 100 µM in isopropanol, stored at −20°C), and 5 mM ε-aminocaproic acid (prepared from 5 M stock solution in water, stored at −20°C).
2. Gel running buffer: Tris(hydroxymethyl)aminomethane 1.5 M, 0.1% sodium lauryl sulfate, adjust pH 8.8 using concentrated hydrochloric acid. Store the buffer at 4°C.
3. Glycerol (50%), dilute glycerol in water, and store at 4°C.
4. Acryl Amide/bis solution 30% (Bio-Rad).
5. Ammonium persulfate; stock solution of 10% is freshly prepared.
6. *N*,*N*,*N*′,*N*′-Tetramethylethylenediamine (TEMED, Bio-Rad).

7. Stacking gel buffer: 0.5 M (hydroxymethyl)aminomethane, sodium lauryl sulfate 0.1%, adjust pH 6.8 using concentrated hydrochloric acid. Store the buffer at 4°C up to 6 months.
8. Running buffer (×10): 0.25 M Trizma base, 1.92 M glycine, 0.1% sodium lauryl sulfate. Do not adjust the pH and dilute in water before use.
9. Blotting buffer (×10): 0.25 M Trizma base, 1.92 M glycine. Do not adjust the pH, dilute in water before use.
10. Ponceau S solution: Ponceau S 0.1%(w/v), in acetic acid 5%(v/v). Store at 4°C.
11. Blocking buffer; bovine serum albumin (Fraction V) 1% (w/v), Tween 20 0.05% (v/v) in PBS. Keep at 4°C to prevent bacterial contamination.
12. Antibodies, dilute in bovine serum albumin (Fraction V) 0.1% (w/v), 0.05% Tween 20 (v/v) in PBS. Antibodies used: rabbit anti-PDX-1 1:5000 (a gift from C.V.E. Wright, USA), mouse anti-NKX6.1, mouse anti-Ngn3, mouse anti-Pax6 all from Developmental Studies Hybridoma Bank (University of Iowa, 1:1000), rabbit anti-VP16 (1:1000, Sigma), and rabbit anti-MafA (1:1000, Abcam). Secondary antibodies, goat anti-Rabbit-Peroxidase conjugated and anti-mouse-Peroxidase conjugated both from Jackson ImmunoResearch (1:10000).
13. Wash buffer: 0.05% Tween 20 (v/v) in PBS.

2.11. Immunofluorescence Analyses

1. Paraformaldehyde: prepare a 4% (w/v) solution in PBS. The solution may need to be carefully heated (use a stirring hot-plate in a fume hood) to dissolve, and then adjust it to room temperature before use; store individual use aliquots at −20°C.
2. Permeabilization solution: 0.1% (v/v) Triton X-100 in PBS.
3. Blocking solution: 3% (w/v) bovine serum albumin in PBS.
4. Antibody dilution buffer: 3% (w/v) bovine serum albumin in PBS.
5. Nuclear stain: 300 nM 4,6-diamidino-2-phenylindole (DAPI) in water.
6. Mounting medium: Fluoromount-G (SouthernBiotech).

2.12. Insulin and C-Peptide Content and Glucose Regulated Secretion

1. Hormones extraction buffer: 35% ethanol in 0.18 N HCl.
2. Krebs–Ringer Buffer medium from KRB stock solutions (stored at 4°C for up to 1 year): Solution 1: 0.46 M NaCl (13.44 g NaCl in 500 mL final volume); Solution 2: 20 mM KCl (0.746 g), 40 mM $NaHCO_3$ (1.68 g); 10 mM $MgCl_2{\cdot}6H_2O$ (1.1017 g) in 500 mL final volume; Solution 3: 10 mM $CaCl_2{\cdot}2H_2O$ (0.735 g) in 500 mL final volume. On

day of use, mix 20 mL of each three solutions above in a 100-mL flask with stir bar. Adjust pH for 20 min by bubbling CO_2 into the solution, then add 0.3813 g HEPES (final HEPES concentration = 20 mM) and 0.08 g bovine serum albumin (fraction V, ultrapure). Fully dissolve solids, and adjust volume to 80 mL with water and the pH to 7.4.

2.13. Amelioration of Hyperglycemia Upon Implantation in Diabetic SCID–NOD Mice

1. Matrigel Matrix (BD) forms a gel above 10°C, thus Matrigel Matrix solution should be kept at low temperatures, and all equipment and reagents (syringes, needles, Matrigel Matrix solution, etc.) should be chilled on ice throughout the process. The Matrigel Matrix solution should be thawed on ice, aliquots, and stored at −20°C. An aliquot should be thawed on ice. Cells pellet is resuspended in 50 μL medium and kept on ice. Mix 50 μL of the matrix with the cold cells and immediately draw in 1 mL cold syringe.
2. NOD/SCID mice are bred and housed under pathogen-free conditions under a 12-h light/dark cycle. Experiments are carried out under the supervision and guidelines of the Institutional Animal Welfare Committee. Seven to eight weeks old males were used for the implantation experiments (18–19 g.)
3. Streptozotocin (STZ, Sigma) is dissolved in Na–Citrate Buffer: dissolve 1.47 g of Na–Citrate in 50 mL water, if necessary adjust buffer to 4.5 pH, using monohydrate Na–Citrate solution. The buffer should be made freshly. Streptozotocin should be stored at −20°C. Weigh the appropriate amount of STZ so your final concentration in the Na–Citrate Buffer will be 7.5 mg/mL and place this into an Eppendorf tube; cover with aluminum foil (light sensitive).
4. Blood glucose was measured by glucometer using disposable test strips (Accutrend® GC, Roche Applied Science).
5. Anesthesia ketamine/xylazine; from commercial stocks of 100 mg/mL ketamine and 100 mg/mL xylazine, make a final concentration of 18 mg/mL ketamine and 1.8 mg/mL xylazine in PBS and use freshly.

3. Methods

3.1. Adult Liver Cells (see Note 1)

3.1.1. Liver Cells Isolation

1. A liver biopsy (at least 2 g) is stored under sterile conditions in a sterile container with cold HBSS (solution must cover the tissue). Store the container at a 4°C refrigerator until isolation (up to 6 h) or on ice (see Note 2).
2. Clean container with 70% ethanol and precede isolation in a laminar flow cabinet.

3. Take tissue out from the container and transfer into a sterile culture plate, in a laminar flow cabinet.
4. Wash tissue three times with cold HBSS on a culture plate, to get rid of blood.
5. Transfer tissue to a new culture plate and cut into 1–2 mm pieces with sterile scalpels (until goes through a 25-mL pipette).
6. Transfer into 50-mL conical tubes with 18 mL HBSS.
7. Add 4.5 mL EGTA 5 mM and gently pipette up and down (>15 times) for Ca^{2+} chelation and increase Collagenase function (below in 10).
8. Centrifuge for 2 min, 660 × *g*, 4°C.
9. Aspirate and discard supernatant.
10. Add collagenase, resuspend with the liver pellet, and transfer to a presterilized Erlenmeyer with a magnetic bar (25 mL of collagenase will be enough for a 2-g tissue).
11. Transfer Erlenmeyer to a magnetic platform in a water bath prewarmed at 37°C. Stir gently for 20 min, or until tissue dissociates (see Note 3).
12. Stop digestion by transferring the Erlenmeyer to an ice bucket and add 2 volumes of Isolation Stop Solution.
13. Transfer into sterile 50 mL conical tubes and centrifuge for 5 min, 815 × *g*, 4°C. Wash cell pellet again with Isolation Stop Solution, centrifuge for 5 min, 815 × *g*, 4°C.
14. Resuspend the cell in medium.
15. Plate the cells on fibronectin-coated tissue culture plates, and incubate at 37°C in a humidified atmosphere of 5% CO_2 95% air.
16. Five to six hours after plating, collect supernatant.
17. Wash original plates twice with HBSS and refeed with fresh medium.
18. Centrifuge collected unattached supernatant (from step 16) for 5 min, 660 × *g*, 4°C (see Note 4).
19. Plate on new fibronectin-coated plates.
20. At the following morning, wash all plates twice with HBSS and replace with new medium.
21. Repeat the wash and medium replacement every day for the first week or until culture reaches confluency.

3.1.2. Primary Cultures of Liver Cells: Maintenance and Treatment (see Note 5)

Liver cells are passage when approaching confluence using trypsin/EDTA.

1. Wash the cells with PBS.

2. Add trypsin/EDTA to the plates (2 mL for 100 mm dish; enough to cover the cells), incubate for 5 min at 37°C in a humidified atmosphere of 5% CO_2 95% air.
3. Add an equal volume of medium to stop the trypsin enzymatic activity.
4. Collect the cells by centrifugation for 5 min, 660 × *g*, at room temperature (RT).
5. Aspirate and discard supernatant, resuspend the cell pellet in 1 mL of medium.
6. Count the cells number; take 20 μL of cells suspension, add equal volume of Trypan blue solution (0.5%), and mix thoroughly. Allow to stand for 5 min at 15–30°C (room temperature).
7. Fill a hemocytometer for cell counting. And under a microscope count the cell number, excluding nonviable (blue stained) cells.
8. Plate 250,000 cells per 100 mm dish, add 10 mL of DMEM and incubate at 37°C in a humidified atmosphere of 5% CO_2 95% air.
9. *Adenovirus infection*: we found that to achieve maximal infection adenoviruses infection should be performed when cells are split. Add the viruses directly to the culture medium when platting the cells. Infection is preformed in viral concentration raging between 1 and 1,000 multiplicity of infection (moi). Multiplicity of infection represents the number of virions per cell.
10. *Growth factors and soluble factors treatment*: All growth factors and soluble factors are added directly to the culture medium. The factors can be added before, with or after the viral infection, alone or in combinations (see Figs. 2 and 4). The recommended final concentrations for human liver cells in vitro are: Nicotinamide, 10 mM; Epidermal Growth Factor, 20 ng/mL; Exendin-4, 5 nM; Betacellulin, 4 nM; Activin-A, 4 nM or 100 moi Ad-CMV-Activin-A; and HGF 10 nM or 5 moi Ad-CMV-HGF.
11. *Serum deprivation*, removal of the serum restricts the rate of proliferation and may promote differentiation. Thus, serum deprivation for transdifferentiated liver cells (after treatment with adenoviruses) increased in our experiments the maturation along the pancreatic lineage and increased the functional outcome. Replace the culture medium with serum-free medium when cells reach 80% confluency. Do not use serum-free medium while platting the cells, wait at least 24 h after plating; anti-trypsin inhibitor present in the serum is needed to neutralize the trypsin function.

3.2. Adenoviruses

For inducing developmental redirection, all recombinant adenoviruses are replication deficient. Since the ectopic transcription factor plays a short-term role as a trigger, both the nonintegrating recombinant adenovirus and the ectopic gene expression vanish as the cells replicate (31–33).

There are several types of recombinant adenoviruses that are used for gene delivery, they are distinctly generated and stored and similarly propagated in 293 cells that supply the viral "missing" viral gene products, needed for propagation:

1. First generation E1-deleted recombinant adenovirus (FGAD), following delivery, transgene expression is at a very high level. But, it decreases rapidly after several days, being low or undetectable after several weeks. The desired genes are subcloned in pAC-plasmid which then undergoes homologous recombination in 293 cells with another plasmid JM-17 which contains most of the viral genome except E-1. The generated virions which contain now the subcloned desired genes propagate in the 293 cells (34).
2. The Ad-Easy system was developed by Vogelstein and colleagues (35). The Ad-Easy system simplifies and speeds up the process of generating a recombinant adenovirus, compared with traditional methods of preparation. The vector contains most the genome of the human adenovirus serotype 5 and is deleted from the E1 and E3 genes.
3. Helper-dependent adenoviral vectors (HdAd) lacking all viral coding sequences display only minimal immunogenicity and negligible side-effects, allowing long-term transgene expression and negligible side-effects (36).

   In our study, we are using both the first generation E1-deleted recombinant adenovirus as Ad-CMV-GFP, Ad-CMV-PDX-1, Ad-RIP-GFP, and Ad-CMV-NKX6.1 (3, 5) and Ad-Easy adenoviruses as Ad-CMV-PDX-1-VP16, Ad-CMV-NeuroD1, and Ad-CMV-Pax4 (12, 24). The propagation process of these adenoviruses although similar requires some modifications.

3.2.1. Propagation of First Generation Recombinant Adenovirus

1. Viral starter preparation: Culture 293 cells in 2 × 100 mm culture dishes, and grow them to 80–90% confluence.
2. Add 25 μL of purified virus (concentration 10^{11}–10^{12} pfu) to each dish and incubate for 48 h.
3. Collect the cells and medium in a 50-mL conical tube, and centrifuge at 4°C 500 × *g* for 10 min.
4. Collect the supernatant which is then used as a starter. The starter can be used freshly, kept in 4°C for up to 2 weeks or −70°C for long periods of time.
5. Culture 293 cells in 24 × 14 cm^2 culture dishes or 18 × 75 cm^2 flasks to 80–90% confluence.
6. Dilute the starter from step 4 in 240 mL of medium (or 180 for flasks).

7. Aspirate the medium and add 10 mL of the medium + starter to each dish/flask.
8. Incubate the cells at 37°C for 90 min.
9. Add additional 30 mL of medium and culture the cells until all the cells round up and detach (about 48 h).
10. Collect the cells and medium in 50-mL conical tubes, and centrifuge at 4°C 800 × *g* for 10 min.
11. Collect the medium in 75 cm^2 flasks (up to 100 mL medium per flask) and discard the cells.
12. Add PEG/NaCl, 50 mL for each 100 mL of medium, rock the flask (300 RPM) at 4°C for 12–18 h (avoid foaming).
13. Centrifuge 16,000 × *g* for 15 min at 4°C in Sorval centrifuge.
14. Aspirate/discard the medium and collect the viral pellet (you may reuse the same tubes to collect all viral stock).
15. Resuspend the viral stock in Adenoviruses in Saline (8 mL for the whole preparation; 24 plates); leave the saline with the viral pellet to stand in 4°C for 24 h until full resuspension.
16. Aliquot the virus and keep in 4°C for 3–4 months or –70°C for longer periods.
17. Before use analyzed for:
    a. Viral particles titration – see below.
    b. Contamination: add 10 μL of the new stock of viruses to culture dish containing medium (with or without cells), and leave for a week. Use only if the viruses are sterile.

3.2.2. Propagation of Ad-Easy Adenoviruses

1. Culture 293 cells in 2 × 100 mm culture dishes, and grow them to 80–90% confluence.
2. Add 25 μL of purified virus (concentration 10^{11}–10^{12} pfu) to each dish and incubate for 48 h (the medium should be yellow and the cells detach).
3. Collect the cells and the medium in a 50-mL conical tube.
4. Perform four freeze–thaw cycles: freeze the cells and medium in dry ice or –70°C and then thaw in 37°C.
5. Centrifuge 4,000 × *g*, 20 min.
6. Collect the supernatant as starter. The starter can be used freshly, or keep at –70°C for long periods of time.
7. Culture 293 cells in 50 × 14 cm^2 culture dishes or 36 × 75 cm^2 flasks and grow them to 70–80% confluence.
8. Dilute the starter from step 6 in 500 mL of medium (or 360 mL for flasks).
9. Aspirate the medium and add 10 mL of the medium + starter to each dish/flask.
10. Incubate the cells at 37°C for 90 min.

11. Add 30 mL of medium and culture the cells until a prominent cytopathic effect (CPE) was reached in cells (48–72 h).
12. Collect the cells and medium in a 50-mL conical tube, and centrifuge at 4°C 800 × *g* for 10 min.
13. Aspirate medium and resuspend in 8 mL 293 medium.
14. Perform four freeze–thaw cycles as in step 4.
15. Centrifuge 4,000 × *g*, 20 min, harvest the supernatant, save at −70°C for long periods.
16. Set the $CsCl_2$ gradient: (two tubes) add 4 mL of the light CsCl to the bottom of the tube, then add 4 mL of the heavy CsCl to the bottom; the light $CsCl_2$ will then move up.
17. Add 4 mL of virus on top of the gradient.
18. Ultracentrifuge 100,000 × *g* at 4°C for 3 h.
19. Harvest the lower band with side puncture needle; add 5 mM HEPES pH 7.8 to reach 4 mL final volume.
20. Set up another CsCl gradient (same as above) and load the viruses or 5 mM HEPES (as balance).
21. Ultracentrifuge 100,000 × *g* at 4°C overnight >18 h.
22. Extract the viral band as in step 19, keep the virus on ice.
23. Clean the viral particles using the PD10 column (GE life science, according to manufacture protocol).
24. Cut the column's tip.
25. Equilibrate the column by adding 25 mL of PBS, let the PBS to drip.
26. Adjust the viral stock volume (from step 22) to 2.5mL with PBS.
27. Load the viruses on the column and let the liquid to drip.
28. Add 3.5 mL of PBS to elute the viruses.
29. Add glycerol to final concentration of 10% (approximately 380 μL).
30. Aliquot the virus stock and store in −70°C.
31. Before use analyzed for:
    a. Viral particles titration – see below.
    b. Contamination: add 10 μL of the new stock of viruses to culture dish containing medium (with or without cells), and leave for week. Use only if no bacterial or fungi contamination was found (see Note 6).

3.2.3. Viral Particle Titration: TCID50 Methods (37)

1. Collect and count 293 cells as previously described (Subheading 3.2.1).
2. Prepare 20 mL of cell suspension at 10^5 cells/mL in medium (containing 2% FBS).

3. Dispense 100 µL (10^4 cells) per well in 2 × 96-well flat bottom plates, using a 12-channel pipette and a sterile reservoir. Allow cells to adhere for 6 h.
4. Label in duplicate eight 5-mL sterile disposable tubes for serial dilutions.
5. Dispense 0.9 mL of the medium into the first tube, and 1.8 mL in all others. Add 0.1 mL of the viral stock into the first tube (this will be 10^{-1} dilution tube).
6. Pipette up and down five times to mix.
7. Take 0.2 mL of 10^{-1} dilution and transfer to the second tube.
8. Pipette up and down five times to mix.
9. Repeat dilutions up to the highest dilution desired.
10. Perform the second series of dilutions from the same viral stock.
11. For each row of the 96-well plate, dispense 0.1 mL/well in the wells #1 to #10 (10 wells per dilution) for the 8 highest dilutions. Columns #11 and #12 are used for the negative control. Add 0.1 mL/well of DMEM 2% to each well in columns #11 and #12 to test the cell viability. When distributing the dilutions, always start with the highest dilution in the top row.
12. Incubate the cells at 37°C in a humidified atmosphere of 5% CO_2 95% air for 10 days.
13. After 10 days read the plates using an inverted microscope. Observe the wells and count observable CPE per row. A well is counted as positive even if only a small spot or a few cells show CPE. If in doubt between CPE and dead cells, compare with the negative control.
14. Determine the ratio of positive wells per row. The test is valid if the negative controls do not show any CPE or cell growth problems and the lowest dilution shows 100% infection (10/10) while the highest dilution shows 0% infection (0/10).
15. The titer determined using the KÄRBER statistical method: for 100 µL of dilution, the titer is $T = 10^{1+d(S-0.5)}$, where d is the Log 10 of the dilution (=1 for a tenfold dilution) and S is the sum of ratios (always starting from the first 10^{-1}dilution).

3.3. Activation of the Pancreatic Lineage in Adult Human Liver Cells, In Vitro

3.3.1. Determine the Efficiency of Adenoviruses Infection of Adult Human Liver Cells, In Vitro

1. Prepare serial dilution of Ad-CMV-GFP in duplicate; use adenoviruses concentration of 10, 100, and 1,000 moi. Multiplicity of infection represents the number of virions per cell. Since we are infecting 100,000 cells for 10 moi used 10^6 virions, for 100 moi used 10^7 virions, and for 1,000 moi used 10^8 virion. Prepare the dilutions in 1 mL of medium.
2. Collect and count liver cells as previously described (Subheading 3.1.2).
3. Plate 100,000 cells in 8 × 60 mm dishes in 4 mL of Medium.

4. Immediately add to six dishes the serial dilution of Ad-CMV-GFP. To the two additional dished add 1 mL of Medium without viruses, the control dishes.
5. Incubate the cells at 37°C in a humidified atmosphere of 5% CO_2 95% air.
6. After 24 and 48 h observed the cultures using inverted microscope using FITC/GFP filter (Excitation 455–495 Emission 510–520) as demonstrated in Fig. 6 (see Note 7).

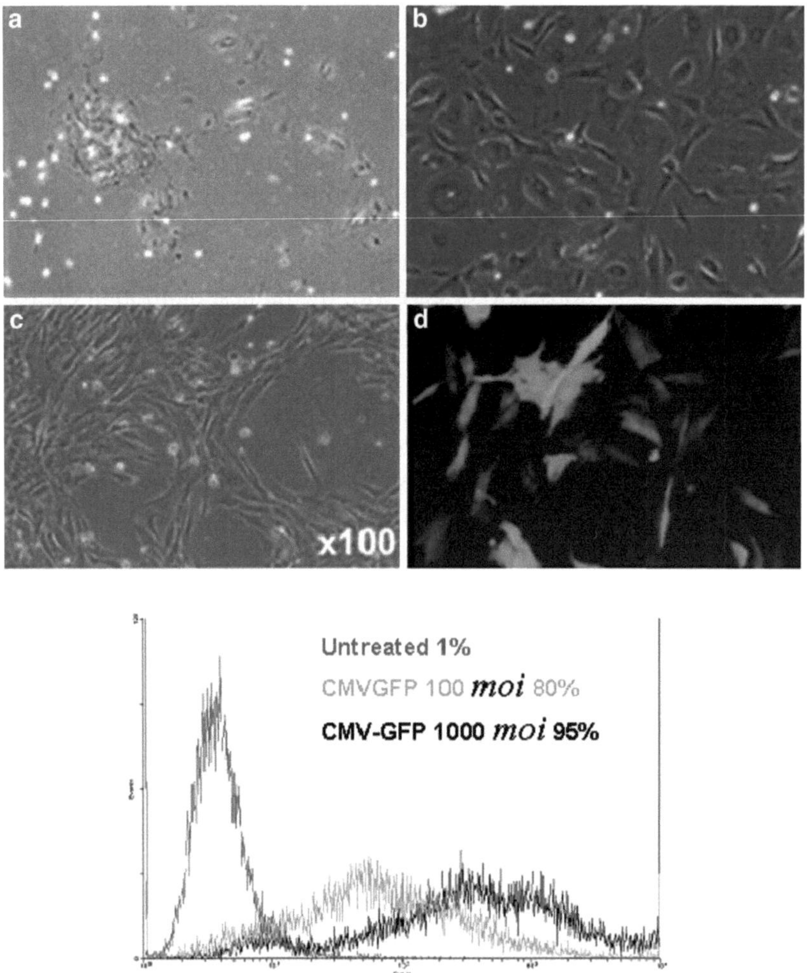

Fig. 6. Primary cultures of adult human liver cells are highly infected by recombinant adenoviruses. Human liver cells were isolated and cultured as described (Subheading 3.1). Representative phase contrast morphology (**a–c**), at passage 0 (3 days postisolation), (**a**), at passage 1 (10 days postisolation), (**b**) and at passage 5 (**c**). Cells at passage 5 (**c–d**) were infected with 1,000 moi Ad-CMV-GFP as described (Subheading 3.1), (**d**) GFP fluorescent of the same field of cells as (**c**). Quantification of the percentile of Ad-CMV-GFP infected cells using fluorescence-activated cell sorting (FACS). **a** and **b**, reproduced with permission from Sapir et al. (3)

7. Observe the number of cells, their overall condition and calculate the percentile of cells that expressed the GFP reporter.
8. Choose the optimal concentration of viruses to further analyze the transdifferentiation process. Take in account the percentile of infection and the survival rate. This concentration should be used in all the experiments that preformed with the analyzed primary liver cells culture.

3.3.2. Activation of Ectopic Insulin Promoter

The activation of an ectopic insulin promoter indicates the first and least stringent levels of β-cell lineage activation.

3.3.2.1. Using Ad-RIP-GFP

1. Collect and count liver cells as previously described (Subheading 3.1.2).
2. Plate 100,000 cells in 8 × 60 mm dishes in 4 mL of medium containing growth factors as Nicotinamide, 10 mM; Epidermal Growth Factor, 20 ng/mL to improve the transdifferentiation process.
3. Infect the cells as previously described using the optimal concentration of viruses (Subheading 3.3.1), prepare two dishes infected with Ad-RIP-GFP and Ad-CMV-β-gal (negative control, for leaky activity of RIP), two dishes coinfected with both Ad-RIP-GFP and Ad-PDX-1 (the experiment) or infected with Ad-RIP-GFP-CMV-PDX-1 (a bifunctional recombinant adenovirus that carries the expression of both genes under the control of distinct promoter), two dishes infected with Ad-CMV-GFP as positive control for the percent of infected cells, and leave two dishes uninfected as additional negative controls.
4. Incubate the cells at 37°C in a humidified atmosphere of 5% CO_2 95% air.
5. After 48 and 72 h analyze the cultures using inverted fluorescent microscope using FITC/GFP filter (Excitation 455–495 Emission 510–520).
6. Determine the number of cells, their overall condition and calculate the percent of cells that expressed the GFP reporter. The percent of cells capable of activating the insulin promoter is calculated from the number of GFP positive cells after RIP-GFP and PDX-1 treatment divided by the number of cells that are positive to GFP upon CMV-GFP treatment multiplied by 100. Usually liver cells treated by RIP-GFP alone do not express fluorescence. A small correction should by applied for double infection (percent of GFP positive cells when infected by similar moi of CMV-β-gal and CMV-GFP, as PDX and RIP-GFP).
7. To calculate the percent of GFP positive cells, a fluorescence-activated cell sorting (FACS) analysis is used.

8. Collect the cells using trypsin as previously described (Subheading 3.2.1) to 15 mL conical tubes.
9. Wash the cells twice with Dulbecco's Phosphate Buffered Saline (without Calcium or Magnesium). Add 5 mL of PBS to each tube, centrifuge for 5 min, 660 × *g*, 4°C, aspirate, and discard supernatant.
10. Resuspend the cell pellet in 0.5 mL PBS. Transfer the cells to specific FACs tubes (polystyrene round bottom 12 × 75 mm tubes).
11. Analyze the GFP florescence of the cells using the FACs and the software available in your institute as FACS Calibur (Becton Dickinson, Heidelberg, Germany) using the CellQuest program.

3.3.2.2. Using Ad-RIP-Luciferase

Using the reporter gene luciferase under the control of the ectopic insulin promoter, gives a quantitative representation of the insulin promoter activity upon distinct treatments of transcription and soluble factors.

1. Collect and count liver cells as previously described (Subheading 3.1.2).
2. Plate 100,000 cells in 8 × 60 mm dishes in 4 mL of medium containing growth factors as Nicotinamide, 10 mM; Epidermal Growth Factor, 20 ng/mL to improve the trans-differentiation process.
3. Infect the cells as previously described using the optimal concentration of viruses (Subheading 3.3.1), prepare two dishes infected with Ad-RIP-Luciferase, two dishes coinfected with both Ad-RIP-Luciferase and Ad-PDX-1, two dishes infected with Ad-CMV-Luciferase as positive control, and leave two dishes uninfected as negative controls and baseline determination.
4. Incubate the cells at 37°C in a humidified atmosphere of 5% CO_2 95% air for 48–72 h.
5. Carefully remove the growth medium from cells to be assayed. Rinse cells with PBS, being careful to not dislodge attached cells. Remove as much of the PBS rinse as possible.
6. Add 400 μL of Luciferase Cell Culture Lysis Reagent (×1).
7. Rock culture dishes several times to ensure complete coverage of the cells with lysis buffer. Scrape attached cells from the dish. Transfer cells and all liquid to a microcentrifuge tube. Place the tube on ice.
8. Vortex-mix the microcentrifuge tubes for 10–15 s, then centrifuge at 12,000 × *g* for 15 s (at room temperature), or up to 2 min (at 4°C).
9. Transfer the supernatant to a new tube.

10. Use the cell lysate fresh or store at −70°C. If the samples are stored, they should be at ambient temperature prior to performing the luciferase activity assay.
11. Dispense 100 μL of the Luciferase Assay Reagent into luminometer tubes, one tube per sample.
12. Program the luminometer to perform a 2-s measurement delay followed by a 10-s measurement read for luciferase activity. The reading time may be shortened if sufficient light is produced.
13. Add 20 μL of cell lysate to a luminometer tube containing the Luciferase Assay Reagent (substrate). Mix by pipetting 2–3 times or vortex briefly.
14. Place the tube in the luminometer and initiate reading.

3.3.3. Molecular Analyses of the Developmental Redirection Process, Quantitative RT-PCR (see Note 8)

3.3.3.1. RNA isolation

Isolation of RNA from human liver cells using TRI Reagent.

1. Plate 250,000 liver cells in 10 cm dishes, treat the cells with recombinant adenoviruses and supplant the medium with growth factors as previously described (Subheading 3.3.1).
2. Incubate the cells at 37°C in a humidified atmosphere of 5% CO_2 95% air for 72–96 h.
3. Collect the cells using trypsin as previously described (Subheading 3.1.2) into 15 mL conical tubes.
4. Wash the cells with Dulbecco's Phosphate Buffered Saline (without Calcium or Magnesium). Add 5 mL of PBS to each tube, centrifuge for 5 min, 660 × *g*, 4°C, aspirate and discard supernatant.
5. Resuspend the pellet in 1 mL of PBS, transfer the cell suspension to sterile RNase-free 1.5-mL microcentrifuge tubes.
6. Centrifuge for 1 min to pellet the cells, and discard supernatant.
7. Add 1 mL of TRI reagent to the tubes.
8. Lyse cells by repetitive pipetting, centrifuge homogenate at 12,000 × *g* for 10 min at 4°C.
9. Transfer the homogenate to a sterile microcentrifuge tube.
10. Incubate samples for 5 min at room temperature.
11. Add 0.1 mL of 1-bromo-3-chloropropane (BCP) to each tube; shake samples vigorously for 15 s.
12. Incubate samples for 5 min at room temperature.
13. Centrifuge samples for 15 min at 12,000 × *g* at 4°C. The phases will be separated by the centrifugation.
14. Transfer the upper aqueous phase to a fresh tube.
15. Add 0.5 mL of isopropyl alcohol to precipitate RNA.
16. Incubate for 5–10 min at room temperature.

17. Centrifuge for 10 min at 12,000 × *g* at 4°C. The RNA will form a pellet on the side or bottom of the tube.
18. Discard the supernatant.
19. Wash pellet with 1 mL 75% ethanol.
20. Mix sample by vortexing. The RNA pellet may float.
21. Centrifuge at 12,000 × *g* for 5 min at 4°C. Repeat steps 18–21 twice.
22. RNA pellet may be stored in ethanol at −70°C for months.
23. For dissolving the RNA, remove supernatant.
24. Air-dry the pellet for 5–10 min. Do not completely dry out the pellet.
25. Dissolve pellet in 30–60 μL RNase-free water by passing the solution through a pipette tip and incubating for 10 min at 55–60°C.
26. RNA should be stored at −70°C, and will be stable for long periods of time.

3.3.3.2. DNase Treatment (DNA-free™, Ambion)

1. Use fresh isolated RNA or thaw on ice RNA that was kept in −70°C. Use 50 μL of the isolated RNA (add RNase-free water to adjust the volume if necessary).
2. Add 5 μL (0.1 volume) 10× DNase I Buffer and 1 μL rDNase I to the RNA, and mix gently.
3. Incubate at 37°C for 20–30 min.
4. Add 5.5 μL (0.1 volume) of DNase Inactivation Reagent and mix well.
5. Incubate 2 min at room temperature, mixing occasionally.
6. Centrifuge at 10,000 × *g* for 1.5 min and transfer the supernatant containing the RNA to a new tube.
7. Measure RNA concentration in spectrometer using absorption of light at 260 and 280 nm (A260/280). An A_{260} reading of 1.0 is equivalent to ~40 μg/mL single-stranded RNA. The A_{260}/A_{280} ratio is used to assess RNA purity and should be close to 2 for high-quality nucleic acid.
8. The RNA should be stored at −70°C, and will be stable for long periods of time.

3.3.3.3. Reverse Transcriptase Reaction

1. Thaw RNA and the kit components on ice (or use fresh).
2. Dilute 1 μg of total RNA in RNase-free water to total volume of 10 μL.
3. Calculate the volume of components needed to prepare the required number of reaction use: 2 μL RT buffer (×10), 0.8 μL dNTPs mix (100 mM), 2 μL RT Random primers (×10), 1 μL multiScribe reverse transcriptase, 1 μL RNase inhibitor, and 3.2 μL RNase-free water; total 10 μL per reaction.

4. Prepare the mix on ice and mix gently.
5. Pipette 10 μL RT mixes into each well of 96-well reaction plate or individual tube.
6. Pipette 10 μL of RNA sample (1 μg) into each well or tube. Pipette up and down twice to mix.
7. Seal the tube.
8. Briefly centrifuge the tube to spin down and to eliminate any air bubbles.
9. Load the thermal cycler. Programmed the conditions as 10 min at 25°C, 120 min at 37°C, 5 s at 85°C, and 4°C until removal.
10. The cDNA can be used directly or stored at −20°C or −70°C for long time.

3.3.3.4. Quantitative RT-PCR Analysis Using SYBR® Green

1. Each reaction (total volume 20 μL): 10 μL SYBR Green Master Mix (×2), 0.5 μL primer F (from 10 μM stock), 0.5 μL primer R (from 10 μM stock), cDNA (1 μL for high copy transcripts as β-actin up to 3 μL for low number copy transcripts as the pancreatic hormones), and RNase-free water (8 for high copy transcripts as β-actin or 6 μL for low number copy transcripts as the pancreatic hormones).
2. Make a β-actin reaction mix for all the reaction needed by mixing the SYBR Green Master Mix with the β-actin primers and RNase-free water. Always add positive control cDNA, and no-template control.
3. Pipette 19 μL reaction mix into each well of 96-well reaction plate or 8-well strips.
4. Prepare the reaction mix for each of the pancreatic hormone genes for all the reactions needed by mixing the SYBR Green Master Mix with the appropriate primers and RNase-free water. Add positive control cDNA, and no-template control.
5. Pipette 17 μL reaction mix into each well of 96-well reaction plate or 8-well strips.
6. Thaw the cDNA samples on ice.
7. Pipette the cDNA sample (1 μL for β-actin or 3 μL for pancreatic hormones) into each well or tube, pipette up and down twice to mix.
8. Seal the plate with Adhesive Film or use Ultra Clear flat Caps.
9. Briefly centrifuge the plate or the strips to spin down and to eliminate any air bubbles.
10. Set the plate to the real-time PCR. The reaction contains: initiation step at 50°C for 10 min, Taq activation at 95°C (2–15 min according to manufacture protocol), 40 cycles of denaturation step for 15 s in 95°C followed by Annealing and Extending for 1 min in 60°C. The last step is dissociation step

where the double-stranded DNA product is melted into single-stranded DNA by raising the temperature successively through 1°C step. As different amplicons melt at different temperature, this step allows the detection of nonspecific products that may contaminate the reaction.

11. Relative quantitative analysis was performed according to the comparative C_T method using the arithmetic formula $2^{-(\Delta\Delta Ct)}$. The cDNA levels were normalized to human β-actin cDNA.

3.3.3.5. Quantitative RT-PCR Analysis Using TaqMan® Probes

1. Each reaction of total volume 20 μL: 10 μL TaqMan Master Mix (×2), 1 μL TaqMan gene expression assay (contains both primers and probe), cDNA (1 μL for high copy transcripts as β-actin up to 3 μL for low number copy transcripts as the pancreatic hormones), and RNase-free water (9 for high copy transcripts as β-actin or 7 μL for low number copy transcripts as the pancreatic hormones).
2. Make a β-actin reaction mix without the cDNA for all the reactions needed, add positive control cDNA, and no-template control.
3. Pipette 19 μL reaction mix into each well of 96-well reaction plate or 8-well strips.
4. Make a reaction mix without the cDNA for each of the pancreatic hormone genes for all the reaction, add positive control cDNA, and no-template control.
5. Pipette 17 μL reaction mix into each well of 96-well reaction plate or 8-well strips.
6. Thaw the cDNA samples on ice.
7. Pipette the cDNA sample (1 μL for β-actin or 3 μL for pancreatic hormones) into each well or tube, pipette up and down twice to mix.
8. Seal the plate with Adhesive Film or use Ultra Clear flat caps.
9. Briefly centrifuge the plate or the strips to spin down and to eliminate any air bubbles.
10. Set the plate to the equipment.
11. The reaction contains: initiation step at 50°C for 10 min, Taq activation at 95°C (2–15 min according to manufacture protocol), 40 cycle of denaturation step for 15 s in 95°C followed by Annealing and Extending for 1 min in 60°C.
12. Relative quantitative analysis was performed according to the comparative C_T method using the arithmetic formula $2^{-(\Delta\Delta Ct)}$. The cDNA levels were normalized to human β-actin cDNA.

3.3.4. Cellular Analysis of the Developmental Redirection Process

3.3.4.1. Western Blot Analysis

3.3.4.1.1. Preparation of Western blot

1. Plate 250,000 liver cells in 10 cm dishes, treat the cells with recombinant adenoviruses and supplant the medium with growth factors as previously described (Subheading 3.2.1).
2. Incubate the cells at 37°C for 72–96 h. Collect by trypsinization as previously described (Subheading 3.2.1).
3. Lyse the cell pellet with 100 μL lysis buffer on ice for 10 min.
4. Spin at 16,000 × *g* in an Eppendorf microfuge for 10 min at 4°C.
5. Transfer the supernatant to a new tube and discard the pellet.
6. Determine the protein concentration using Bradford assay (Bio-Rad).
7. Take 40–50 μg (the volume should be less than the 60 μL for 8 lanes gels or less than the 35 μL for 15 lanes gels) and mix with 8–10 μL of Lane Marker Reducing Sample Buffer (×5, Pierce).
8. Boil for 5 min.
9. Cool at RT for 5 min.
10. Flash spin to bring down condensation prior to loading gel. *Polyacrylamide gel using* Mini-PROTEAN 3 Electrophoresis System (Bio-Rad).
11. Assemble the Mini-PROTEAN 3 Electrophoresis System; assemble the glass plates and spacers (1.5 mm thick) using the casting stand and casting frames, according to manufactures protocol.
12. Prepare the 10 mL of running gel; 2.3 mL water, 2 mL Glycerol (50%), 2.5 mL running gel buffer, 3.3 mL Acryl amide 30%, 120 μL ammonium persulfate (10%), and 15 μL TEMED.
13. Pour the running gel to about 1 cm below the wells of the comb (~7.5 mL).
14. Seal with 1 mL water.
15. When gel has set, pour off the water (the gel can be left at 4°C up to 48 h wrapped in plastic sheet to prevent dehydration).
16. Prepare the 3 mL of stacking gel; 6.3 mL water, 2.5 mL stacking gel buffer, 1.2 mL Acryl amide 30%, 100 μL ammonium persulfate (10%), and 10 μL TEMED.
17. Pour the stacking gel (~2.5 mL) and insert the comb immediately, avoid trapped air bubbles between the comb and the gel.
18. When the stacking gel has set, place in gel in the electrophoresis module.

19. Fill the electrophoresis module with running buffer, remove the comb, and flush the wells out thoroughly with running buffer.
20. Load the samples into the wells; load Prestained Protein Standards in the first lane.
21. Run with constant voltage (150 V) with current set at >200. Usual running time is about 1.5 h.
22. Prewet the sponges, filter papers (slightly bigger than gel) and Pure Nitrocellulose membrane (0.45 μM size of membrane 7 × 8.4 cm), in 1× Blotting buffer.
23. Assemble "sandwich" for Bio-Rad's Trans-blot, using the gel holder cassettes; Sponge – filter paper – gel – membrane – filter paper – sponge. Put attention to the direction of the cassette in the Trans-blot, putting the cassette in the wrong direction will cause the protein to run away from the gel to the filter paper and out to the blotting buffer. Add the cold pack before starting the blotting.
24. Transfer with constant current (350 mA) with voltage set at >200, run for 1 h.
25. When finished, immerse membrane in Ponceau S solution to identify the presence of the protein on the membrane.
26. Rinse the blot well with water until all the color is washed from the membrane.
27. Incubate for western wash buffer for 5 min (the blot can be dried and stored at 4°C for long period of time).

3.3.4.1.2. Antibodies and Protein Detection

28. Incubate the membrane in blocking buffer while rocking for 1 h at room temperature or overnight at 4°C.
29. Incubate with primary antibody diluted in blocking buffer for 90 min at room temp or overnight at 4°C.
30. Wash 3 × 10 min with western wash buffer.
31. Incubate with secondary antibody diluted in blocking buffer for 60 min at room temperature.
32. Wash 3 × 10 min with western wash buffer.
33. Detect with ECL kit (Amersham) according to manufacturer protocol.

3.3.4.2. Immunofluorescence Analyses

1. Plate 250,000 liver cells in 10 cm dishes, treat the cells with recombinant adenoviruses, and supplement the medium with growth factors as previously described.
2. Incubate the cells at 37°C in a humidified atmosphere of 5% CO_2 95% air for 72 h.
3. Detach cells using trypsin and collect them as previously described.

4. Add sterile cover-glasses into each well of a multiwell plate, use 13 mm cover-glasses for 24-well plates or 18 mm cover-glasses for 12- or 6-well plates.
5. Plate 10,000 cells per well in 24-well plates or 50,000 cells per well in 6-well plates.
6. Incubate at 37°C in a humidified atmosphere of 5% CO_2 95% air for 48 h or until the cells reach subconfluence.
7. Wash twice with PBS.
8. Fix with 4% Paraformaldehyde for 20 min at 4°C.
9. Wash three times with PBS.
10. Permeabilize cells with 0.1% Triton X-100, for 4 min, and wash three times with PBS.
11. Block by 3% bovine serum albumin for 20 min at room temperature.
12. Incubate with the first antibody in 3% bovine serum albumin for 2 h at room temperature or overnight at 4°C in a humid chamber.
13. Wash three times with PBS.
14. Incubate with second antibody coupled to a fluorochrome in 3% bovine serum albumin for 1–2 h at room temperature in a humid chamber.
15. Wash three times with PBS.
16. Stain the nuclei with DAPI, add DAPI (10 μg/mL), and incubate for 5 min in room temperature.
17. Add one drop of mounting medium (Fluoromount-G, SouthernBiotech) to the slide.
18. Remove excess of liquid from the side of the cover-glass with a filter paper, place cover-glass carefully cell-face-down without pressure, and wait until mounting medium has polymerized before you move the slides.
19. Store your samples protected from light at 4°C, and analyze results under a fluorescent microscope or fluorescent confocal microscope.

3.3.4.3. Insulin Content

1. Plate 250,000 liver cells in 10 cm dishes, treat the cells with recombinant adenoviruses and supplement the medium with growth factors as previously described.
2. Incubate the cells at 37°C in a humidified atmosphere of 5% CO_2 95% air for 72 h.
3. Replace medium to fresh and add Diazoxide (500 μM), which inhibit hormone secretion.
4. Incubate the cells at 37°C in a humidified atmosphere of 5% CO_2 95% air for 12–18 h.

5. Detach cells by trypsin and collect them as previously described.
6. Count the cell number in each sample.
7. Wash the cell pellet twice with 4 mL of PBS.
8. Freeze the pellet at −70°C or extract immediately.
9. Add 1.5 mL 35% ETOH/0.18 N HCl to the cell pellet, sonicate the sample or freeze–thaw three times.
10. Incubate 1 h at 4°C with continuously shaking.
11. Centrifuge at 6,000 × *g* for 5 min at 4°C.
12. Transfer the supernatants to a new 2 mL tubes (caps-lock). With hot needle make three small holes in the capes.
13. Freeze the samples in −70°C or liquid Nitrogen.
14. Lyophilize the sample until all the samples are dry. Keep the dry samples at −20°C.
15. Resuspend the samples in 0.5 mL RIA buffer, add protease inhibitors cocktail to the buffer. Incubate for 2 h 37°C with continuous shaking.
16. Centrifuge at 2,000 × *g* for 5 min at 4°C.
17. Transfer the supernatants to a new 2-mL tubes.
18. Measure insulin content immediately or freeze in −20°C.

3.3.4.3.1. RadioImmuno Assay (RIA) Analyses for Insulin and C-Peptide

1. Thaw the frozen samples on ice, add protease inhibitors cocktail to each sample.
2. Measure insulin or C-peptide using RIA kits, according to manufacture protocol.
3. Calculate the amount of insulin/C-peptide in each sample.
4. Use the measurement of cell number to calculate the amount of insulin produced per cell or per protein concentration.

3.3.5. Functional Analysis of the Developmental Redirection Process In Vitro

3.3.5.1. Insulin and C-Peptide Secretion

1. Plate 50,000 cells per well in 6-well plates, treat the cells with recombinant adenoviruses, and supplement the medium with growth factors as previously described (Subheading 3.1.2).
2. Incubate the cells at 37°C in a humidified atmosphere of 5% CO_2 95% air for 72–96 h.
3. Wash the cells twice with PBS.
4. Prepare KRB solution with 2 mM glucose; calculate 4 mL of this solution to each well.
5. Add 2 mL of KRB solution with 2 mM glucose to each well and incubate at 37°C for 2 h.
6. Remove the KRB and add 2 mL of KRB solution with 2 mM glucose to each well and incubate at 37°C for 15–60 min.
7. Take a 1-mL sample of the KRB buffer for insulin and C-peptide secretion; keep the sample at −20°C until further analyses.

8. Remove the KRB and add 2 mL of KRB solution with 17.5 mM glucose to each well and incubate at 37°C for 15–60 min.
9. Take a 1-mL sample of the KRB buffer for insulin and C-peptide secretion; keep the sample at −20°C until further use.
10. Wash the cell with PBS.
11. Collect cells after trypsinization as previously described (Subheading 3.1.2).
12. Wash the cell pellet with PBS.
13. Keep the sample at −20°C for protein determination or RNA isolation.
14. Measure the amount of insulin/C-peptide secreted using kits (Linco Research, Inc., Subheading 3.3.4).
15. Calculate the amount of insulin/C-peptide secreted in 1 h per million cells or per amount of protein in the sample.

Amelioration of hyperglycemia upon implantation in diabetic SCID–NOD mice.

3.3.5.2. Preparation of Transdifferentiated Liver Cells for Transplantation

1. Plate 750,000 liver cells in 6 × 140 mm dishes per mouse, treat the cells with recombinant adenoviruses, and supplement the medium with growth factors as previously described (Subheading 3.1.2).
2. Incubate the cells at 37°C for 96 h.
3. Wash the cell with PBS.
4. Collect by trypsinization as previously described (Subheading 3.1.2), and count the cells number.
5. Transfer 5×10^6 cells in a new tube. Prepare a separate tube for each mouse.
6. Wash the cell pellet twice with PBS.
7. Resuspend the cell pellet in 100 μL of 1:1 with medium and Matrigel matrix.
8. Draw the cells in the Matrigel up into a 1-mL syringe and keep on ice until the implantation.

3.3.5.3. Induction of Diabetes in SCID–NOD Mice

1. Mice should be fasted prior to injection; 4 h is usually sufficient.
2. Inject appropriate amount of the STZ solution intraperitoneally so the final dosage is 180 mg/kg mouse (STZ should be dissolved right before the application, separately for each mouse). You may need to anesthetize, upon your local IACUC procedures.
3. Supply mice with 10% sucrose water, if necessary, to avoid sudden hypoglycemia postinjection.

4. Mice should be tested for their blood glucose levels every second day starting 3 days post-STZ injection. A small drop blood is drawn from the tail for determination of glucose levels. Mice consider diabetic when his blood glucose levels were ≥300 mg/dL on two consecutive measurements.

3.3.5.4. Implantation of Cells Under the Kidney Capsule of the Diabetic Mice

1. The mice are handled using sterile technique and anesthetized by intraperitoneal injection of ketamine (75 mg/kg). The mice are placed on a heating pad during the procedure.
2. Use a clipper to shave the abdomen, then wash the abdomen sequentially with: (1) iodine-based solution and (2) 70% ethanol solution.
3. The mouse is placed on its right side (left side up). Make a 1-cm (approximately) flank incision with scissors, just below the costal margin on the left side.
4. Deliver the kidney into the wound using two pairs of forceps.
5. Inject the cell suspension under the renal capsule.
6. Deliver the kidney back into the abdomen, close the abdominal wall in one layer, take peritoneum, muscle, and skin in that layer. Use a running 4.0 vicryl suture.
7. Inject the animal with analgesic buprenorphine (900 μg Buprenex) intraperitoneal.
8. Let the animal recover for 20–30 min; utilize a heating pad and/or light to ensure that the mice are not rendered hypothermic. Observe the animals until they have regained consciousness and are walking around the cage.
9. Test the mice glucose levels every second day starting 3 days postimplantation. A small drop blood is drawn from the tail for determination of glucose levels.

4. Notes

1. All material and equipment should be sterile for tissue culture. Make sure to use sterile material or filter or autoclave before use.
2. Liver cell isolation from the biopsy is best no longer than 12 h after harvest.
3. Collagenase digestion time can vary upon different tissues depending on the age of the donor and the composition of the tissue. Avoid over-digesting!
4. If large pieces of undigested liver remain after first collagenase digestion, an additional digestion with collagenase or trypsin can be preformed, only on these chunks.

5. Liver cells proliferate and transdifferentiate efficiently in culture for up to 20 passages. However, if the rate of liver cell proliferation or adenoviruses infection capacity is reduced, the culture must be replaced.
6. Adenoviruses do not contain a lipid envelop; therefore, organic solvent as chloroform (1% total volume) can be used to disinfect the viruses. However, it is not recommended to use these viruses without further analyses of the virus activity.
7. Recombinant adenoviruses delivered transgene is starting to be expressed within 17 h from the time of infection. Protein accumulation as detected by GFP fluorescent occurs 24–48 h postinfection depending on the cell type, the promoter activity, and the multiplicity of infection.
8. All material and equipment should be nuclease-free throughout the procedure.

References

1. Ferber, S., Halkin, A., Cohen, H., Ber, I., Einav, Y., Goldberg, I., Barshack, I., Seijffers, R., Kopolovic, J., Kaiser, N., and Karasik, A. (2000) Pancreatic and duodenal homeobox gene 1 induces expression of insulin genes in liver and ameliorates streptozotocin-induced hyperglycemia. *Nat Med* **6,** 568–72.
2. Ber, I., Shternhall, K., Perl, S., Ohanuna, Z., Goldberg, I., Barshack, I., Benvenisti-Zarum, L., Meivar-Levy, I., and Ferber, S. (2003) Functional, persistent, and extended liver to pancreas transdifferentiation. *J Biol Chem* **278,** 31950–7.
3. Sapir, T., Shternhall, K., Meivar-Levy, I., Blumenfeld, T., Cohen, H., Skutelsky, E., Eventov-Friedman, S., Barshack, I., Goldberg, I., Pri-Chen, S., Ben-Dor, L., Polak-Charcon, S., Karasik, A., Shimon, I., Mor, E., and Ferber, S. (2005) Cell-replacement therapy for diabetes: generating functional insulin-producing tissue from adult human liver cells. *Proc Natl Acad Sci U S A* **102,** 7964–9.
4. Shternhall-Ron, K., Quintana, F. J., Perl, S., Meivar-Levy, I., Barshack, I., Cohen, I. R., and Ferber, S. (2007) Ectopic PDX-1 expression in liver ameliorates type 1 diabetes. *J Autoimmun* **28,** 134–42.
5. Meivar-Levy, I., Sapir, T., Gefen-Halevi, S., Aviv, V., Barshack, I., Onaca, N., Mor, E., and Ferber, S. (2007) Pancreatic and duodenal homeobox gene 1 induces hepatic dedifferentiation by suppressing the expression of CCAAT/enhancer-binding protein beta. *Hepatology* **46,** 898–905.
6. Meivar-Levy, I., and Ferber, S. (2006) Regenerative medicine: using liver to generate pancreas for treating diabetes. *Isr Med Assoc J.* **8,** 430–4.
7. Meivar-Levy, I., and Ferber, S. (2003) New organs from our own tissues: liver-to-pancreas transdifferentiation. *Trends Endocrinol Metab* **14,** 460–6.
8. Cao, L. Z., Tang, D. Q., Horb, M. E., Li, S. W., and Yang, L. J. (2004) High glucose is necessary for complete maturation of Pdx1-VP16-expressing hepatic cells into functional insulin-producing cells. *Diabetes* **53,** 3168–78.
9. Horb, M. E., Shen, C. N., Tosh, D., and Slack, J. M. (2003) Experimental conversion of liver to pancreas. *Curr Biol* **13,** 105–15.
10. Imai, J., Katagiri, H., Yamada, T., Ishigaki, Y., Ogihara, T., Uno, K., Hasegawa, Y., Gao, J., Ishihara, H., Sasano, H., Mizuguchi, H., Asano, T., and Oka, Y. (2005) Constitutively active PDX1 induced efficient insulin production in adult murine liver. *Biochem Biophys Res Commun* **326,** 402–9.
11. Kaneto, H., Matsuoka, T. A., Nakatani, Y., Miyatsuka, T., Matsuhisa, M., Hori, M., and Yamasaki, Y. (2005) A crucial role of MafA as a novel therapeutic target for diabetes. *J Biol Chem* **280,** 15047–52.
12. Kaneto, H., Nakatani, Y., Miyatsuka, T., Matsuoka, T. A., Matsuhisa, M., Hori, M., and Yamasaki, Y. (2005) PDX-1/VP16 fusion

protein, together with NeuroD or Ngn3, markedly induces insulin gene transcription and ameliorates glucose tolerance. *Diabetes* **54,** 1009–22.

13. Kojima, H., Fujimiya, M., Matsumura, K., Younan, P., Imaeda, H., Maeda, M., and Chan, L. (2003) NeuroD-betacellulin gene therapy induces islet neogenesis in the liver and reverses diabetes in mice. *Nat Med* **9,** 596–603.
14. Li, W. C., Horb, M. E., Tosh, D., and Slack, J. M. (2005) In vitro transdifferentiation of hepatoma cells into functional pancreatic cells. *Mech Dev* **122,** 835–47.
15. Miyatsuka, T., Kaneto, H., Kajimoto, Y., Hirota, S., Arakawa, Y., Fujitani, Y., Umayahara, Y., Watada, H., Yamasaki, Y., Magnuson, M. A., Miyazaki, J., and Hori, M. (2003) Ectopically expressed PDX-1 in liver initiates endocrine and exocrine pancreas differentiation but causes dysmorphogenesis. *Biochem Biophys Res Commun* **310,** 1017–25.
16. Zalzman, M., Gupta, S., Giri, R. K., Berkovich, I., Sappal, B. S., Karnieli, O., Zern, M. A., Fleischer, N., and Efrat, S. (2003) Reversal of hyperglycemia in mice by using human expandable insulin-producing cells differentiated from fetal liver progenitor cells. *Proc Natl Acad Sci U S A* **100,** 7253–8.
17. Muniappan, L., and Ozcan, S. (2007) Induction of insulin secretion in engineered liver cells by nitric oxide. *BMC Physiol* **7,** 11.
18. Jin, C. X., Li, W. L., Xu, F., Geng, Z. H., He, Z. Y., Su, J., Tao, X. R., Ding, X. Y., Wang, X., and Hu, Y. P. (2008) Conversion of immortal liver progenitor cells into pancreatic endocrine progenitor cells by persistent expression of Pdx-1. *J Cell Biochem* **104(1),** 224–36.
19. Wang, A. Y., Ehrhardt, A., Xu, H., and Kay, M. A. (2007) Adenovirus transduction is required for the correction of diabetes using Pdx-1 or Neurogenin-3 in the liver. *Mol Ther* **15,** 255–63.
20. Yatoh, S., Akashi, T., Chan, P. P., Kaneto, H., Sharma, A., Bonner-Weir, S., and Weir, G. C. (2007) NeuroD and reaggregation induce beta-cell specific gene expression in cultured hepatocytes. *Diabetes Metab Res Rev* **23(3),** 239–49
21. Yamada, S., Yamamoto, Y., Nagasawa, M., Hara, A., Kodera, T., and Kojima, I. (2006) In vitro transdifferentiation of mature hepatocytes into insulin-producing cells. *Endocr J* **53,** 789–95.
22. Fodor, A., Harel, C., Fodor, L., Armoni, M., Salmon, P., Trono, D., and Karnieli, E. (2007) Adult rat liver cells transdifferentiated with lentiviral IPF1 vectors reverse diabetes in mice: an ex vivo gene therapy approach. *Diabetologia* **50,** 121–30.
23. Song, Y. D., Lee, E. J., Yashar, P., Pfaff, L. E., Kim, S. Y., and Jameson, J. L. (2007) Islet cell differentiation in liver by combinatorial expression of transcription factors neurogenin-3, BETA2, and RIPE3b1. *Biochem Biophys Res Commun* **354,** 334–9.
24. Tang, D. Q., Cao, L. Z., Chou, W., Shun, L., Farag, C., Atkinson, M. A., Li, S. W., Chang, L. J., and Yang, L. J. (2006) Role of Pax4 in Pdx1-VP16-mediated liver-to-endocrine pancreas transdifferentiation. *Lab Invest* **86,** 829–41.
25. Tang, D. Q., Lu, S., Sun, Y. P., Rodrigues, E., Chou, W., Yang, C., Cao, L. Z., Chang, L. J., and Yang, L. J. (2006) Reprogramming liver-stem WB cells into functional insulin-producing cells by persistent expression of Pdx1- and Pdx1-VP16 mediated by lentiviral vectors. *Lab Invest* **86,** 83–93.
26. Zalzman, M., Anker-Kitai, L., and Efrat, S. (2005) Differentiation of human liver-derived, insulin-producing cells toward the beta-cell phenotype. *Diabetes* **54,** 2568–75.
27. Koizumi, M., Doi, R., Toyoda, E., Tulachan, S. S., Kami, K., Mori, T., Ito, D., Kawaguchi, Y., Fujimoto, K., Gittes, G. K., and Imamura, M. (2004) Hepatic regeneration and enforced PDX-1 expression accelerate transdifferentiation in liver. *Surgery* **136,** 449–57.
28. Nakajima-Nagata, N., Sakurai, T., Mitaka, T., Katakai, T., Yamato, E., Miyazaki, J., Tabata, Y., Sugai, M., and Shimizu, A. (2004) In vitro induction of adult hepatic progenitor cells into insulin-producing cells. *Biochem Biophys Res Commun* **318,** 625–30.
29. Ferber, S. (2000) Can we create new organs from our own tissues? *Isr Med Assoc J* **2,** 32–6.
30. Tang, D. Q., Cao, L. Z., Chou, W., Shun, L., Farag, C., Atkinson, M. A., Li, S. W., Chang, L. J., Yang, L. J., Lu, S., Sun, Y. P., Rodrigues, E., and Yang, C. (2006) Role of Pax4 in Pdx1-VP16-mediated liver-to-endocrine pancreas transdifferentiation. *Lab Invest* **86,** 829–41.
31. Breyer, B., Jiang, W., Cheng, H., Zhou, L., Paul, R., Feng, T., and He, T. C. (2001) Adenoviral vector-mediated gene transfer for human gene therapy. *Curr Gene Ther* **1,** 149–62.
32. Lai, C. M., Lai, Y. K., and Rakoczy, P. E. (2002) Adenovirus and adeno-associated virus vectors. *DNA Cell Biol* **21,** 895–913.

33. Vorburger, S. A., and Hunt, K. K. (2002) Adenoviral gene therapy. *Oncologist* **7,** 46–59.
34. Becker, T. C., Noel, R. J., Coats, W. S., Gomez-Foix, A. M., Alam, T., Gerard, R. D., and Newgard, C. B. (1994) Use of recombinant adenovirus for metabolic engineering of mammalian cells. *Methods Cell Biol* **43,** 161-89.
35. He, T. C., Zhou, S., da Costa, L. T., Yu, J., Kinzler, K. W., and Vogelstein, B. (1998) A simplified system for generating recombinant adenoviruses. *Proc Natl Acad Sci U S A* **95,** 2509–14.
36. Morral, N., O'Neal, W., Rice, K., Leland, M., Kaplan, J., Piedra, P. A., Zhou, H., Parks, R. J., Velji, R., Aguilar-Cordova, E., Wadsworth, S., Graham, F. L., Kochanek, S., Carey, K. D., and Beaudet, A. L. (1999) Administration of helper-dependent adenoviral vectors and sequential delivery of different vector serotype for long-term liver-directed gene transfer in baboons. *Proc Natl Acad Sci U S A* **96,** 12816–21.
37. LaBarre, D. D., and Lowy, R. J. (2001) Improvements in methods for calculating virus titer estimates from TCID50 and plaque assays. *J Virol Methods* **96,** 107–26.

Chapter 17

In Vitro Reprogramming of Pancreatic Cells to Hepatocytes

Daniel Eberhard, Kathy O'Neill, Zoë D. Burke, and David Tosh

Abstract

Transdifferentiation is defined as the conversion of one cell type to another. One well-documented example of transdifferentiation is the conversion of pancreatic cells to hepatocytes. Here we describe a robust in vitro model to study pancreas to liver transdifferentiation. It is based on the addition of the synthetic glucocorticoid dexamethasone to the rat pancreatic exocrine cell line AR42J. Following glucocorticoid treatment, cells resembling hepatocytes are induced. Transdifferentiated hepatocytes express many of the properties of *bona fide* hepatocytes, e.g. production of albumin and ability to respond to xenobiotics. These hepatocytes can be used for studying liver function in vitro as well as studying the molecular basis of transdifferentiation.

Key words: Transdifferentiation, AR42J, Exocrine pancreas, Hepatocytes, Dexamethasone, Oncostatin M

1. Introduction

During development, multipotent progenitors undergo a progressive restriction of cell fate, eventually giving rise to mature cell types. This process is usually irreversible, permanently fixing the morphological and molecular characteristics of terminally differentiated cells. However, the interconversion of differentiated cell types, or transdifferentiation, has been described both in animal models and in human pathology (1). Studies of transdifferentiation not only extend our understanding of the differentiated state, but also help to identify key transcription factors required for normal development. In addition, reprogrammed cells may be used in cell replacement therapy, thus avoiding the need for donor organs and immunosuppression.

Transdifferentiation usually occurs between tissues that arise from neighboring regions of the embryo, which may be distinguished by

S. Ding (ed.), *Cellular Programming and Reprogramming: Methods and Protocols*, Methods in Molecular Biology, vol. 636, DOI 10.1007/978-1-60761-691-7_17, © Springer Science+Business Media, LLC 2010

the expression of just one or a few transcription factors (1). The pancreas and liver, for example, arise from a bipotential precursor in the anterior endoderm (2), and interconversion between the two cell populations is well documented. Perhaps the most widely used model of pancreas to liver reprogramming is the copper depletion–repletion model (3, 4), in which adult rats are maintained on a copper deficient diet (in the presence of a copper chelator) for 7–9 weeks, and then returned to a normal diet. Following 6–8 weeks of recovery, liver cells can occupy up to 60% of the pancreatic volume. Hepatic foci are also found in: the pancreas of rats fed with the peroxisome proliferator ciprofibrate (5), hamsters fed with *N*-nitrosobis (2-oxopropyl) amine (6), and in transgenic mice expressing keratinocyte growth factor (KGF) under the control of the insulin promoter (7).

The demonstration of transdifferentiation requires the following criteria to be fulfilled: (a) extensive morphological, biochemical, and molecular characterization of the two populations of cells and (b) lineage tracing analysis to establish a direct ancestor–descendant relationship between the two cell types (8). To overcome the difficulties of achieving these aims in vivo, we have developed an in vitro model of pancreas to liver transdifferentiation using the amphicrine (exocrine and neuroendocrine) AR42J-B13 (B13) cell line.

The AR42J-B13 (B13) cell line is a subclone of the parental AR42J cell line, originally derived from an azaserine-treated rat (9). Culture of AR42J cells with activin A and HGF, or with betacellulin, results in transdifferentiation to insulin-expressing beta cells (10). Similarly, GLP1 induces a mixture of beta cells and glucagon-expressing alpha cells (11). The B13 subclone was isolated on the basis of an increased tendency to acquire a beta-cell phenotype in the presence of activin A and HGF (10). Short-term (48 h) exposure of B13 cells to 10 nM dexamethasone (a synthetic glucocorticoid) increases expression of the exocrine enzyme amylase (12). Treatment with higher doses of dexamethasone (1 μM) over a 2-week period gives rise to cells with a hepatocyte-like morphology: the cells are larger and flatter and contain cellular features characteristic of hepatocytes, e.g., the presence of bile canaliculi. Moreover, the hepatocytes express markers representing a variety of key liver functions: Plasma protein secretion (albumin and transferrin), gluconeogenesis (glucose-6-phosphatase), biotransformation (Cyp2E1 and Cyp3A1), phase I and II detoxification, ammonia detoxification (CPS1 and GS), and xenobiotic metabolism (13–15).

Key liver enriched transcription factors, such as hepatocyte nuclear factor 1 (HNF1α), Foxa2 (HNF3β), Foxa3 (HNF3γ), and HNF6 (Onecut1), are expressed in pancreas, and are not upregulated during transdifferentiation of B13 cells to hepatocytes. However, the basic region/leucine zipper transcription factor

CCAAT-enhancer binding protein beta (C/EBPbeta) is absent in B13 cells and induced after 3 days of dexamethasone treatment (16). C/EBPbeta is also induced during transdifferentiation of embryonic pancreas and primary exocrine cells (16, 17). In B13 cells transfected with C/EBPbeta, amylase is downregulated and liver markers such as transferrin, glucose-6-phosphatase, and transthyretin are expressed (16). Conversely, transfection of B13 cells with liver inhibitory protein (LIP), a dominant negative form of C/EBPbeta, blocks transdifferentiation (16). C/EBPbeta therefore appears to act as the master regulator of pancreas to liver transdifferentiation, perhaps reflecting a role in distinguishing pancreas and liver during normal development.

Here we describe a protocol for the conversion of the pancreatic exocrine cell line AR42J-B13 to hepatocytes with dexamethasone treatment (16). Using the described protocol, transdifferentiation of pancreatic cells to hepatocytes can also be studied in organ cultures of pancreatic buds from mouse embryos (16, 18, 19) and primary pancreatic exocrine cells (17).

2. Materials

1. The parental AR42J cell line can be purchased from the ATCC or the ECACC. The experiments in our lab were largely performed using the AR42J-B13 subclone (provided by Dr. Itaru Kojima, Gunma University, Japan). Both the parent cell line and the subclone can be induced to convert to hepatocytes, but there are differences in the ability to form hepatocytes: the B13 subclone is more effective than the parental cell line at producing hepatocytes (16).
2. Culture medium: Dulbecco's modified Eagle's medium (DMEM) with 2 mM L-glutamine, 0.5 U/ml penicillin, 500 ng/ml streptomycin (10,000 U/ml, 10 mg/ml stock), and 10% fetal bovine serum (FBS).
3. Trypsin-EDTA solution.
4. Culture dishes (35 mm).
5. Glass coverslips (22 × 22 mm) (SLS).
6. Dexamethasone is dissolved in ethanol at a stock concentration of 1 mM.
7. Recombinant human Oncostatin M (OSM, R&D System). OSM is dissolved to a stock concentration of 10 μg/ml in phosphate-buffered saline (PBS) containing 0.1% bovine serum albumin.
8. RU-486 (also known as Mifepristone).

Table 1
Antibodies used to investigate transdifferentiation of pancreatic cells to hepatocytes. See ref.13 for additional markers

Name	Company/origin	Dilution	Observation of positive cells
Primary antibodies			
Rb anti-Transferrin	Dako, A0061	1/200	>d3
Rb anti-Albumin	Sigma, A0433	1/500	>d9
Mouse anti-ApoB	Chemicon	1/100	>d3
CYP2E1	Cypex, Dundee, UK	1/300	>d3
Sheep anti-UGT	Cypex, Dundee, UK	1/300	>d5
Rb anti-Amylase	Sigma, A8273	1/300	Decrease
Secondary antibodies			
Goat anti-rabbit IgG FITC	Vector labs, FI-1000	1/200	
Rb anti-sheep IgG FITC	Vector labs, FI-6000	1/200	
Horse anti-mouse IgG FITC	Vector labs, FI-2001	1/200	

Rb rabbit, *ApoB* apolipoprotein B, *CYP2E1* Cytochrome P450 2E1, *UGT* UDP-glucuronosyltransferase, *FITC* fluorescein isothiocyanate

9. Antisera to pancreatic and hepatic proteins (see Table 1).
10. Triton X-100.
11. Blocking reagent for immunostaining (Roche Applied Science).
12. 4′,6-diamidino-2-phenylindole (DAPI) nuclear stain.
13. Aqueous-based mounting medium for mounting, e.g. Gelmount or Fluoromount.
14. Cell freezing media: FBS containing 10% dimethyl sulphoxide.
15. 4% paraformaldehyde (pFA) in PBS.
16. Phosphate-buffered saline, pH 7.4.

3. Methods

Transdifferentiation of pancreatic AR42J cells to hepatocytes can be induced by culture with 1 μM dexamethasone (Dex) (16) but lower concentrations (1 nM) have been used to induce the hepatic phenotype (16). Transdifferentiation can be followed under

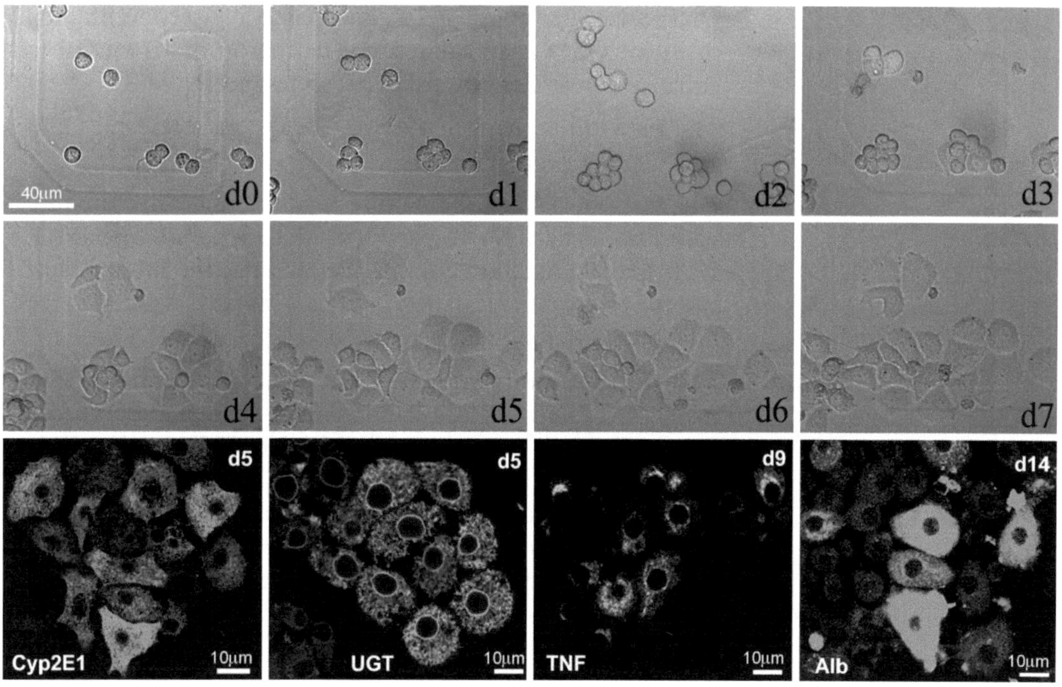

Fig. 1. Time course of B13 cells during days 0–7 of Dex treatment. Small and round B13 cells change their morphology and gradually convert to hepatocytes (large and flattened cells). *Lower row* (adapted from (13)), immunostaining of liver markers expressed in Dex-treated B13 cells using antibodies described in Table 1

transmitted light microscopy. Hepatocytes induced from AR42J cells are morphologically distinct and are larger and flatter than their control pancreatic counterparts. Morphological changes are observed between days 3 and 5 after Dex addition (Fig. 1). The proportion of transdifferentiated cells can be increased by coculture with 1 μM and 10 ng/ml OSM. Oncostatin M promotes liver maturation (20). At least 80% of the pancreatic cells can be converted to hepatocytes after 14 days of combined Dex and OSM treatment. The conversion to a liver phenotype can be confirmed by immunostaining for exocrine pancreatic (amylase) and hepatocyte (transferrin and albumin) proteins (Table 1).

1. The AR42J cell line is maintained in DMEM containing 10% FBS, 0.1 mg/ml penicillin-streptomycin, and 2 mM L-Glutamine (see Note 1). The medium is changed every 2–3 days and the cells are subcultured every 4–6 days at a ratio of 1:7. The cells can be passaged from 1:4 to 1:10 once 80% confluent.
2. AR42J-B13 cells are seeded at low density (10–20% confluent) in 35 mm dishes on sterile, noncoated glass coverslips, and cultured in 1.5–2.0 ml of medium (see Note 2).

3. On the day after splitting, Dex is added to a final concentration of 1 μM. OSM is added at a concentration of 10 ng/ml to enhance conversion toward hepatocytes (see Note 3).
4. The medium is changed every 2–3 days and Dex and OSM added as appropriate.
5. For fixation, discard the culture medium and rinse the dish with PBS. Fix cells in 4% paraformaldehyde for 20 min at RT. Cells may be fixed on days 3–7 to observe the intermediate stages of transdifferentiation or after 7–14 days to analyze the fully transdifferentiated hepatocytes (Fig. 1, Table 1).
6. The cells are immunostained for pancreatic or hepatocyte markers (Table 1) using a standard immunostaining protocol. Briefly, the cells are permeabilized in 0.1% Triton X-100 in PBS for 30 min and then incubated in 2% blocking buffer to block nonspecific binding sites (see Note 4). The primary and secondary antibodies are added sequentially (Table 1). The primary antibody is added overnight at 4°C. The following day, the primary antibody is removed and the cells washed with PBS to remove excess antibody. The wash is repeated a total of three times. The secondary antibody (usually fluorescently conjugated) is added for 3 h at room temperature. The cells are then counterstained with DAPI (500 ng/ml) and mounted on a slide with mounting medium. The fluorescently labeled cells can be viewed by conventional fluorescence microscopy (Fig. 1).
7. Transdifferentiation can be inhibited by addition of RU486 (which inhibits glucocorticoid receptor activity). RU486 can be added at a concentration of 2.5 μM with the treatment commencing 1 h before dexamethasone addition (see Note 5).

4. Notes

1. The ATCC suggests formulated F12-K (30-2004) medium to maintain AR42J cells. We commonly use DMEM.
2. The cell density is critical for efficient transdifferentiation. Cell growth and conversion rates are low in cultures below 20% cell density. AR42J cell cultures also resist efficient conversion to hepatocytes if too confluent.
3. Prepare a stock of 1 mM dexamethasone and add 1 μl/ml of culture medium. Dexamethasone is prepared in 100% ethanol and can be stored at −20°C for several months. Dexamethasone is added to the medium, mixed and then added to the pancreatic cells.

4. Due to cross-reactivity, it is not recommended to use a blocking buffer containing bovine serum albumin when staining for albumin. Alternatively, BSA-free blocking reagent (Roche) is very effective.
5. RU-486 binds to the glucocorticoid receptor and inhibits transdifferentiation (16).

Acknowledgments

We wish to thank the Medical Research Council and Wellcome Trust for funding.

References

1. Tosh, D., and Slack, J. M. (2002) How cells change their phenotype. *Nat Rev Mol Cell Biol.* **3,** 187-94.
2. Deutsch, G., Jung, J., Zheng, M., Lora, J., and Zaret, K. S. (2001) A bipotential precursor population for pancreas and liver within the embryonic endoderm. *Development.* **128,** 871-81.
3. Rao, M. S., Dwivedi, R. S., Subbarao, V., Usman, M. I., Scarpelli, D. G., Nemali, M. R., Yeldandi, A., Thangada, S., Kumar, S., and Reddy, J. K. (1988) Almost total conversion of pancreas to liver in the adult rat: a reliable model to study transdifferentiation. *Biochem Biophys Res Commun.* **156,** 131-6.
4. Rao, M. S., Subbarao, V., and Reddy, J. K. (1986) Induction of hepatocytes in the pancreas of copper-depleted rats following copper repletion. *Cell Differ.* **18,** 109-17.
5. Reddy, J. K., Rao, M. S., Qureshi, S. A., Reddy, M. K., Scarpelli, D. G., and Lalwani, N. D. (1984) Induction and origin of hepatocytes in rat pancreas. *J Cell Biol.* **98,** 2082-90.
6. Scarpelli, D. G., and Rao, M. S. (1981) Differentiation of regenerating pancreatic cells into hepatocyte-like cells. *Proc Natl Acad Sci U S A.* **78,** 2577-81.
7. Krakowski, M. L., Kritzik, M. R., Jones, E. M., Krahl, T., Lee, J., Arnush, M., Gu, D., and Sarvetnick, N. (1999) Pancreatic expression of keratinocyte growth factor leads to differentiation of islet hepatocytes and proliferation of duct cells. *Am J Pathol.* **154,** 683-91.
8. Eguchi, G., and Kodama, R. (1993) Transdifferentiation. *Curr Opin Cell Biol.* **5,** 1023-8.
9. Longnecker, D. S., Lilja, H. S., French, J., Kuhlmann, E., and Noll, W. (1979) Transplantation of azaserine-induced carcinomas of pancreas in rats. *Cancer Lett.* **7,** 197-202.
10. Mashima, H., Ohnishi, H., Wakabayashi, K., Mine, T., Miyagawa, J., Hanafusa, T., Seno, M., Yamada, H., and Kojima, I. (1996) Betacellulin and activin A coordinately convert amylase-secreting pancreatic AR42J cells into insulin-secreting cells. *J Clin Invest.* **97,** 1647-54.
11. Zhou, J., Wang, X., Pineyro, M. A., and Egan, J. M. (1999) Glucagon-like peptide 1 and exendin-4 convert pancreatic AR42J cells into glucagon- and insulin-producing cells. *Diabetes.* **48,** 2358-66.
12. Logsdon, C. D., Moessner, J., Williams, J. A., and Goldfine, I. D. (1985) Glucocorticoids increase amylase mRNA levels, secretory organelles, and secretion in pancreatic acinar AR42J cells. *J Cell Biol.* **100,** 1200-8.
13. Burke, Z. D., Shen, C. N., Ralphs, K. L., and Tosh, D. (2006) Characterization of liver function in transdifferentiated hepatocytes. *J Cell Physiol.* **206,** 147-59.
14. Kurash, J. K., Shen, C. N., and Tosh, D. (2004) Induction and regulation of acute phase proteins in transdifferentiated hepatocytes. *Exp Cell Res.* **292,** 342-58.
15. Tosh, D., Shen, C. N., and Slack, J. M. (2002) Differentiated properties of hepatocytes induced from pancreatic cells. *Hepatology.* **36,** 534-43.
16. Shen, C. N., Slack, J. M., and Tosh, D. (2000) Molecular basis of transdifferentiation of pancreas to liver. *Nat Cell Biol.* **2,** 879-87.

17. Lardon, J., De Breuck, S., Rooman, I., Van Lommel, L., Kruhoffer, M., Orntoft, T., Schuit, F., and Bouwens, L. (2004) Plasticity in the adult rat pancreas: transdifferentiation of exocrine to hepatocyte-like cells in primary culture. *Hepatology.* **39,** 1499-507.

18. Percival, A. C., and Slack, J. M. (1999) Analysis of pancreatic development using a cell lineage label. *Exp Cell Res.* **247,** 123-32.

19. Shen, C. N., Seckl, J. R., Slack, J. M., and Tosh, D. (2003) Glucocorticoids suppress beta-cell development and induce hepatic metaplasia in embryonic pancreas. *Biochem J.* **375,** 41-50.

20. Kamiya, A., Kinoshita, T., Ito, Y., Matsui, T., Morikawa, Y., Senba, E., Nakashima, K., Taga, T., Yoshida, K., Kishimoto, T., and Miyajima, A. (1999) Fetal liver development requires a paracrine action of oncostatin M through the gp130 signal transducer. *Embo J.* **18,** 2127-36.

Chapter 18

Generation of Novel Rat and Human Pluripotent Stem Cells by Reprogramming and Chemical Approaches

Wenlin Li and Sheng Ding

Abstract

Although embryonic stem cells (ESCs) have been established from mice since 1981, attempts to derive its counterparts from various other mammals, including rats, have not succeeded. Recently, induced pluripotent stem cells (iPSCs) have been generated from both mouse and human somatic cells by genetic transduction. We had successfully established novel rat iPSCs (riPSCs), which can be homogenously maintained by LIF and a cocktail of ALK5 inhibitor, GSK3 inhibitor and MEK inhibitor. riPSCs share conventional mouse ESC characteristics and most importantly can contribute extensively to chimeras. We also generated novel human iPSCs (hiPSCs) with "mouse ESC-like" characteristics, which can be surprisingly maintained in culture in the presence of MEK inhibitor and ALK5 inhibitor.

Key words: Embryonic stem cells, Pluripotent stem cells, iPS cells, Inhibitors

1. Introduction

Recently, pluripotent stem cells were derived from the postimplantation egg cylinder stage epiblasts of mouse and rat (1, 2). These novel stem cells were named epiblast stem cells (EpiSCs). EpiSCs seem to correspond very closely to human embryonic stem cells (hESCs) in the colony morphology and culture/signaling requirements for maintaining pluripotency, but exhibit a range of significant phenotypic and signaling response differences from the conventional mouse ES cells (mESCs). Leukemia inhibitory factor (LIF) is essential for maintaining the pluripotency of mESCs in the presence of serum through JAK-STAT3 pathway (3). However, in serum-free medium, BMP4 is also required, together with LIF, to sustain mESC self-renewal by inducing inhibitor of differentiation (Id) protein expression (4) and inhibiting ERK activation (5).

S. Ding (ed.), *Cellular Programming and Reprogramming: Methods and Protocols*, Methods in Molecular Biology, vol. 636, DOI 10.1007/978-1-60761-691-7_18, © Springer Science+Business Media, LLC 2010

In contrast to mESCs, LIF cannot support EpiSCs/hESCs, which typically require basic fibroblast growth factor (bFGF) and Activin A for long-term self-renewal. Undifferentiated hESCs display high-level basal activity of ERK through bFGF signaling (6). BMP4 doesn't support EpiSC/hESC self-renewal either, but instead induces EpiSC/hESC to differentiate into trophoblasts or primitive endoderm (1, 2, 7). In addition to bFGF, Activin A/Nodal signaling has been shown to support the undifferentiated state of hESCs/EpiSCs (1, 2, 8), while it is dispensable for mESCs. These results strongly support the notion that EpiSCs and hESCs are intrinsically similar and raise an attractive hypothesis that mESCs and EpiSCs/hESCs represent two distinct pluripotent states: the mESC-like state representing the preimplantation inner cell mass (ICM) and EpiSC-like state representing later epiblast cells, respectively.

mESCs can be usually derived from certain mouse strains using feeder layer-based cell culture conditions (9). However, it has been proven difficult to derive authentic ES cells from rats under similar conditions (10–13). Similarly, although (in vitro) pluripotent rat EpiSCs had been derived, both rat and mouse EpiSCs show little or no ability to be reincorporated into the preimplantation embryo and contribute to chimeras (1, 2). Recently, induced pluripotent stem cells (iPSCs) generated from both mouse and human somatic cells by defined genetic transduction have attracted enormous interests (14–19). Based on the cell-signaling differences for sustaining the pluripotency of mESC or EpiSC/hESC, the novel "mESC-like" rat and human pluripotent cells could be captured and maintained by combining genetic reprogramming and cell-signaling-based selection using small molecules.

2. Materials

2.1. Cell Culture

1. Diploid rat WB-F344 cells (Grisham et al., 1993), a kind gift from Prof. William B. Coleman at University of North Carolina. Human fibroblasts IMR90 were from ATCC. Viral package cell lines: Plat-E and 293T.
2. Rat iPS cell growth medium: Knockout™ Dulbecco's Modified Eagle's Medium (DMEM), 20% Knockout serum replacement, 1% Glutamax, 1% Nonessential amino acids, 1% penicillin/streptomycin, 0.1 mM β-mercaptoethanol, and 10^3 U/ml mLIF (Millipore). Human iPS cell growth medium: Knockout™ DMEM, 20% Knockout serum replacement, 1% Glutamax, 1% Nonessential amino acids, 1% penicillin/streptomycin, 0.1 mM β-mercaptoethanol, and 10 μg/ml hLiF. Viral pack-

age cell line medium: DMEM with 10% fetal bovine serum. All cell culture products were from Invitrogen/Gibco BRL except where mentioned.

2.2. Viral Package and Cell Transduction

1. Plasmids: pMXs-based retroviruses for mouse Oct4, Klf4, and Sox2 (Addgene); pSin-EF2-Puro-based lentiviruses for human Oct4, Sox2, Nanog, and Lin28; (Addgene) and lentiviral package plasmids pMD 2G and psPAX2 were prepared using Qiagene Hi-Speed kit.
2. Polybrene (10 mg/ml) (Millipore) was used as 4 μg/ml during viral transduction.

2.3. Cytochemistry and Immunofluorescence Assay

1. Alkaline Phosphatase staining was performed using the Alkaline Phosphatase Detection Kit (Millipore).
2. Fixation buffer for immunofluorescence: 4% paraformaldehyde in PBS.
3. Washing buffer: PBS containing 0.1% Triton X-100.
4. Blocking buffer: 0.1% Triton X-100 and 10% normal donkey serum (Jackson ImmunoResearch Laboratories Inc) in PBS.
5. The primary antibodies: Mouse anti-Oct4 antibody (1:250) (Santa Cruz), rabbit anti-Sox2 antibody (1:2,000) (Chemicon), mouse anti-SSEA1 antibody (1:250) (Santa Cruz), rabbit anti-Nanog antibody (1:500) (Abcam), rat anti-SSEA3 antibody (1:1,000) (Chemicon), mouse anti-SSEA4 antibody (1:1,000) (Chemicon), mouse anti-TRA-1-81 antibody (1:1000) (Chemicon), rabbit anti-Pdx1 (1:1,500), a gift from Dr. C. Wright (Vanderbilt University, TN), mouse anti-βIII-Tubulin (Tuj1) antibody (1:1,000) (Covance Research Products), rabbit anti-albumin antibody (1:1,000) (DAKO), rabbit anti-Brachyury antibody (1:200) (Santa Cruz).
6. Secondary antibodies: Alexa Fluor 486/555 donkey anti-mouse, anti-rat, anti-goat, or anti-rabbit IgG (1:500) (Invitrogen).
7. DAPI (Sigma) is dissolved in tissue-culture water at 1 mg/ml, stored in aliquots at –20°C, and used at 1 μg/ml.
8. Nikon Eclipse TE2000-U microscope.

2.4. Teratoma Assay

Mice: 4–6 week-old male SCID-Beige mice from Charles River Breeding Laboratories.

3. Methods

3.1. Generation of Rat Induced Pluriopotent Stem Cells (riPSCs)

3.1.1. Retroviral Package and Target Cell Transduction

1. Thaw a vial of frozen Plat-E cells from the liquid nitrogen, transfer the cell suspension to a tube with 10 ml DMEM medium containing 10% FBS and seed the cells in Poly-lysine-coated 100-mm dish. Four hours later, replace the medium with fresh DMEM medium containing 10% FBS and incubate the cells in a 37°C, 5% CO_2 incubator. The Plat-E cells are passaged when approaching 70–80% confluence with trypsin/EDTA, and then replate them to new 100-ml dishes at 1:6 dilution.
2. Before transduction, dissociate the Plat-E cells by trypsin/EDTA, suspend the cells in appropriate amount of DMEM medium with 10% FBS by gently pipetting. Count the cell number and adjust the concentration to 7×10^5 cells per ml with fresh medium. Seed cells at 100-mm Poly-lysine coated culture dishes (12 ml per dish), and incubate overnight at 37°C, 5% CO_2. Replace the medium with antibiotics free DMEM, 10% FBS (7 ml/dish). The Plat-E cells are ready to be transduced.
3. Transfer 0.45 ml of OPTI-MEM into a 1.5-ml tube. Add 9 μg of pMXs plasmid DNA (encoding Oct3/4, Sox2, Klf4, and c-Myc), mix gently by finger tapping. Add 30 μl of Fugene HD transfection reagent into each of the above tubes, mix gently by finger tapping, and then incubate for 15 min at room temperature. Add the DNA/Fugene HD complex dropwise into the Plat-E dish, and incubate overnight at 37°C, 5% CO_2. Replace the transfection reagent-containing medium with 7 ml of fresh DMEM, 10% FBS medium, and incubate the cells for another 24 h. Collect the medium from the transduced Plat-E dishes and filter it through a 0.45 μm pore size cellulose acetate filter.
4. After adding polybrene into the filtrated virus-containing medium at 4 μg/ml final concentration, the medium containing Oct-3/4, Sox2, Klf4 retroviruses was mixed equally and used to transduce 20–30% confluence diploid rat WB-F344 cells at passage 7 for overnight.

3.1.2. The Establishment and Culture of riPSCs

1. Twenty-four hours later after retroviral transduction, 1×10^5 transduced WB-F344 cells were seeded on the X-ray-inactivated CF1 MEFs in 100-mm dish and incubated with mESC growth medium: Knockout™ DMEM, 20% Knockout serum replacement, 1% Glutamax, 1% Nonessential amino acids, 1% penicillin/streptomycin, 0.1 mM β-mercaptoethanol, and 10^3 U/ml mLIF. The medium was changed every another day.

2. After 10 days, the riPSC colonies were visualized and picked up under a microscope using a Pipetman. The individual colonies were trypsinized and seed on MEF feeder cells in 24-well plates for expansion in mESC growth medium with MEK inhibitor PD0325901 (0.5 μM), ALK5 inhibitor A-83-01 (0.5 μM), and GSK3β inhibitor CHIR99021 (3 μM). When the cells reach 80–90% confluency, they were passaged into six-well plates and subcultured regularly with the presence of above small molecules (see Note 1).

3.1.3. The Pluripotency of riPSCs

1. To analyze the expression of pluripotent markers by riPSCs, cells were fixed in 4% paraformaldehyde for 10 min and washed with PBS containing 0.1% Triton X-100. The fixed cells were then incubated in blocking buffer, 0.1% Triton X-100 and 10% normal donkey serum in PBS, for 30 min at room temperature. The cells were then incubated with primary antibody overnight at 4°C in blocking buffer. The day after, cells were washed with PBS three times, 5 min per time and incubated with secondary antibody in PBS containing 0.1% Triton X-100 for 1 h at room temperature. Nuclei were stained by 1 μg/ml DAPI for 30 s. Then the cells were washed three times, 5 min per time with PBS. Images were captured using a Nikon Eclipse TE2000-U microscope. riPSCs should be positive to typical mESC markers, such as Oct4, Sox2, SSEA-1, Nanog, but are negative to the hESC markers, such as SSEA3, SSEA4, and TRA-1-81.
2. For Alkaline Phosphatase detection, riPSCs were fixed by 4% Paraformaldehyde in PBS for 1–2 min. Aspirate fixative and rinse with PBS. Stain the cells by mixing Fast Red Violet (FRV) with Naphthol AS-BI phosphate solution and water in a 2:1:1 ratio (FRV:Naphthol:water) in dark at room temperature for 15 min. Aspirate staining solution and rinse with PBS. Red colonies are riPSCs (see Note 2).
3. For in vitro differentiation of riPSCs, the cells were dissociated by 0.05% Trypsin-EDTA and cultured in ultra-low attachment 100-mm dish in DMEM medium supplemented with 10% FBS to form embryoid bodies (EBs). The medium was changed every another day. One week later, the EBs were harvested and transferred into Matrigel-coated six-well plate in DMEM medium with 10% FBS. Three to seven days later, the cells were fixed for immunocytochemistry analysis as described above. Mesoderm marker Brachyury should be detected at 3 days after transferring EBs on Matrigel-coated plate. Endoderm and ectoderm markers, such as Albumin, Pdx-1, and βIII-Tubulin, should be detected 7 days later.
4. For teratoma formation, 10^6 riPSCs were resuspended in PBS and injected into subrenal capsule of immune-compromised

SCID mice. Xenografted masses formed within 4–6 weeks, and paraffin sections were stained with hematoxylin and eosin for all histological determinations.

3.2. Generation of Mouse ES Cell-Like Human Induced Pluripotent Stem Cells

3.2.1. Lentiviral Package and Target Cell Transduction

1. Thaw a vial of frozen 293T cells from the liquid nitrogen, culture and split the cells exactly following the protocol of culturing Plat-E cells.
2. One day before transduction, dissociate the 293T cells by trypsin/EDTA, seed cells at 100-mm Poly-lysine-coated culture dishes (12 ml per dish), and incubate overnight at 37°C, 5% CO_2. Replace the medium with antibiotics-free DMEM, 10% FBS (7ml/dish). The 293T cells are ready to be transduced.
3. Transfer 0.45 ml of OPTI-MEM into a 1.5-ml tube. Add 4 μg of pSin-EF2-Puro-based lentiviral plasmid DNA (encoding human Oct4, Sox2, Nanog, and Lin28), 3.5 μg psPAX2, and 0.5 μg pMD2.G, mix gently by finger tapping. Add 30 μl of Fugene HD transfection reagent into each of the above tubes, mix gently by finger tapping, and then incubate for 15 min at room temperature. Add the DNA/Fugene HD complex dropwise into the 293T dishes, and incubate overnight at 37°C, 5% CO_2. Replace the transfection reagent-containing medium with 7 ml of fresh DMEM, 10% FBS medium, and incubate the cells for another 24 h. Collect the medium from the transduced 293T dishes and filter it through a 0.45 μm pore size cellulose acetate filter.
4. After adding polybrene into the filtrated virus-containing medium at 2 μg/ml final concentration, the medium containing human Oct4, Sox2, Nanog, and Lin28 lentiviruses was mixed equally and used to transduce 40–50% confluence diploid human fibroblast IMR-90 cells for 4 h. After incubated with fresh medium over night, IMR-90 cells were transduced again for 4 h by the lentiviruses.

3.2.2. The Establishment and Culture of Mouse ES Cell-Like Human Induced Pluripotent Stem Cells

1. Twenty-four hours later, 1×10^5 transduced IMR90 cells were seeded on the X-ray-inactivated CF1 MEFs in 100-mm dish and incubated with human ES cell growth medium: Knockout DMEM, 20% Knockout serum replacement, 1% Glutamax, 1% Nonessential amino acids, 1% penicillin/streptomycin, 0.1 mM β-mercaptoethanol, and 10 μg/ml hLiF. The medium was changed every another day.
2. After 3 weeks, the mouse ES cell-like colonies were visualized. The colonies were picked up under a microscope using a Pipetman. The individual colonies were dissociated by Accutase and seed on MEF feeder cells in 24-well plates for expansion in human ES cell growth medium with MEK

inhibitor PD0325901 (0.5 μM), ALK5 inhibitor A-83-01 (0.25 μM), and GSK3β inhibitor CHIR99021 (3 μM). When the cells reach ~50% confluency, they were passaged into six-well plates by Accutase and subcultured regularly (see Note 3).

3.2.3. The Pluripotency of Mouse ES Cell-Like Human Induced Pluripotent Stem Cells

1. To analyze the expression of pluripotent markers by Human Induced Pluripotent Stem Cells (hiPSCs), immunostaining was carried out as described above to detect the expression of pluripotent markers by riPSCs. These hiPSCs should homogenously express typical pluripotency markers, such as Oct4, Sox2, Nanog, TRA-1-81, SSEA3, and SSEA-4. hiPSCs should also express Alkaline Phosphatase by cytochemistry assay.
2. For in vitro differentiation of mouse ES cell-like hiPSCs, the cells were dissociated by Accutase and cultured in ultra-low attachment 100-mm dish in DMEM medium supplemented with 10% FBS to form embryoid bodies (EBs). The medium was changed every another day. One week later, the EBs were harvested and transferred into Matrigel-coated six-well plate in DMEM medium with 10% FBS. Three to seven days later, the cells were fixed for immunocytochemistry analysis as described above. Mesoderm marker Brachyury should be detected at 3 days after transferring EBs on Matrigel-coated plate. Endoderm and ectoderm markers, such as Albumin, Pdx-1, and βIII-Tubulin, should be detected 7 days later.
3. For teratoma formation, 10^6 hiPSCs were resuspended in PBS and injected into subrenal capsule of immune-compromised *SCID* mice. Xenografted masses formed within 4–6 weeks, and paraffin sections were stained with hematoxylin and eosin for all histological determinations.

4. Notes

1. riPSCs were very easy to detach from feeders. So, always passage the cells before riPSC colonies become too big and always dissociate riPSC colonies into single cells by trypsin.
2. Do not overfix cells. Fixing cells longer than 2 min will result in inactivation of alkaline phosphatase.
3. It would be helpful to include 5 μg/ml Rock inhibitor (Y-27632) in the medium, then seed the picked-up colonies dissociated by Accutase.

References

1. Tesar, P. J., Chenoweth, J. G., Brook, F. A., Davies, T. J., Evans, E. P., Mack, D. L., et al. (2007) New cell lines from mouse epiblast share defining features with human embryonic stem cells. Nature. 448, 196–199.
2. Brons, I. G., Smithers, L. E., Trotter, M. W., Rugg-Gunn, P., Sun, B., Chuva de Sousa Lopes, S. M., et al. (2007) Derivation of pluripotent epiblast stem cells from mammalian embryos. Nature. 448, 191–195.
3. Niwa, H., Burdon, T., Chambers, I., and Smith, A. (1998) Self-renewal of pluripotent embryonic stem cells is mediated via activation of STAT3. Genes Dev. 12, 2048–2060.
4. Ying, Q. L., Nichols, J., Chambers, I., and Smith, A. (2003) BMP induction of Id proteins suppresses differentiation and sustains embryonic stem cell self-renewal in collaboration with STAT3. Cell. 115, 281–292.
5. Qi, X., Li, T. G., Hao, J., Hu, J., Wang, J., Simmons, H., et al. (2004) BMP4 supports self-renewal of embryonic stem cells by inhibiting mitogen-activated protein kinase pathways. Proc Natl Acad Sci USA. 101, 6027–6032.
6. Dvorak, P., Dvorakova, D., Koskova, S., Vodinska, M., Najvirtova, M., Krekac, D., et al. (2005) Expression and potential role of fibroblast growth factor 2 and its receptors in human embryonic stem cells. Stem Cells. 23, 1200–1211.
7. Xu, R. H., Chen, X., Li, D. S., Li, R., Addicks, G. C., Glennon, C., et al. (2002) BMP4 initiates human embryonic stem cell differentiation to trophoblast. Nat Biotechnol. 20, 1261–1264.
8. Sato, N., Sanjuan, I. M., Heke, M., Uchida, M., Naef, F., and Brivanlou, A. H. (2003) Molecular signature of human embryonic stem cells and its comparison with the mouse. Dev Biol. 260, 404–413.
9. Martin, G. R. (1981) Isolation of a pluripotent cell line from early mouse embryos cultured in medium conditioned by teratocarcinoma stem cells. Proc Natl Acad Sci USA. 78, 7634–7638.
10. Ueda, S., Kawamata, M., Teratani, T., Shimizu, T., Tamai, Y., Ogawa, H., et al. (2008) Establishment of rat embryonic stem cells and making of chimera rats. PLoS ONE. 3, e2800.
11. Demers, S. P., Yoo, J. G., Lian, L., Therrien, J., and Smith, L. C. (2007) Rat embryonic stem-like (ES-like) cells can contribute to extraembryonic tissues in vivo. Cloning Stem Cells. 9, 512–522.
12. Schulze, M., Ungefroren, H., Bader, M., and Fandrich, F. (2006) Derivation, maintenance, and characterization of rat embryonic stem cells in vitro. Methods Mol Biol. 329, 45–58.
13. Ruhnke, M., Ungefroren, H., Zehle, G., Bader, M., Kremer, B., and Fandrich, F. (2003) Long-term culture and differentiation of rat embryonic stem cell-like cells into neuronal, glial, endothelial, and hepatic lineages. Stem Cells. 21, 428–436.
14. Takahashi, K., Tanabe, K., Ohnuki, M., Narita, M., Ichisaka, T., Tomoda, K., et al. (2007) Induction of pluripotent stem cells from adult human fibroblasts by defined factors. Cell. 131, 861–872.
15. Yu, J., Vodyanik, M. A., Smuga-Otto, K., Antosiewicz-Bourget, J., Frane, J. L., Tian, S., et al. (2007) Induced pluripotent stem cell lines derived from human somatic cells. Science. 318, 1917–1920.
16. Takahashi, K., and Yamanaka, S. (2006) Induction of pluripotent stem cells from mouse embryonic and adult fibroblast cultures by defined factors. Cell. 126, 663–676.
17. Dimos, J. T., Rodolfa, K. T., Niakan, K. K., Weisenthal, L. M., Mitsumoto, H., Chung, W., et al. (2008) Induced pluripotent stem cells generated from patients with ALS can be differentiated into motor neurons. Science. 321, 1218–1221.
18. Han, J., and Sidhu, K. S. (2008) Current concepts in reprogramming somatic cells to pluripotent state. Curr Stem Cell Res Ther. 3, 66–74.
19. Park, I. H., Zhao, R., West, J. A., Yabuuchi, A., Huo, H., Ince, T. A., et al. (2008) Reprogramming of human somatic cells to pluripotency with defined factors. Nature. 451, 141–146.

Chapter 19

Small Molecule Screen in Zebrafish and HSC Expansion

Eirini Trompouki and Leonard I. Zon

Abstract

The zebrafish (*Danio rerio*) has emerged as a valuable model organism that is amenable for large-scale chemical and genetic screens. The ability of zebrafish to produce large quantities of synchronized, externally fertilized, transparent embryos makes them ideal for screens, which often are not possible in mammalian models. Signaling pathways important for hematopoiesis are well conserved between zebrafish and mammals, making many targets identified in zebrafish screens applicable to mammals. Hematopoiesis in zebrafish occurs in two waves: the primitive or embryonic wave and the definitive or adult wave. Definitive hematopoietic stem cells arise in the aorta-gonad-mesonephros region (AGM) and express conserved markers such as *runx1* and *c-myb* that allow for the detection of stem cells by whole-mount in situ hybridization (WISH). In this protocol, we will discuss a chemical screen in zebrafish embryos to detect compounds that expand or deplete hematopoietic stem cells (HSCs) in vivo. This type of screen represents a powerful tool to study HSCs in zebrafish.

Key words: Zebrafish hematopoiesis, HSCs: hematopoietic stem cells, Chemical screen, Chemical compound library, Chemoinformatics

1. Introduction

Zebrafish is an attractive model to study vertebrate hematopoiesis (1, 2) and has been used extensively for the study of many human hematological disorders (3). It shares characteristics of invertebrate models such as generation of large numbers of progeny, short generation time, and relatively small size that make it ideal for large-scale studies. In addition, hematopoiesis is well conserved between zebrafish and mammals (4). Zebrafish embryos are transparent and therefore ideal for fluorescent microscopy studies. Screens in whole organisms offer many advantages in comparison to tissue culture screens since the cells are in their physiological environment where it is easy to monitor in vivo in the context of adjacent tissues (5, 6).

S. Ding (ed.), *Cellular Programming and Reprogramming: Methods and Protocols*, Methods in Molecular Biology, vol. 636, DOI 10.1007/978-1-60761-691-7_19, © Springer Science+Business Media, LLC 2010

As in all vertebrates zebrafish hematopoiesis occurs in two waves, the primitive or embryonic hematopoietic wave and the definitive or adult wave. In zebrafish the primitive wave occurs in the intermediate cell mass (ICM), which gives rise to primitive erythroblasts, and the rostral blood island (RBI), which predominantly generates macrophages. The primitive erythrocytes are morphologically distinct from the adult erythrocytes and they enter the circulation around 24 h postfertilization (hpf). Several transcription factors, which are homologous to mammalian transcription factors, are expressed during the primitive wave of hematopoiesis and allow for the study of this event including *scl, fli1, gata2, lmo2, and tif1γ*. Definitive hematopoiesis gives rise to long-term HSCs, which maintain all mature hematopoietic lineages for the lifetime of the animal. This initially takes place in the aorta-gonad-mesonephros (AGM) region that in zebrafish is found in the ventral wall of the dorsal aorta around 36–40 hpf. Markers of the definitive HSCs include *c-myb* and *runx1*. These runx1+ HSCs then translocate to the kidney that becomes the major source of hematopoiesis during the lifetime of the zebrafish (7–9).

Large-scale screens are a major tool in zebrafish biology. Since HSCs are well defined in zebrafish it is expected that screens have been performed in order to find chemical factors and signaling pathways that can expand the HSC pool. In this chapter, we will describe a chemical screen in zebrafish that has been performed successfully and led to the discovery of chemical compounds that affect HSCs (10).

Zebrafish embryos, as already mentioned, are ideal for small molecule screens (11,12). Chemical screens in zebrafish can lead to the discovery of molecules that affect certain pathways or developmental procedures, providing information that can finally lead to drug discovery. Zebrafish is also being used successfully to study potential safety liabilities of novel drugs (13).Chemical screens can provide information about toxicity of specific chemicals as well as tissue specificity since the screen is performed in whole organisms. A live organism may metabolize the chemical that is being tested, leading to activation of a pro-drug or an inactive metabolite. These effects cannot be studied in vitro and are studied much more accurately in vivo in a whole organism. Chemical screens can be used also for the discovery of modifiers in zebrafish mutants. If the mutation is embryonic lethal, these screens can be more labor intensive as heterozygous adults must be mated to generate mutant embryos. Since only 25% of the clutch will be mutants, fewer chemicals can be screened each week (14). Although small molecule screening in zebrafish is beneficial some potential problems have been identified. First, since the chemicals are added to the water, there is a question as to whether a given chemical can penetrate the chorion and permeate the embryo. Warfarin and dexamethasone are two examples of chemicals that have been shown to permeate into the embryo (15, 16). Colorful chemicals can provide direct evidence for the reagent permeability, for example, geimsa stain in a chemical library turns the embryos blue, while beta-carotene turns them yellow. Another major concern is the possibility that small molecules discovered by zebrafish screens will not be active in a mammalian system although some important drug-metabolizing enzymes studied in zebrafish seem to be conserved across species (17).

A small molecule screen to evaluate the effect of chemicals on HSCs has been performed successfully in zebrafish (10). The general design of this screen, described in detail below, is to collect thousands of zebrafish embryos, treat them with chemicals over the course of HSC development and perform whole-mount in situ hybridization for two HSC markers, *runx1* and *c-myb*. Chemicals that augment or decrease HSCs in the AGM can be recognized, confirmed, and further tested in a mammalian system (Fig. 1).

Different screening procedures can be adapted depending on the desired result. For HSC screening in zebrafish, whole-mount in situ hybridization (see Note 1) (18) has been successfully employed as a screening modality. As mentioned earlier, two markers for the definitive stem cells of zebrafish are *runx1* and *c-myb*.(see Note 2)

2. Materials

2.1. Embryo Preparation

1. E3 embryo medium: 5 mM NaCl, 0.17 mM KCl, 0.33 mM $CaCl_2$, 0.33 mM $MgSO_4$, and 10^{-5} methylene blue (see Note 3).

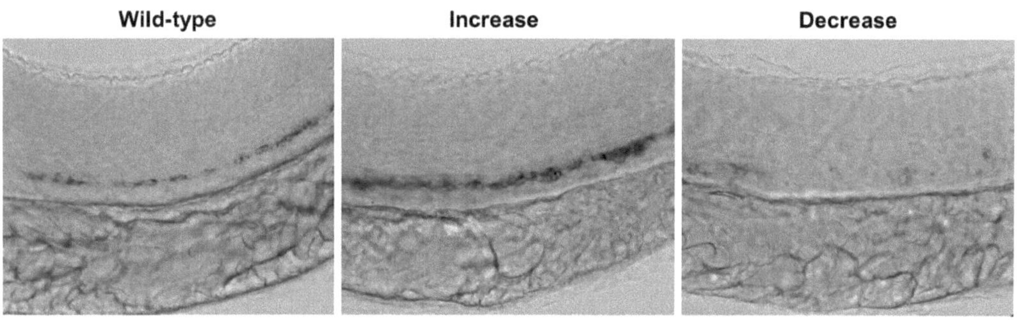

Fig. 1. Runx1/c-myb staining of zebrafish embryos at 36 hpf. The first picture represents the staining in a wild-type embryo. The second is a picture of an embryo treated with a chemical that increases HSC population whereas the third picture represents an embryo that has been treated with a chemical that decreases HSC population in the AGM. (The pictures in this figure are kindly provided by Trista E. North)

2. 1,000× stock methylene blue.
3. Plastic pipettes and glass pasteur pipettes.
4. Tissue culture dishes.
5. Watchmaker's forceps (Dumont).
6. Incubator 28°C.
7. Cages with dividers for timed embryo production.

2.2. Chemical Libraries and Chemical Screen

E3

E3+: E3 with 1% DMSO and 1% Pen/Strep

There are numerous chemical libraries commercially available. Chemicals are usually stored in 96- or 384-well plates at −20°C, −80°C, or at room temperature under 100% nitrogen. The libraries differ in the number and complexity of compounds as well as the purity of them. High purity of the compounds is essential since impurities can give false-positive or false-negative results on a screen. In the best available libraries the quality control of the compounds is performed by NMR. Chemical libraries can contain known or unknown chemical substances or they may be compiled from naturally occurring chemical compounds.

Some examples of the available chemical libraries are given below

- *Libraries that contain biologically active compounds with defined biological activity.* These libraries are the best solution for dissecting a biological pathway, most importantly, they offer good coverage over several biologically important pathways and the chemicals are usually provided in high purity. The disadvantage of some of these libraries is that usually they include only one or two chemicals per pathway. An example of this kind of library is the "ICCB known bioactive library" by Biomol.
- *Natural product libraries.* These libraries are the best choice when the screen is seeking a novel clinical or pharmaceutical product. Their major disadvantage is that they usually contain chemicals with low purity 50–80%. That means that a possible active biochemical needs to be biochemically purified and retested in order to be verified, and to ensure that the impurities don't interfere with the action of the chemical. Many companies offer natural product libraries.
- *Functional libraries.* These libraries contain a collection of chemicals against a very specific target or with a specific action. Examples include a neurotransmitter library, or kinase inhibitor library.

- *Commercial libraries.* These libraries are in general huge and contain tens of thousands of compounds. The advantage is maximal chemical diversity, while the disadvantage is that most compounds will not have any biological activity. These libraries are most useful when the screen is very high throughput.

Many companies offer the ability to create a custom chemical library (Chembridge http://www.chembridge.com/collected-screening-libraries.html). The National Institute of Health offers a vast array of chemicals available for screens free of charge. Sigma Aldrich also offers a variety of chemical libraries, including the LOPAC1280 collection of pharmaceutically active compounds.

The libraries offered can be broad in order to cover a great extent of the available chemical compounds, or can be more biased according to the chemical identities of the compounds, the molecular weight of the compounds, the medicinal use, or other.

1. 96- or 48-well plates.
2. TECAN liquid handling robot to easily transfer chemicals from 96 to 48-well plates if desired (Tecan, Durham, NC) (see Note 4)

2.3. Embryo Fixation, Dechorionation, Bleaching, Rehydration, and Proteinase K Treatment

1. Phosphate buffer saline (PBS): 0.14 M NaCl, 0.003 M KCl, 0.002 M KH_2PO_4, 0.01 M Na_2HPO_4, pH 7.2.
2. 4%PFA in PBS (see Note 5).
3. Pronase: 50 mg/ml in E3, aliquot, and store at −20°C (see Note 6).
4. Bleach solution: 0.8% KOH, 0.9% H_2O_2, 0.1%Tween in deionized water. Should be prepared fresh just before use (see Note 7).
5. PBST: PBS, 0.1%Tween-20.
6. Proteinase K: 10 μg/ml in PBST, aliquot and store at −20°C (see Note 8).
7. Watchmaker's forceps (Dumont).
8. 48-well mesh plates.

2.4. Hybridization

1. Deionized formamide: formamide should be aliquoted and stored at −20°C. Formamide is toxic and must be handled with care. (see Note 9).
2. Saline sodium citrate (SSC): 20×: 3M NaCl, 300 mM trisodium citrate, pH 6.0.
3. Hybridization solution (hybe solution): 50% formamide, 5×SSC, 0.1%Tween-20. Can be prepared and stored at −20°C.
4. PBST: PBS, 0.1%Tween-20.

5. Probe: dilute 100 ng of probe in 100 µl of Hybe solution.
6. 70°C oven.
7. Blocking solution: 2% heat-inactivated lamb serum 0.2%BSA in PBST.
8. Anti-digoxigenin alkaline phosphatase (AP) Fab fragments (Roche) (see Note 10).
9. Antibody solution: 1 µl of antibody in 5 ml of blocking solution.

2.5. DIG Staining (see Note 11)

1. PBST: PBS, 0.1%Tween-20.
2. Staining buffer: 100 mM Tris-base, pH 9.5, 50 mM $MgCl_2$, 100 mM NaCl, 0.1% Tween-20. Add Tween-20 before use. The rest of the solution can be kept at 4°C.
3. 4-Nitroblue tetrazolium (NBT) (50 mg/ml) in 70% dimethylformamide.
4. 5-Bromo-4-chloro-3-indolyl-phosphate (BCIP) (50 mg/ml) in 100% dimethylformamide.
5. Staining solution: dilute 35 µl of 50 mg/ml BCIP and 45 µl of 50 mg/ml NBT in 10 ml of staining buffer.

2.6. Probe Preparation

1. Template DNA in appropriate vector.
2. Enzyme suitable for vector linearization.
3. DIG-labeling kit (Roche): DIG-11-UTP NTP mix, 5× Transcription buffer, SP6, T3 or T7 polymerase, RNAse inhibitor, RNAse-free DNAse, and RNAse-free ddH_2O.
4. RNeasy Mini kit or some other kit for purification of RNA.

2.7. Observation and Photography

1. Coverslips and microscope slides.
2. PBST.
3. Glycerol.
4. Microscope (we use Nikon Eclipse E600).
5. Camera (we use CoolPix 4500 4.0 Mega Pixel).

3. Methods

3.1. Embryo Preparation

1. Set up the desired adult pairs overnight (male and female separated by divider) in mass mating cages 2–3 females and 2–3 males per cage, or as single pairs.
2. Pull out the dividers to combine fish around 8–9 a.m. to best synchronize fertilization timing.

3. Collect different clutches into individual petri dishes using E3 water and methylene blue and stage the embryos to verify that they are synchronized.
4. At about 5 hpf clean out dead embryos and debris.
5. When the embryos have reached the desired stage check them again to confirm that they are synchronized and pool all the embryos together (synchronized staging of the embryos is really crucial!).
6. For a screen in HSCs the treatment with the chemical can start as early as 5 somites stage or as late as 24 hpf depending on the specificity of the target desired. For different kind of screens it may be necessary to choose an earlier or a later stage. (see Note 12)

3.2. Chemical Libraries

1. Preparation of chemical plates.
2. Get the desired chemical library, for example, the ICCB.
3. Add 150 μl E3+ into each well of the desired number of 96-well plates and add 1 μl of chemical in each well. (see Note 13)
4. Transfer chemicals to 48-well plates using Tecan robot or manually. Final volume is 300 μl per well (drugs are used at an approximate concentration of ~20 μM).
5. Pool synchronized embryos together to avoid clutch bias and remove as much of the E3 as possible. Scoop desired number of embryos with a small spatula and flick briskly to place embryos into each well. Embryo number can be adjusted according to the demands of each screen. For screens of wild-type embryos 5–10 embryos per well is an acceptable number. (see Note 14)
6. Incubate embryos in the chemicals at 28.5°C. The time can be adjusted according to the demands of each screen. For an HSC screen embryos can be incubated from 5 somites stage till 36 h postfertilization. (see Note 15)
7. It is crucial that untreated, negative and positive controls (if available) are placed in every single plate of the chemical screen. The positive control can be used to verify the stopping point for the in situ hybridization staining and can validate the success of the whole procedure for each plate.

3.3. In Situ Hybridization (see Note 16)

3.3.1. Embryo Dechorionation, Fixation, Bleaching and Rehydration, and Proteinase K Treatment

1. Dechorionate the embryos. Dechorionation of embryos younger than 20 somites should be done manually with two fine watchmaker's forceps. The user can grab the chorion with the forceps and gently pull in opposite directions. The procedure should be performed under a dissecting microscope. After the dissection the yolk will appear dark but this will not affect the outcome of in situ hybridization. For an HSC screen the embryos will be collected at 36 hpf. In this

case, the dechorionation can be performed with the addition of pronase. Add 10 μl of 50 mg/ml pronase per well and leave for 5 min. The chorion should be very easily detachable. Prolonged incubation with the pronase may result in embryo dissociation.

2. After dechorionation with pronase the embryos should be washed four times in E3 to remove pronase and the chemical.
3. Fix the embryos overnight in 4% PFA.
4. After fixation, embryos that are 36 hpf or older need bleaching, otherwise, the optical clarity of the embryos, a great advantage of zebrafish, will be severely diminished. To bleach the embryos remove PFA and wash twice with PBST twice. Put 300 μl of bleaching solution in each well and incubate 10 min at RT for embryos 36 hpf to 2 dpf, 30 min for 3 dpf embryos and 45 min for older embryos. Embryos after bleach are generally fragile so they must be handled with care.
5. Remove bleach and wash twice for 5 min with PBST.
6. Fix again in 4% PFA for at least 2 h at RT or overnight at 4°C (PFA should be fresh or recently defrozen because embryos are fragile after bleaching).
7. Wash in MeOH three times for 5 min and fix in MeOH o/n at 4°C or −20°C for at least 2 h. Embryos can be stored at −20°C for up to 2 weeks.
8. Rehydrate the embryos. From this step onward you can transfer embryos to 48-well mesh screen trays, which facilitate the procedure or for use with the BioLane machine if available.
9. For each of the solutions mentioned use 500 μl per well unless otherwise indicated. If you use the mesh 48-well plates, use 50 ml of solution per plate.
10. Wash once in MeOH:PBST 2:1 for 5 min RT.
11. Wash once in MeOH:PBST 1:2 for 5 min RT.
12. Wash four times in PBST for 5 min each RT.
13. Dilute pK in PBST to a final dilution of 10 μg/ml and add 200 μl per well. This treatment can be hard for the embryos, so timing is really crucial. Another important point is that every batch of pK may differ slightly, as it is also possible for older versus newer stocks.
14. Suggesting timing for pK treatment
    - Embryos younger that "bud": treat them for 30 s at RT.
    - Early somitogenesis embryos: treat them for 1 min at RT.
    - Late somitogenesis: treat them for 2 min at RT.
    - Embryos 24 hpf: treat them for 5 min at RT.

- Embryos 36/48 hpf: treat them for 10 min at RT.
- Zebrafish 5–6 dpf: treat them for 20 min at RT.

15. Wash once quickly with PBST. Washing should start immediately to prevent prolonged treatment with pK. In mesh-48-well plates the solution can be changed quickly. In the case of regular plates it is advisable to dilute the pK solution with PBST before trying to remove it, so that the time of exposure to pK is not prolonged.
16. Fix in 4% PFA for 20 min at RT.
17. Wash four times in PBST for 5 min each at RT.

3.3.2. Hybridization

1. Prehybridize the embryos in hybe solution at 70°C for 30 min up to 3 h.
2. Dilute the probe in hybe solution at a final concentration of 1 ng/μl and prewarm it at 70°C.
3. Replace hybe solution with hybe solution+probe (250 μl) and incubate o/n at 70°C.
4. Next day prepare the wash solutions and preheat the appropriate ones at 70°C. Use at least 500 μl of washing solution for each well or 50 ml if you are using the mesh plates.
5. Hybe/2×SSC 3:1 for 15 min at 70°C.
6. Hybe/2×SSC 1:1 for 15 min at 70°C.
7. Hybe/2×SSC 1:3 for 15 min at 70°C.
8. 2×SSC for 15 min at 70°C.
9. 0.2×SSC for 30 min at 70°C.
10. 0.2×SSC for 30 min at 70°C.
11. 0.2×SSC/PBST 3:1for 10 min at RT.
12. 0.2×SSC/PBST 1:1 for10 min at RT.
13. 0.2×SSC/PBST 1:3 for 10 min at RT.
14. PBST for 10 min at RT.
15. Block the embryos in blocking solution 500 μl per well for at least 30 min at RT.
16. Dilute the antibody (0.15 U/ml, 1 μl of antibody per 5 ml of blocking solution) and incubate o/n at 4°C (see Note 17).

3.3.3. DIG Staining

1. Wash once quickly with PBST to remove the antibody solution.
2. Wash five times with PBST for 15 min RT on a rocker at 40 rpm.
3. Prepare staining solution and vortex. In 10 ml of staining buffer add 35 μl of BCIP stock (50 mg/ml) and 45 μl of NBT stock (50 mg/ml).

4. Wash once with staining buffer and then add 500 µl of staining solution and 500 µl of staining buffer. This reaction is light sensitive so the plates must be kept in the dark. Monitor the reaction every 30–60 min to prevent high background staining. If *runx1* and *c-myb* are used as probes for staining of the AGM staining should be stopped after several hours or it can be left o/n.
5. When the staining has reached the desired level remove the staining solution and wash multiple times with PBST. Stop the reaction by adding PFA 4%, and store at 4°C. Observe the embryos under a microscope for changes in the staining. In a screen, it is crucial that the same person evaluates all the plates of the screen so that there is consistency. The best solution is for two independent observers to judge the results and compare.

3.3.4. Probe Preparation

Probes for in situ hybridization can be made from a variety of vectors that contain a T3, T7, or SP6 polymerase site and can be linearized opposite that site with a restriction enzyme. In general, any in vitro transcription kit that adds digoxigenin-labeled UTP into antisense RNA can be used. The probe is crucial for the outcome of the screen so it is better to use a probe that gives a strict pattern and that is easily recognizable. The probes should be around 1 kb in length but longer or shorter probes or partial probes can work equally well.

1. Digest 10 µg of DNA with the appropriate enzyme. Digestion should be complete so longer than usual incubation times or even o/n incubation may be needed.
2. Purify the DNA and measure the OD with nanodrop. Run a small sample on an agarose gel to check whether the digestion is complete. Dilute the DNA to a final concentration of 1 µg/µl.
3. For the transcription reaction use
   - 1 µg of linearized DNA
   - Transcription buffer (T3, T7 or SP6 polymerase) 2 µl
   - NTP-DIG-RNA 2 µl
   - RNAse inhibitor (35 units/µl) 1 µl
   - T3/T7/SP6 polymerase (20 units/µl) 1 µl
   - ddH_2O to 20 µl total volume
4. Incubate for 1 h at 37°C.
5. Digest the DNA template by adding 1 µl RNAse-free DNAse for 30 min at 37°C.
6. Purify the RNA with RNeasy MINI kit from Qiagen or some other purification method.

7. Measure the optical density of the RNA and run a gel to check for the integrity and the size of the RNA fragment. Aliquot probe and store at −80°C.

3.3.5. Observation and Photography

Observe the screen plates under the microscope for possible "hits." In case you have a "hit," the embryos should be photographed. Embryos that are 36 hpf are more easily photographed using a depression or a bridge slide to keep the embryos properly oriented. Coverslip slides by Nomarsky are also appropriate for taking pictures of the embryos. To make a bridge slide, stack and glue 3–4 coverslips on each side of a normal slide. The embryo can be put in the middle in a drop of glycerol, positioned as desired, then covered with another coverslip that bridges the two stacks. The embryos should be transferred from PFA 4% to PBST and finally to glycerol so that they can be photographed (Fig. 1).

3.4. Chemoinformatics (19, 20)

Performing a chemical screen can lead to the identification of many chemicals that produce a desired effect. It is possible that these chemicals have no known biological activity. In this case, there are tools to help the researcher to identify the potential activity of the unknown chemical by comparison to known chemicals. It is also possible to identify a known and FDA-approved drug that may exhibit the same activity as the unknown compound. The computational tool that gives the researcher these possibilities is chemoinformatics.

Chemoinformatics is the use of information technology to manage chemical information and solve chemical problems. Usually the chemicals that are included in a certain library, even though they may not have a known biological activity, are represented as molecules in different formats. The researcher can use these formats to compare to chemicals with known biochemical activity. Many types of formats exist. One of the most commonly used is the Simplified Molecular Input Line Entry Specification (SMILES) format (useful information about SMILES can be found in the following link http://www.daylight.com/dayhtml/doc/theory/theory.smiles.html). Smiles is a very powerful format because it is simple and can be read rapidly by a computer. SMILES represent only the fundamental property of the molecule so they provide information that is not arbitrary. Another standard format is the Chemical Markup Language (CML) and the GROMACS file format family which is used in combination with the molecular simulation software GROMACS. Other formats are the CHARMM format, the Chemical File Format, the SYBYL Line Notation. It is also possible to convert between different formats. Two open source tools that can be used to convert between formats are the OpenBabel and JOELib.

The researcher can use these formats to search available chemical databases, using algorithms somewhat analogous to a

BLAST search. Examples of such databases includePubChem, Wombat, MDL Drug Data Report (MDDR), Thor, CrossFire Beilstein, ChemBank, and many more. These repositories contain unique features but most of them enable the user to identify analog chemical substance to the one that they are interested in and they provide a link to 3D models and Safety Data Sheets as well as information about the purity of each chemical or the availability. MDL ACD provides also a number of suppliers for different chemicals. MDL ACD can be reached online via Discovery Gate where the researcher can execute structural or text-based searches. Most of the databases provide also links to literature and patents. The methods to query each database vary, but include the commonly used Tanimoto similarity score, which allows the researcher to determine how similar a given compound is to others in the literature with n dimensions by finding the angle between them. Alternatives include the SEA algorithm (http://shoichetlab.compbio.ucsf.edu/~keiser/sea/) or Tversky similarity (http://chembank.broad.harvard.edu/welcome.htm) These tools, although not always successful, may sometimes point toward a class of similar compounds that can be tested for their ability to phenocopy the "hit" from a given screen.

4. Notes

1. In situ hybridization is an established readout for a screen in zebrafish. Another powerful readout is imaging of fluorescent fish, and the particular screen that we describe in this protocol could be performed using fish that express *c-myb GFP*(AGM staining), *or c-myb GFP* and *lmo2-dsRED*(AGM and vessel staining). The techniques for such a screen are still evolving. There are three main issues to contend with: (1) Sibling fish tend to exhibit significant variability in the strength of fluorescence. For that case, one should perform a prescreen and isolate some zebrafish pairs that produce embryos with similar fluorescent levels. (2) High-throughput fluorescent imaging needs to be rapid so that the embryos in the first well are at a similar stage to those in the last well screened on a given day. (3) Fluorescent visualization in typical 48-well plates tends to be of low quality, but this can be improved with the use of optical prism plates that allow for much improved image quality.
2. In situ hybridization on a large scale can be carried out using a robot. We use the Biolane HTI robot. This robot consists of two individual trays that can be each run independently with separate in situ protocols. The whole machine can be

programmed to perform all the steps of the in situ hybridization. We usually use the robot from the rehydration step of the protocol until the washes, whereas the staining is performed manually. For the robot we use mesh plates provided by the same company (Hölle & Hüttner AG).

3. E3 can be made and kept at room temperature as a 50× stock. Working stock is 1× and can be kept up to 2 weeks at RT. Methylene blue is usually added to 1× E3 as a fungicide.
4. Chemicals can be transferred to the wells manually, but an automated system ensures a much faster and more accurate system avoiding pipetting errors.
5. Dilute PFA in PBS and heat the solution to 65°C to dissolve. If the powder is not completely dissolved add some drops of 1 N NaOH to the solution till it is clear (pH ~7.5). PFA should be used fresh but it can be stored in aliquots and kept at −20°C. Do not freeze and thaw PFA.
6. Pronase is a mixture of a broad spectrum of proteases. Its activity is extended to both denatured and native proteins which can be digested into individual aminoacids. The mixture contains various types of endopeptidases (serine and metalloproteases) and exopeptidases (carboxypeptidases and aminopeptidases) as well as neutral and alkaline phosphatases. It is used here for the dechorionation of zebrafish embryos. It is obvious that prolonged exposure of embryos to pronase will affect their integrity so if the embryos appear fragile reduce the time of exposure to pronase.
7. Another common way to produce unpigmented embryos is the use of Phenylthiouria (PTU). PTU should be used to treat embryos younger than 24 hpf but it has shown to cause delayed hatching, retardation, and embryo malformation in some cases although these effects are very mild at the working dose of 0.003% in fish water. PTU has also been shown to block thyroid hormone expression (21). For older embryos, bleaching is recommended.
8. Proteinase K is a broad spectrum serine protease. The predominant site of cleavage is the peptide bond adjacent to the carboxyl group of aliphatic, aromatic, or hydrophobic amino acids. Prolonged exposure to proteinase K may damage the embryos. The bleached embryos are especially fragile so a shorter treatment with proteinase K is recommended for bleached embryos.
9. Formamide lowers the melting point of nucleic acids and as a result the strands can be separated more easily. Formamide is highly corrosive in contact with skin or eyes and may be deadly if ingested. Formamide should never be handled without proper safety attire including gloves and goggles.

10. Instead of digoxigenin other molecules can be also used in in situ hybridization probes, for example, fluoroscein or its derivatives such as fluoroscein isothyocyanate (FITC).
11. These solutions are toxic and light sensitive. The user should wear gloves while handling these solutions. BCIP and NBT are chemical compounds used for the sensitive detection of alkaline phosphatase (AP). BCIP is the alkaline phosphatase substrate and NBT acts as the oxidant. These two compounds together naturally form a purple precipitate but alkaline phosphatase act as a catalyst in this reaction and accelerates it approximately 1,000 fold. BCIP binds to the alkaline phosphatase active site but upon interaction with NBT, BCIP is released from the enzyme and the colorful precipitate is formed. In case the reaction is very slow, the addition of polyvinyl alcohol (PVA) accelerates the reaction by 20-fold. In that case, the user should add PVA 10%w/v to the staining buffer that contains no Tween-20, boil the solution, let it cool down, and finally add Tween-20 (0.1%) and use the solution. The use of PVA is recommended for less robust chromogenic substrates such as Fast Red. Levamisole can be used to block high endogenous alkaline phosphatase activity, and should be diluted in the staining buffer at a final concentration of 1 mM.
12. If a positive control for the screen is available, a time course with the known chemical should be performed in order to check the optimal initiation time for the screen (primary screen). Embryos should be treated at different stages and for different time points and a check should be made for the best possible readout. Earlier stages, for example, the start of gastrulation (50% epiboly stage) may cause greater toxicity than choosing a stage after completion of gastrulation. The positive control is also essential to verify that the chemicals are able to penetrate the chorion since the treatment is performed before dechorionation. Important evidence regarding the penetrance of the chemicals is the fact that some chemicals are colored so they can be easily visualized to color the embryo. A negative control can also provide the time window, dose and different types or levels of nonspecific effects, but the positive control will also provide a picture of a potential positive regulator.
13. In a chemical screen it is generally too labor intensive to use different concentrations of the chemicals. This can result in toxicity from chemicals that should have been used in lower concentrations. It is also possible to miss a "hit" because a chemical needed to be more concentrated in order to have an effect. So, if the user has identified some chemicals that affect a certain pathway but missed others in the same pathway, we

recommend retesting these chemicals and performing a dose–response curve to find out the optimal dilution for each chemical.

14. The number of embryos depends on the kind of the screen. If the screen is performed on wild-type embryos, five embryos per well are generally enough to detect the desired result. When screens are performed on homozygous mutants that are embryonic lethal, the embryos must be generated by in-crossing heterozygous adults so that only 25% of the total embryo population will be homozygous mutants. In this case, at least 20 embryos per well should be used to ensure enough numbers of the mutant embryos are tested with each chemical.
15. Pooling of chemicals is another suggestion in case large chemical libraries are screened. This is recommended for natural product libraries or commercial libraries that contain thousands of chemicals most of which will have no biological activity. Depending on the format of the library and the screening plate, the number of compounds per well can vary. The advantage of this method is that it reduces greatly the number of embryos requested for the screen. If the pooling strategy allows for each chemical to be represented twice in a plate, individual hits can be identified. The major drawback for the pooling strategy is the increased toxicity that can be very high, especially in libraries with known bioactive compounds.

    Another issue is the incubation temperature. The suggested temperature can be modified according to the needs of the screen. Lower temperatures, for example, 25˚C or 21˚C will delay the development of the embryos if it is necessary.
16. As a general instruction for the whole in situ protocol, use large volumes of liquid for the washes. Smaller volumes are required for the Proteinase K treatment, bleaching, and the hybridization with the probe.
17. The antibodies can be absorbed against whole zebrafish embryos or zebrafish powder but we don't find this step necessary if we use dilution of the antibody at 1:5,000.

Acknowledgments

We would like to thank Trista E North for providing Fig. 1. We would also like to thank Trista E North, Jill L.O. de Jong, Xuining Le, and Richard M. White for providing their expertise in chemical screening and in particular Richard M.White for the chemoinformatics part of the protocol. Finally, we would like to thank Trista E North, Jill L.O. de Jong, and Richard M. White for careful review of the manuscript. Eirini Trompouki is supported by Jane Coffin Childs Memorial Fund.

References

1. Amatruda, J. F., and Zon, L. I. (1999) Dissecting hematopoiesis and disease using the zebrafish. *Dev Biol* **216**, 1–15.
2. Dooley, K., and Zon, L. I. (2000) Zebrafish: a model system for the study of human disease. *Curr Opin Genet Dev* **10**, 252–6.
3. Shafizadeh, E., and Paw, B. H. (2004) Zebrafish as a model of human hematologic disorders. *Curr Opin Hematol* **11**, 255–61.
4. Galloway, J. L., and Zon, L. I. (2003) Ontogeny of hematopoiesis: examining the emergence of hematopoietic cells in the vertebrate embryo. *Curr Top Dev Biol* **53**, 139–58.
5. Kari, G., Rodeck, U., and Dicker, A. P. (2007) Zebrafish: an emerging model system for human disease and drug discovery. *Clin Pharmacol Ther* **82**, 70–80.
6. Lally, B. E., Geiger, G. A., Kridel, S., Arcury-Quandt, A. E., Robbins, M. E., Kock, N. D., Wheeler, K., Peddi, P., Georgakilas, A., Kao, G. D., and Koumenis, C. (2007) Identification and biological evaluation of a novel and potent small molecule radiation sensitizer via an unbiased screen of a chemical library. *Cancer Res* **67**, 8791–9.
7. Davidson, A. J., and Zon, L. I. (2004) The 'definitive' (and 'primitive') guide to zebrafish hematopoiesis. *Oncogene* **23**, 7233–46.
8. de Jong, J. L., and Zon, L. I. (2005) Use of the zebrafish system to study primitive and definitive hematopoiesis. *Annu Rev Genet* **39**, 481–501.
9. Hsia, N., and Zon, L. I. (2005) Transcriptional regulation of hematopoietic stem cell development in zebrafish. *Exp Hematol* **33**, 1007–14.
10. North, T. E., Goessling, W., Walkley, C. R., Lengerke, C., Kopani, K. R., Lord, A. M., Weber, G. J., Bowman, T. V., Jang, I. H., Grosser, T., Fitzgerald, G. A., Daley, G. Q., Orkin, S. H., and Zon, L. I. (2007) Prostaglandin E2 regulates vertebrate haematopoietic stem cell homeostasis. *Nature* **447**, 1007–11.
11. Murphey, R. D., and Zon, L. I. (2006) Small molecule screening in the zebrafish. *Methods* **39**, 255–61.
12. Mandrekar, N., and Thakur, N. L. (2008) Significance of the zebrafish model in the discovery of bioactive molecules from nature. *Biotechnol Lett.*
13. Barros, T. P., Alderton, W. K., Reynolds, H. M., Roach, A. G., and Berghmans, S. (2008) Zebrafish: an emerging technology for in vivo pharmacological assessment to identify potential safety liabilities in early drug discovery. *Br J Pharmacol* **154**, 1400–13.
14. Murphey, R. D., Stern, H. M., Straub, C. T., and Zon, L. I. (2006) A chemical genetic screen for cell cycle inhibitors in zebrafish embryos. *Chem Biol Drug Des* **68**, 213–9.
15. Jagadeeswaran, P., and Sheehan, J. P. (1999) Analysis of blood coagulation in the zebrafish. *Blood Cells Mol Dis* **25**, 239–49.
16. Langenau, D. M., Ferrando, A. A., Traver, D., Kutok, J. L., Hezel, J. P., Kanki, J. P., Zon, L. I., Look, A. T., and Trede, N. S. (2004) In vivo tracking of T cell development, ablation, and engraftment in transgenic zebrafish. *Proc Natl Acad Sci U S A* **101**, 7369–74.
17. Tseng, H. P., Hseu, T. H., Buhler, D. R., Wang, W. D., and Hu, C. H. (2005) Constitutive and xenobiotics-induced expression of a novel CYP3A gene from zebrafish larva. *Toxicol Appl Pharmacol* **205**, 247–58.
18. Thisse, B., Heyer, V., Lux, A., Alunni, V., Degrave, A., Seiliez, I., Kirchner, J., Parkhill, J. P., and Thisse, C. (2004) Spatial and temporal expression of the zebrafish genome by large-scale in situ hybridization screening. *Methods Cell Biol* 77, 505–19.
19. Parker, C. N., and Schreyer, S. K. (2004) Application of chemoinformatics to high-throughput screening: practical considerations. *Methods Mol Biol* **275**, 85–110.
20. Brown, F. (2005) Editorial opinion: chemoinformatics - a ten year update. *Curr Opin Drug Discov Devel* **8**, 298–302.
21. Elsalini, O. A., and Rohr, K. B. (2003) Phenylthiourea disrupts thyroid function in developing zebrafish. *Dev Genes Evol* **212**, 593–8.

Chapter 20

Zebrafish Small Molecule Screen in Reprogramming/Cell Fate Modulation

Jing-Ruey J. Yeh and Kathleen M. Munson

Abstract

Embryonic zebrafish have long been used for lineage-tracing studies. In zebrafish embryos, the cell fate identities can be determined by whole-mount in situ hybridization, or by visualization of live embryos if using fluorescent reporter lines. We use embryonic zebrafish to study the effects of a leukemic oncogene AML1-ETO on modulating hematopoietic cell fate. Induced expression of AML1-ETO is able to efficiently reprogram hematopoietic progenitor cells from erythroid to myeloid cell fate. Using the zebrafish model of AML1-ETO, we performed a chemical screen to identify small molecules that suppress the cell fate switch in the presence of AML1-ETO. The methods discussed herein may be broadly applicable for identifying small molecules that modulate other cell fate decisions.

Key words: Chemical screen, Zebrafish, Hematopoiesis, AML, Leukemia, Reprogramming, Cell fate, In vivo, Erythroid, Myeloid

1. Introduction

Many leukemic oncogenes, including AML1-ETO, contribute to leukemogenesis by modulating hematopoietic stem/progenitor cell differentiation. Embryonic zebrafish is a powerful model to study hematopoietic cell fate. Within the first day postfertilization, zebrafish embryos develop two localized pools of hematopoietic progenitor cells (HPCs). The anterior blood island expresses *pu.1*, and will give rise to the myeloid cells (1). On the other hand, the posterior blood island expresses *gata1*, and will differentiate into the erythroid cells (1). These synchronized pools of HPCs are useful for studying the signaling pathways that underlie or affect hematopoietic cell fate determination.

We have shown that expressing AML1-ETO, an oncogene frequently associated with acute myelogenous leukemia, leads to

S. Ding (ed.), *Cellular Programming and Reprogramming: Methods and Protocols*, Methods in Molecular Biology, vol. 636, DOI 10.1007/978-1-60761-691-7_20,

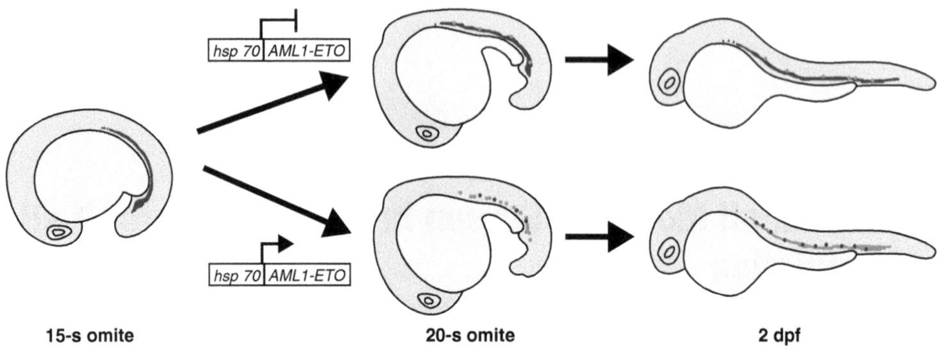

Fig. 1. *Expression of AML1-ETO in embryonic zebrafish reprograms hematopietic cell fate.* At 15-somite stage, erythroid cell markers such as *gata1* are strongly expressed in the intermediate cell mass (ICM) of zebrafish embryos (indicated as a red stripe in the ventral side of the trunk). In the absence of AML1-ETO (top scheme), the hematopoietic progenitor cells in the ICM will continue to express erythroid cell markers and will eventually become red blood cells. Only a small number of cells in the ICM express myeloid cell markers (indicated as *green dots*). However, induced expression of AML1-ETO (bottom scheme) leads to rapid downregulation of *gata1* in the ICM and eventually the accumulation of myeloid cells at 2 days postfertilization (dpf). Red stripes and red dots, cells expressing erythroid cell markers; *green* stripe and green dots, cells expressing myeloid cell markers

a rapid and efficient cell fate switch in the posterior blood island of zebrafish embryos (Fig. 1). This fate switch is characterized by downregulation of *gata1*, suggesting suppression of erythropoiesis, and upregulation of *mpo*, indicating conversion into the granuocytic cell fate (2). Furthermore, to identify candidate small molecules that can reverse AML1-ETO's effects and the mechanisms by which AML1-ETO reprograms hematopoietic cell fate, we conducted a chemical suppressor screen using zebrafish embryos. From this screen we have identified several classes of compounds that restored *gata1* expression in the presence of AML1-ETO (3). The chemical suppressors of AML1-ETO identified from the in vivo zebrafish screen may provide not only new insights into AML1-ETO-mediated hematopoietic differentiation but also new means to block AML1-ETO's effects in the clinical settings.

For the chemical suppressor screen of AML1-ETO we used a transgenic zebrafish line, Tg (*hsp:AML1-ETO*), in which AML1-ETO expression is controlled by a zebrafish heat-shock *hsp70* promoter. Thus, AML1-ETO expression can be induced by incubating zebrafish embryos at 37–42°C as compared to the regular embryo culture temperature at 24–28.5°C. Tg(*hsp:AML1-ETO*) embryos were arrayed into 96-well screening plates. Compounds from the chemical library were also added to the screening plates, and the plates were subjected to the heat treatment to induce AML1-ETO expression. Subsequently, the embryos were fixed and processed for whole-mount in situ hybridization of *gata1* through both manual and automated procedures.

Conceivably, the present method could also be adapted for investigation of other cell fate decisions by using other zebrafish lines and cell markers. The combination of facile detection of cell fates and high-throughput in vivo small molecule screening surely will make the embryonic zebrafish a unique model system to study reprogramming/cell fate modulation.

2. Materials

2.1. Zebrafish

1. Adult wild-type and transgenic Tg (*hsp:AML1-ETO*) zebrafish, males and females (see Note 1).
2. Mating cages for the zebrafish that have divider slots in the middle.
3. E3 buffer: 5 mM NaCl, 0.17 mM KCl, 0.33 mM $CaCl_2$, 0.33 mM $MgSO_4$.
4. Petri dishes.
5. Egg strainers for collecting zebrafish embryos.
6. Incubators at 24°C and 28.5°C.

2.2. Chemical Screening in Zebrafish Embryos

1. A dissecting microscope.
2. A desiccator.
3. Pronase, which is a mixture of proteinases available from various commercial sources.
4. Pipette Pump™, 10 ml (#378980000, Scienceware®).
5. Glass Pasteur pipettes with a large bore size.
6. MultiScreen-Mesh Filter plates with 96-well receiver plates (MANMN6010, Millipore). Each set of the plates includes a cover, a 96-well nylon mesh (60 µm) plate, a 96-well receiver plate, and a single-well reservoir tray.
7. A multichannel pipette.
8. Chemical libraries – Any small molecule libraries may be used. The compounds in the chemical libraries are generally dissolved in dimethyl sulfoxide (DMSO) and stored in a 96- or 384-well format.
9. Dimethyl sulfoxide (DMSO).
10. Foil plate seals.

2.3. Incubation and Fixation of Compound-Treated Zebrafish Embryos

1. A water bath at 39°C for heat-shock treatment.
2. 4% paraformaldehyde in 1× phosphate-buffered saline (PBS), (4% PFA/PBS): 4 g PFA in 100 mL 1× PBS. Dissolve

at 65°C. Alternatively, dilute 37% formaldehyde solution to 4% with 1× PBS. Store in foil-wrapped Falcon tubes at 4°C.

2.4. Digoxigenin Probe Labeling

1. Plasmid DNA pBS-ZG1 (4).
2. *Xba*I restriction enzyme.
3. QIAquick PCR Purification kit (#28104, Qiagen).
4. T7 RNA polymerase (#P2075, Promega).
5. 0.1 M dithiothreitol (DTT).
6. RNasin® Ribonuclease Inhibitor (#N2115, Promega).
7. DIG RNA Labeling Mix (#11277073910, Roche).
8. RNase-free DNase (#M6101, Promega).
9. Mini Quick Spin RNA Columns (#11814427001, Roche).
10. A 37°C incubator.

2.5. Whole-Mount In Situ Hybridization

1. 10× PBS: 800 mL RNase-free water, 80 g NaCl, 2 g KCl, 14.4 g Na_2HPO_4, 2.4 g KH_2PO_4, adjust pH to 7.4 with HCl and add RNase-free water to 1 L. Store at room temperature.
2. 1× PBS: Dilute 1 part of 10× PBS with 9 parts of RNase-free water.
3. PBST: 1× PBS with 0.1% Tween-20. Store at room temperature.
4. 20× SSC: 175.3 g NaCl, 88.2 g sodium citrate, adjust pH to 7.0, and add RNase-free water to 1 L. Store at room temperature.
5. Hyb(−) solution: 50% formamide, 5× SSC, 0.1% Tween-20. Use RNase-free water. Add 1 ml of 1 M citric acid per 100 ml of hyb(−) solution. Store at −20°C.
6. Hyb(+) Solution: Hyb(−) solution with 500 μg/mL Torula Yeast RNA and 50 μg/mL heparin. Store at −20°C.
7. Maleic acid buffer: 0.1 M maleic acid, pH 7.5, 0.15 M NaCl.
8. Blocking solution: Maleic acid buffer with 10% calf serum.
9. Antibody solution: Diluted anti-digoxigenin-AP antibody (#11093274910, Roche) 1:5,000 in the blocking solution.
10. NTMT: 0.1 M Tris–HCl, pH 9.5, 0.1 M NaCl, 50 mM $MgCl_2$, 0.1% Tween-20.
11. NBT: 4-nitro blue tetrazolium chloride solution (100 mg/ml).
12. BCIP: 5-bromo-4-chloro-3-indolyl-phosphate 4-toluidine salt (50 mg/ml).
13. BioLane™ HTI (Holle & Huttner AG), which is an automated liquid handling platform that can be programmed for various applications.
14. A 68°C incubator.
15. A dissecting microscope.

3. Methods

3.1. Collection of Zebrafish Embryos

1. *Day 1*: Set up matings of wild-type and transgenic fish pairs using the mating cages with dividers in the middle to separate males and females.
2. *Day 2*: Pull the dividers in the morning. Two hours later, collect embryos using the egg strainers and place each clutch of the embryos in a 10-cm Petri dish with E3 buffer. Incubate dishes with the embryos at 28.5°C for 5 h and then at 24°C overnight (see Note 2).

3.2. Arraying the Embryos and Administering the Chemicals

1. *Day 3:* Set water bath temperature at 39°C.
2. Thaw compound plates from −80°C in a desiccator (see Note 3).
3. Discard dead embryos and stage embryos under a dissecting microscope, pooling clutches of similarly staged embryos. Embryos should be younger than 14-somite stage.
4. Reduce the E3 volume in the Petri dish just to the point at which the embryos can move freely in the dish when swirled. Add 0.5 μg of pronase to each ml of E3 buffer. Swirl Petri dishes to mix. Monitor the embryos, swirling often to promote the dechorionation of the embryos. Once all of the embryos come out of the chorions, rinse at least five times in E3 to remove chorion debris and any remaining pronase (see Note 4).
5. Assemble the screening plate with one 96-well mesh plate, one 96-well receiver plate, and one single-well reservoir tray. Add 250 μl of E3 to each well of the screening plate using a multichannel pipette. Make sure that the solution flows through the mesh (see Note 5).
6. Using a Pipette Pump with a glass Pasteur pipette, transfer five dechorionated embryos to each well of the screening plate (see Note 6):
   (a) In many chemical libraries, Columns 1 and 12 of the compound plates are empty. Thus, these wells can be used for positive and negative controls.
   (b) Add wild-type embryos to wells in Column 1 to serve as positive controls. These embryos should stain positive with *gata1* probe.
   (c) Add transgenic embryos to the remaining wells including Column 12.
   (d) Add 0.5 μl DMSO to wells in Column 12. These wells should stain negative with *gata1* probe and will serve as negative controls for the assay.

7. Add 0.5 µl of each individual compound from the compound plate to the corresponding wells in Column 2–11 of the screening plate. Use the pipette tip to gently mix the solution while adding the compound (see Note 7).
8. Reseal the compound plate with foil seal when finished. Store the chemical libraries at –80°C.

3.3. Heat-Shock Treatment and Fixing the Embryos

1. Put the lid back on the screening plate and let stand at 25–28.5°C for at least 1 h before heat shock. The embryos should not exceed 18-somite stage before heat shock.
2. Remove the single-well reservoir tray from the rest of the screening plate and float the screening plate (including the lid, 96-well mesh plate and 96-well receiver plate) in a 39°C water bath for 1 h to heat-shock the embryos.
3. Remove the screening plate from the water bath, replace the reservoir tray to the bottom of the screening plate and incubate embryos at 28.5°C for 90 min.
4. Remove the reservoir tray and pour 30 ml of 4% PFA/PBS solution into it. Lift the mesh plate, let drain of the solution and place the mesh plate directly into the reservoir tray containing the fixative. Make sure all of the embryos are covered in the solution. Store the screening plate overnight at 4°C.
5. Clean the 96-well receiver plate for future use (Subheading 3.5, step 7).

3.4. Digoxigenin Labeling of Antisense RNA Probe

1. Digest 10 µg of pBS-ZG1 with *Xba*I. This will yield linearized plasmid DNA. Run one hundredth of the sample on an agarose gel to confirm that the digestion is complete.
2. Purify the linearized DNA with Qiagen PCR Purification kit. Alternatively, the DNA may be purified by phenol/chloroform 1:1 extraction followed by ethanol precipitation.
3. Mix the following components in an RNase-free microcentrifuge tube:

   1 µg linearized DNA

   8 µl 5× transcription buffer

   4 µl 0.1 M DTT

   4 µl DIG RNA-Labeling Mix

   1 µl RNase inhibitor (40 U/µl)

   2 µl T7 RNA polymerase

Add RNase-free water to 40 µl and incubate in a 37°C water bath for 2 h to overnight.

4. Following the reaction, add 2 µl of RNase-free DNase I to the reaction mix and incubate for additional 30 min in the 37°C water bath.
5. Use Mini Quick Spin RNA Columns from Roche to purify the probe. Follow the manufacturer's protocol (see Note 8).
6. Dilute the purified probe from one 40-µl reaction into 16 ml of hyb(+) solution. This is enough for one 96-well screening plate. Store the probe/hyb(+) solution at −20°C.

3.5. Whole-Mount In Situ Hybridization

1. (Continued from Subheading 3.3) *Day 4*: Lift the mesh plate with the embryos, let drain of the solution, and pour 4% PFA/PBS from the reservoir tray to a waste container. Replace with 30 ml of 1× PBS, put the mesh plate back into the reservoir tray and incubate for 5 min at room temperature.
2. Lift the mesh plate, let the solution drain, and pour out the PBS in the reservoir tray. Replace with 30 ml of methanol, put the mesh plate back into the reservoir tray and incubate for 5 min.
3. Repeat methanol washes three times for 5 min each, leaving final wash. Store embryos for a minimum of 2 h in methanol at −20°C. Alternatively, the plates may be stored long-term at −20°C.
4. *Day 5:* Wash plate as described above for 5 min each of 3:1, 1:1, 1:3 methanol:PBST solutions to rehydrate the embryos at room temperature. Use 30 ml of solution per plate for all washes (see Note 9).
5. Wash plates two times quickly in PBST, followed by four more washes of PBST for 15 min at room temperature. During these washes, heat hyb(−) solution and probe/hyb(+) solution in a 68°C incubator.
6. Replace final PBST wash with warm hyb(−) buffer and incubate at 68°C for 30 min to 2 h.
7. At the end of step 6, add 160 µl of probe/hyb(+) solution to each well of the 96-well receiver plate.
8. After hyb(−) incubation, lift the mesh plate from the reservoir tray and let the solution drain. Place the mesh plate into the 96-well receiver plate. Make sure that all of the embryos are covered with probe/hyb(+) solution. Place the screening plate including the lid, the mesh plate and the receiver plate in a humidified chamber and incubate at 68°C overnight. Clean the reservoir tray for later use (step 10).

9. Keep hyb(−) solution at 68°C. Prepare 3:1, 1:1, 1:3 hyb(−):2× SSC solutions, 80 ml per plate, 2× SSC, 80 ml per plate, and 0.2× SSC, 160 ml per plate. Store at 68°C for *Day 6*.
10. *Day 6*: Pour 30 ml of hyb(−) solution into the reservoir tray. Transfer the mesh plate from the 96-well receiver plate to the reservoir tray and incubate at 68°C for 5 min.
11. Collect the probe/hyb(+) solution from the 96-well receiver plate into a 50-ml Falcon tube. Solution can be used once again. Bring up to 16 ml with hyb(+) and store at −20°C.
12. Pour warm 3:1 hyb(−):2× SSC solution into the tray of the BioLane™ HTI machine. Transfer the mesh plate from the reservoir tray to BioLane™ HTI. Up to four plates may be processed at once using BioLane™ HTI (see Note 10). Clean the reservoir tray for later use (step 16).
13. Set up the program in BioLane™ HTI according to Table 1. Connect the solutions to the corresponding ports of the machine. The volume required per wash per plate is 80 ml. Start the program.
14. *Day 7*: Prepare NTMT solution. Reserve 25 ml NTMT per plate and connect remaining NTMT to the correct port on BioLane™ HTI.
15. Add 56.25 μl of NBT (100 mg/ml) and 87.5 μl of BCIP (50 mg/ml) substrates per 25 ml of reserved NTMT solution.
16. At the end of the final NTMT wash, pour NTMT with substrates into the reservoir tray and remove the mesh plate from the machine into the reservoir tray. Cover the screening plate with the lid, wrap in aluminum foil, and incubate at room temperature.
17. Check staining periodically using a dissecting microscope. Check the staining of the positive and negative control embryos (Fig. 2). The development is complete when the positive control embryos show strong purple staining in the posterior blood island, or the intermediate cell mass (ICM). Stop the reaction if the negative control embryos start to show any staining in the ICM or when the background staining becomes obvious.
18. To stop the reaction, lift the mesh plate and change the solution in the reservoir tray into PBST. Replace the mesh plate and incubate for 5 min at room temperature. Repeat PBST wash once more.
19. Inspect the staining of each embryo under a dissecting microscope and log the results into 8 × 12-grid Excel spread sheets.

Table 1
Program for whole-mount in situ hybridization using BioLane™ HTI

Step	Solution	Time	Temperature
1	3:1 hyb(−):2× SSC	15 min	68°C
2	1:1 hyb(−):2× SSC	10 min	68°C
3	1:3 hyb(−):2× SSC	10 min	68°C
4	2× SSC	10 min	68°C
5	0.2× SSC	30 min	68°C
6	0.2× SSC	30 min	68°C
7	3:1 0.2× SSC:PBST	5 min	22°C
8	1:1 0.2× SSC:PBST	5 min	22°C
9	1:3 0.2× SSC:PBST	5 min	22°C
10	Blocking solution	2 h	22°C
11	Antibody solution	2 h	22°C
12	PBST	10 min	22°C
13	PBST	10 min	22°C
14	PBST	4 h	4°C
15	PBST	4 h	4°C
16	PBST	4 h	4°C
17	PBST	4 h	4°C
18	PBST	10 min	22°C
19	PBST	10 min	22°C
20	PBST	4 h	22°C
21	NTMT	10 min	22°C
22	NTMT	10 min	22°C
23	NTMT	10 min	22°C

Embryos can remain in PBST to be photographed using a light microscope and camera either as separate wells of the mesh plate (Fig. 2) or as individual embryos on the lid of petri dish. The embryos can be stored within the plate in PBST at 4°C for several weeks, although the color of the staining will fade over time.

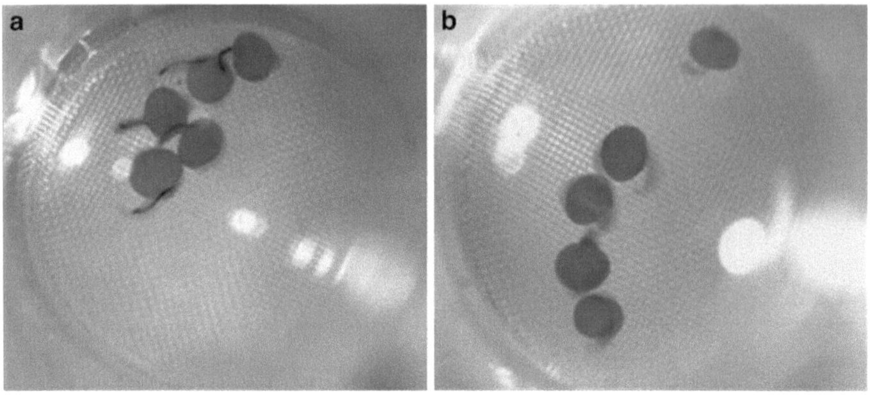

Fig. 2. *The gata1 staining in the positive and negative control embryos.* Strong staining of *gata1* probe appears in the ICM region of the heat-treated wild-type embryos (**a**) but not the heat-treated Tg (*hsp:AML1-ETO*) embryos (**b**). Five embryos were arrayed into each well of the screening plate. The *gata1* staining was performed as described in the Methods

4. Notes

1. In this experiment, we use Tg(*hsp:AML1-ETO*) zebrafish to identify chemical modifiers of AML1-ETO function. Other lines of zebrafish may be used for different experimental designs.
2. The use of dividers ensures the synchrony of embryo stages, which is crucial to this experiment. Transferring embryos from 28.5 and 24°C slows down the embryonic development, so that they will be at the desired stage on the next day.
3. The chemical libraries should be aliquoted into several copies to prevent frequent freeze–thaw cycles.
4. In general, it takes about 15–30 min to dechorionate embryos with pronase. Prolonged incubation with pronase or insufficient rinsing after pronase incubation will cause destruction of the embryo, as evidenced by disintegration of the embryo during later steps of the in situ hybridization protocol. Once dechorionated, embryos are fragile and have a tendency to stick to dry plastic surface. Handle with care.
5. If the solution does not flow through the mesh right away, suck the solution back up with the multichannel pipette and expel the solution again. Once the mesh is wet, the solution should flow through easily.
6. To transfer embryos without increasing the volume of the solution in each well, simply hold the dial on the pump (to prevent expelling the solution) and gently tap the tip of the glass pipette in the solution of each well. The embryos will naturally come out of the glass pipette and sink into the well.

7. The volume of the chemical libraries to use is determined by the concentrations of the compound stocks and the desired effective dose range. We performed this screen at 20 μM concentration. Compounds may be added using a multichannel pipette or a 96-pin transfer device. If using a pin transfer device, the device should be cleaned between different compound plates. This is done by dipping the pins into a DMSO bath and an ethanol bath, and by briefly flaming the device to remove any residual solution.
8. The quality of the RNA probe synthesized may be monitored on a denaturing agarose gel. The yield of the RNA probe may be quantified using a spectrophotometer. We typically obtain around 10–20 μg per 40-μl reaction.
9. The steps from rehydration to hybridization may also be done using BioLane™ HTI. However, the volumes of the solutions required for each step will need to scale up to 80 ml per plate. In addition, all solutions containing Tween-20 should not be stored for long-term use. Stocks may be stored for long term without Tween-20.
10. All of the following steps may also be done manually by changing the solutions in the reservoir tray.

Acknowledgments

The authors would like to thank Dr. Randall T. Peterson for his advice and support during the development of this project. J.-R. J. Yeh is supported by a Career Development Award (AG031300) from the National Institute of Aging. This work was supported by RO1 CA118498 and the Claflin Distinguished Scholar Award.

References

1. Davison, A. J. and Zon, L. I. (2004) The 'definitive' (and 'primitive') guide to zebrafish hematopoiesis. *Oncogene* **23**, 7233–7246.
2. Yeh, J.-R. J., Munson, K. M., Chao, Y. L., Peterson, Q. P., MacRae, C. A., Peterson, R. T. (2008) AML1-ETO reprograms hematopoietic cell fate by downregulating scl expression. *Development* **135**, 401–410.
3. Yeh, J.-R. J., Munson, K. M., Elabgib, K. E., Goldfarb, A. N., Sweetser, D. A., Peterson, R. T. (2009) Discovering chemical modifiers of oncogene-regulated hematopoietic differentiation. Nat. Chem. Biol. 5, 236 –243.
4. Detrich, H. W., Kieran, M. W., Chan, F. Y., Barone, L. M., Yee, K., Rundstadler, J. A., Pratt, S., Ransom, D., Zon, L. I. (1995) Intraembryonic hematopoietic cell migration during vertebrate development. *PNAS* **92,** 10713–10717.

INDEX

MIX
Papier aus verantwortungsvollen Quellen
Paper from responsible sources
FSC® C105338

If you have any concerns about our products,
you can contact us on
ProductSafety@springernature.com

In case Publisher is established outside the EU,
the EU authorized representative is:
Springer Nature Customer Service Center GmbH
Europaplatz 3, 69115 Heidelberg, Germany

Printed by Libri Plureos GmbH
in Hamburg, Germany